T0212439

PERSPECTIVES ON
STRING PHENOMENOLOGY

ADVANCED SERIES ON DIRECTIONS IN HIGH ENERGY PHYSICS

ISSN: 1793-1339

Published

Vol. 1   – High Energy Electron–Positron Physics (*eds. A. Ali and P. Söding*)
Vol. 2   – Hadronic Multiparticle Production (*ed. P. Carruthers*)
Vol. 3   – CP Violation (*ed. C. Jarlskog*)
Vol. 4   – Proton–Antiproton Collider Physics (*eds. G. Altarelli and L. Di Lella*)
Vol. 5   – Perturbative QCD (*ed. A. Mueller*)
Vol. 6   – Quark–Gluon Plasma (*ed. R. C. Hwa*)
Vol. 7   – Quantum Electrodynamics (*ed. T. Kinoshita*)
Vol. 9   – Instrumentation in High Energy Physics (*ed. F. Sauli*)
Vol. 10 – Heavy Flavours (*eds. A. J. Buras and M. Lindner*)
Vol. 11 – Quantum Fields on the Computer (*ed. M. Creutz*)
Vol. 12 – Advances of Accelerator Physics and Technologies (*ed. H. Schopper*)
Vol. 13 – Perspectives on Higgs Physics (*ed. G. L. Kane*)
Vol. 14 – Precision Tests of the Standard Electroweak Model (*ed. P. Langacker*)
Vol. 15 – Heavy Flavours II (*eds. A. J. Buras and M. Lindner*)
Vol. 16 – Electroweak Symmetry Breaking and New Physics at the TeV Scale
           (*eds. T. L. Barklow, S. Dawson, H. E. Haber and J. L. Siegrist*)
Vol. 17 – Perspectives on Higgs Physics II (*ed. G. L. Kane*)
Vol. 18 – Perspectives on Supersymmetry (*ed. G. L. Kane*)
Vol. 19 – Linear Collider Physics in the New Millennium (*eds. K. Fujii, D. J. Miller
           and A. Soni*)
Vol. 20 – Lepton Dipole Moments (*eds. B. Lee Roberts and William J. Marciano*)
Vol. 21 – Perspectives on Supersymmetry II (*ed. G. L. Kane*)
Vol. 22 – Perspectives on String Phenomenology (*eds. B. Acharya, G. L. Kane
           and P. Kumar*)

Forthcoming

Vol. 23 – 60 Years of CERN Experiments and Discoveries (*eds. H. Schopper and
           L. Di Lella*)
Vol. 8   – Standard Model, Hadron Phenomenology and Weak Decays on
           the Lattice (*ed. G. Martinelli*)

Advanced Series on
Directions in High Energy Physics — Vol. 22

PERSPECTIVES ON STRING PHENOMENOLOGY

Editors

Bobby Acharya
King's College London, UK & ICTP, Trieste, Italy

Gordon L Kane
University of Michigan, USA

Piyush Kumar
Yale University, USA

World Scientific

NEW JERSEY · LONDON · SINGAPORE · BEIJING · SHANGHAI · HONG KONG · TAIPEI · CHENNAI

Published by

World Scientific Publishing Co. Pte. Ltd.
5 Toh Tuck Link, Singapore 596224
USA office: 27 Warren Street, Suite 401-402, Hackensack, NJ 07601
UK office: 57 Shelton Street, Covent Garden, London WC2H 9HE

Library of Congress Cataloging-in-Publication Data
Perspectives on string phenomenology / edited by Bobby Acharya, King's College London, UK,
Gordon Kane, University of Michigan, USA, Piyush Kumar, Yale University, USA.
    pages cm. -- (Advanced series on directions in high energy physics ; volume 22)
    Includes bibliographical references.
    ISBN 978-9814602662
    1. String models. I. Acharya, Bobby, editor. II. Kane, G. L., editor. III. Kumar, Piyush, editor.
    QC794.6.S85P47 2015
    539.7'258--dc23
                                2014042636

British Library Cataloguing-in-Publication Data
A catalogue record for this book is available from the British Library.

First published 2015 (Hard cover)
Reprinted 2015 (in paperback edition)
ISBN 978-981-4641-93-7 (pbk)

Printed in Singapore

Contents

Preface

Quantum field theory is a very broad, mathematical framework within which one can successfully describe an incredibly large, rich and diverse plethora of physical systems. There are many examples of such systems ranging from the "microscopic" (e.g. descriptions of elementary particle physics) to the "macroscopic" (e.g. low temperature superconductors). In fact, one of the most celebrated *examples* of a quantum field theory — the Standard Model of Particle Physics — is the most accurate scientific model ever constructed, successfully confronting experimental observations at the level of one part in 100 billion in some cases.

Like quantum field theory, string/M theory is also an extremely broad, mathematical framework which addresses an incredibly rich and diverse plethora of physical systems. However, string/M theory has a distinctive advantage over quantum field theory: *string theory consistently unifies elementary particle forces with gravity*. For instance, in some solutions of string theory, gravitons and photons are simply *different massless states of one and the same quantum string*. Quantum field theory, in this sense, seems incomplete; gravity cannot be incorporated into the framework in any obvious way.

String Phenomenology is usually the name given to the branch of string/M theory devoted to addressing questions in elementary particle physics, physics beyond the Standard Model, dark matter and cosmology and was really born in the mid-eighties shortly after Green and Schwarz discovered "anomaly cancellation" and the existence of unified gauge groups like $E_8 \times E_8$.

Why String Phenomenology? The Standard Model of Particle Physics is based upon relativistic wave equations: Maxwell's equations, the Dirac equation, the Yang-Mills equations and, now that the Higgs boson has been discovered, the Klein-Gordon equation. A mathematical fact is that all of these equations emerge from string theory in the simple, low energy limit.

A bonus is that one also gets Einstein's equations in the same limit! But, you might say, these are very general statements, what about details? Well, one of the most highly cited papers on String Phenomenology was written in the mid-eighties by Candelas, Horowitz, Strominger and Witten. This paper clearly demonstrates that solutions of the low energy limit of the heterotic string theory with three large dimensions of space give rise to a model with non-Abelian gauge symmetry and chiral families of interacting charged fermions. Subsequent work showed that the masses of these fermion families are generically hierarchical. Therefore, the key properties of the Standard Model of Particle Physics clearly emerge from string/M theory in a straightforward fashion.

But, what about the fine details like the electron mass and the W-boson mass? The quick answer to that is that they are a calculation in progress and, in fact, the articles in this volume describe some of the tremendous progress which has been made towards questions like this. In principle, these quantities are calculable in any given solution of string theory, but in practice such calculations are extremely difficult at a precision level and, moreover, such calculations do not seem to provide any conceptual or scientific insights. However, in most other frameworks including quantum field theory, these quantities are not even calculable in principle.

In any case, the framework called string/M theory contains solutions which very plausibly include the Standard Model of particle physics, address physics beyond the Standard Model and address important physics topics like dark matter and dark energy.

This volume contains Perspectives on String Phenomenology by a number of experts in the field but also includes the viewpoint of physicists not normally labelled under that banner. The volume is designed not only to provide a snapshot of the state of the art today; it includes both general, introductory level articles as well as technical material and both reviews as well as articles at the frontier of string phenomenology research. Malcolm Perry has provided an introduction to "the electron" from the string theory perspective and Joe Conlon a basic review of the physics of "moduli fields", which are the low energy remnants of extra dimensions. Mary K. Gaillard has written an introduction to the early heterotic string phenomenology alluded to earlier, whilst Hans Peter Nilles and Patrick Vaudrevange describe the state of the art in that area in a later article. General, personal perspectives on the whole subject are provided separately by Keith Dienes, Mike Douglas and Bert Schellekens. Fernando Quevedo reviews some of the tremendous progress in Type IIB string theory solutions in recent years, whilst Saukura Schafer-Nameki does the same for aspects of F-theory. Piyush Kumar and Gordy Kane have both provided perspectives on both M theory phenomenology and more general aspects of string/M theory phenomenology. A review of the cosmological constant problem and dark energy from the string theory perspective is provided by Brian Greene and Gary Shiu, whereas Alexander Westphal has written a perspective on inflationary cosmology. Finally, Tom Banks, who has argued against the effective field theory approach to the low energy limit of string theory that has been adopted by many of our contributors, describes an alternative approach, dubbed cosmological supersymmetry breaking, which leads to a set of particle physics predictions.

Bobby Samir Acharya
Gordon Kane
Piyush Kumar
September 2014

Acknowledgments

We would like to wholeheartedly thank all of the contributors for their articles, without which this volume would not have been possible. We gratefully acknowledge CERN for the image of the Higgs Event, and Andrew J. Hanson for the image of a Calabi-Yau manifold, on the cover page. We also acknowledge Ran Lu for help with combining the images. The image of the nutcracker on the cover symbolizes the fact that to crack a hard problem and unravel the mysteries within requires probing both from the top-down and bottom-up.

1

What is an Electron?

Malcolm J. Perry

DAMTP, Centre for Mathematical Sciences, Wilberforce Road, Cambridge, CB3 0WA, England

Can we try to say what an electron (or a quark, gluon or photon) is, as opposed to just describing its properties. String theory can address that question.

We all know how to describe an electron and its behavior. Since the discovery of the Dirac equation, we have had a satisfactory theory of electrons. Quantum electrodynamics was developed so that the electron interacting with photons could be described and this led to a theory that has been verified to an accuracy of around one part in 10^9. Let's examine the g-factor for the magnetic dipole moment of the electron. The Dirac equation predicts that $g = 2$, but once virtual photons are taken into account in quantum electrodynmics, we find that $g - 2 = 115\,965\,218\,17.8 \times 10^{-13}$ theoretically, compared to $115\,965\,218\,07.3 \times 10^{-13}$ experimentally. This makes quantum electrodynamics extraordinarily accurate. In quantum electrodynamics, the electron is just a quantized fluctuation in the electron field. This explains why all electrons are identical. Why should one look further than this picture. The answer is because there is more to the world than electrodynamics. There are the other interactions, the strong and weak nuclear forces and gravitation. It is possible to include the strong and weak interactions into a straightforward generalization of quantum electrodynamics. Gravity however cannot be treated in the same way. For this reason, we move on to string theory.

What is string theory and why should one study it? These questions are those that one is always asked by those outside the string theory community. It does not matter who is doing the asking; it could be a condensed matter theorist or an immigration inspector or someone you meet at a cocktail party. In each case, the question is always the same. What follows is an attempt to provide some answers on the route to explaining what an electron is in string theory. For us, normally an electron is a spin-half particle with almost no mass and an electric charge that couples to the electromagnetic field that describes photons. But what is it really?

The Standard Model (SM) of particle physics is perhaps the most successful theory ever invented. It provides us with explanations of most of the phenomena in nature. It is a construction that assumes that point particles are properly described by a normalization relativistic quantum field theory. Contained in it are the basic interactions of nature. It contains three generations of leptons and colored quarks together with gauge bosons that carry the fundamental forces of nature, massless colored gluons that mediate the strong force, the massless photons that are responsible for the electromagnetic interaction and the massive $W^{\pm}$ and Z bosons that mediate the weak interaction. Color is rather like electric charge but is carried only by the quarks and gluons. In addition, there are Higgs scalars which act to provide the mass to those particles that are massive. The whole theory is based on the gauge group $SU(3) \otimes SU(2) \otimes U(1)$. Each of the strong, electromagnetic and weak interactions are associated to a coupling strength.

What the Standard Model does not provide is any insight into dark matter or dark energy which make up around 27% and 68% of the Universe respectively. Also, what goes into the Standard Model does seem rather arbitrary. For example, there is no obvious reason why the gauge group should be what it is, and no obvious reason why there should be three generations. There is however a hint, from a knowledge of the magnitudes of the three coupling strengths, that there should be some deeper picture underlying the standard model. Couplings depend on energy in a way specified by the renormalization group equations. If one extrapolates the three couplings to an energy scale of around 10^{15} GeV, they become numerically similar. If one makes a modest extension of the standard model to include supersymmetry, then this numerical coincidence becomes even stronger. We take this to mean that there is some kind of unification of these three interactions at that scale.

Missing from the Standard Model, and indeed from any conceivable extension of it is any description of the interaction that controls the behavior of the entire Universe; gravitation.

It is often said that the gravitational force is a special case and unlike the other forces. Whilst it is very familiar, at the level of individual subatomic particles it is vastly weaker than the other forces, by about a factor of 10^{39}. This makes it difficult to study, and so the microscopic mechanism behind gravitation is yet to be fully understood. The only reason we notice gravitation is that unlike the other forces, it is purely attractive in nature. The other forces can either be attractive or repulsive; that is to say they have charges of both signs. In electromagnetism, like charges attract and unlike charges repel. That picture holds for all interactions except gravity. Gravity only has one sign of charge. So although it is weak on the atomic scale, its effects build up and become easily observable for large objects. It is for that reason that gravity is most easily observed on astronomical distance scales. There is an excellent theory of gravitation that works at the classical level and that is Einstein's general theory of relativity.

Yet, the unification scale of 10^{15} GeV is so close to the scale of gravitation, the Planck scale of 10^{19} GeV, that to ignore gravitation is surely misconceived. At these energy scales, the strengths of the other couplings become similar to that of gravitation. And since there can hardly be one set of laws for gravitation and a different set for everything else we are driven to try to find a picture in which all of the interactions can be accounted for at once. Superstring theory is the only known picture that includes gravitation that does not suffer from some kind of fatal difficulty. Trying to incorporate gravitation into a quantum field theory of particles, a method that has been astonishingly successful for the other forces, encounters numerous problems.

To do so, one postulates the existence of a graviton that transmits the gravitational force in a way that is similar to how a photon transmits the electromagnetic force. A graviton viewed as an elementary particle must be massless and have a spin of two in contrast to the photon which is massless and has a spin of one. It is the fact that the gravity is universally attractive that requires the graviton to have a spin of two. Perhaps the most serious difficulty that nobody has ever found a way around is the ultra-violet divergence problem. In relativistic quantum field theory, one always encounters infinite quantities. One can think of this as being associated with the fact that the self-energy of a point particle is infinite. In theories without gravitation, it is possible to control these infinities through a process known as renormalization. If one tries to construct a quantum theory of gravity by following the same route, it is impossible to get rid of these infinities without violating some essential physical principle. The principles that one comes into conflict with are either causality, the idea that no information can propagate faster than light or unitarity, the idea that the probability for any event lies in the range zero to one inclusively. These two principles as so basic to our understanding of the universe that it impossible for us to conceive of any theory that does not satisfy these conditions. Such a theory is termed unrenormalizable and as a fundamental physical theory makes no sense.

In fact, it is easy to see why gravity is different. The idea of a point particle in gravitation does not really exist. To see why, consider a spherical body like the earth. To escape from it, one needs to achieve a certain velocity, the escape velocity. If a body of fixed mass gets smaller, its escape velocity increases. If this velocity becomes equal to the speed of light, then nothing can escape from the body. Such an object is a black hole. In general relativity this is the closest one can get to a point particle. Black holes have finite size. For example if the Sun were to be shrunk to form a black hole,m then it would have a radius of around 3 km. One might think that this would require a density that is impossible to obtain, but in fact for the Sun the density would be equivalent to about that of the nucleus. As one considers larger objects, then their density gets lower and if one thinks about the density of black holes in the center of galaxies, their density is about that of water.

The simplest string theory is the bosonic string. It is termed bosonic because it results in objects in spacetime that are all bosons. The bosonic string cannot describe fermions and therefore at best is a toy model for realistic physics. One replaces the "worldline" of a particle, a one-dimensional timelike line in spacetime generated as the particle moves, with a two-dimensional surface with one time direction and one spatial direction. The spatial direction can be either a line segment - the open string - or a circle - the closed string. Thus, a string is a surface in space time. An open string is one that has spatial endpoints that sweep out lines in spacetime. A closed string is a cylinder in spacetime. One can regard a string as being a generalization of a particle.

In classical mechanics, to describe the behavior of a particle, one constructs an action and uses its Euler-Lagrange equations as the particle equations of motion. The action is proportional to the proper length of the worldline. Extremising the action with fixed initial and final positions yields the particle equations of motion. Interactions between particles can now happen but the interactions must be concentrated at points and in practice are restricted by conservation laws that constrain how elementary particles are observed to interact. At the classical level these interactions must be introduced by hand. In quantum field theory, these interactions are introduced naturally as cubic or higher order terms involving the fields corresponding to particular particles. Their diagrammatic treatment is as if the particles had classical interactions involving three or higher point vertices. The nature of these interactions is rather *ad hoc* and is governed by what one has observed experimentally together with some consistency requirements such as gauge invariance, hermiticity and renormalizability.

To describe the behavior of a string, a similar action can be constructed. The action is proportional to the proper area of the string worldsheet. Unlike point particle theory, all string interactions are contained in this action. Whilst free strings are described by tubes or planar worldsheets, there is no restriction on the topology of the worldsheet. The surface that describes the string can have holes or junctions where string meet. The string action can describe how strings interact without having to introduce any assumptions about how interactions take place. A string can in principle live in d spacetime dimensions. A string is described by whether it is open or closed, by its center of mass motion, and how it is vibrating. if one thinks about a point on the surface of the string, it can move in two directions in the plane of the string, or in $d - 2$ directions perpendicular to the surface of the string. Motion in the direction of the string is just moving one point on the worldsheet into another and does not correspond to any physical change in the string, just how it is described. Since the surface of the string has one timelike and one spacelike direction, the physical degrees of freedom all correspond to spacelike displacements and so it is only these $d - 2$ directions that are physical string motions. These vibrations in $d - 2$ directions transverse to the string should really be thought of as waves traveling along the string rather in the same way that vibrations of a string in a musical instrument such as a guitar involve waves causing displacement of the string perpendicular to the string itself.

Lets first consider the closed bosonic string. The spectrum of states of the string consists of a collection of excitations that are classified by their mass and other quantum numbers. States can either be massless or have masses on the order of the string scale. Since the string scale is of the order of the Planck scale, roughly 10^{19} GeV, the only excitations that are directly observable are those that are massless. These massless states can be described using the same language one uses for elementary particles. That is really because on any distance scale we can measure directly, the strings are invisibly small, of the order of the Planck length, roughly 10^{-33} cm. One finds that the spectrum of massless states contains three types of object: something with the quantum numbers of a scalar particle, something that is a bit like a photon and an object with the quantum numbers expected for a graviton, the object that transmits the gravitational force. These massless excitations are universal features of all string theories. It is this graviton like object that is most intriguing as it is something which comes out of the theory automatically rather than something that has to be put in by hand. It is a hint that string theory really does contain gravitation.

An open bosonic string is a little different. Here the waves can be reflected at the endpoints of the string and set up standing waves. An open string has massless excitations which look rather like a photon. A simple modification of the open bosonic string is to place a label at each end of the string which can take integer values from 1 to N. The result is that instead of an abelian gauge theory like a photon, one finds excitations that behave like Yang-Mills particles with gauge group $U(N)$. $U(N)$ is the unitary group of rank N which can be thought of as the set of all N by N unitary matrices. It is rather like the $SU(2)$ or $SU(3)$ groups found in the Standard Model. In a similar way, one can also realise gauge theories of the orthogonal groups $SO(N)$ and the symplectic groups $USp(N)$. Thus strings are represented in low-energy physics, the physics that we see, by fields in quantum field theory and can carry quantum numbers such a spin and charge. One way to think of Yang-Mills particles is that they are generalizations of the photon in which the gauge transformations have become rather more complicated and controlled by a components of a Lie algebra rather than just numbers as would be the case for electromagnetism. Just as photons can couple to objects with electric charge, Yang-Mills particles can couple to objects with generalizations of electric charges.

One can then ask if the string theory is quantum mechanically consistent. This is a rather non-trivial step which is forced on one because string theory has a hidden classical symmetry, called conformal symmetry. The conformal symmetry is essential for the string to work properly. Yet quantum effects can cause conformal symmetry to be broken. This would be an undesirable state of affairs. It turns out that the fields that yield string theories with unbroken conformal symmetry are precisely those that obey the classical Einstein euqations or Yang-Mills equations or for scalars a version of the wave equation. This indicates a deep relationship between the known laws of physics and string theory. In fact, one could say that these laws of physics instead of being postulated, have in fact been derived from just the symmetry of string theory.

There are some difficulties in promoting bosonic string theories into fully sensible models of fundamental physics. The first is that quantum effects result in the string making sense only if $d = 26$, which is rather a long way from the $d = 4$ we observe. A second problem that they do not contain any fermions. A final problem is that all bosonic string theories contain a fatal flaw in the form of a tachyon. For the most part, we not need to worry about massive string states simply because their mass is so high that we cannot observe them. An exception to this is what happens in bosonic string theory, and that is that its spectrum contains a tachyon, that is a particle that has to move faster than light. Such objects as well as being objectionable due to their conflict with causality, also tend to indicate instabilities. Since their masses of the order of the Planck scale, whatever instability they signal will governed by a timescale of order of the Planck time of 10^{-43} seconds, in gross conflict with the observed age of the universe.

However, we have learnt some interesting facts. The first is that one finds a quantum field theory of massless particles as a low energy version of string theory. All other string excitations are not directly observable at energies scales we can access. In amongst these string excitations, there is the graviton and so we have the potential to describe gravity using string theory. One might wonder if such a theory has the same kind of problems with infinites as gravity on its own does. The answer is no. String theory does not have the kind of divergences encountered in quantum field theory. The reason is that when one tries to calculate the types of quantities that diverge in quantum field theory, we find they do not diverge in string theory. Mathematically, this comes about because of the effect of all the massive excitations of the string is to kill off those divergences. We have made progress.

But now we need to fix up the difficulties found in bosonic string theory. The remedy is supersymmetry and the result of incorporating supersymmetry this way yields the superstring. The basic idea of supersymmetry is to introduce a symmetry that exchanges bosons with fermions. For every boson, there will be a fermion. The supersymmetry we are interested in is going to be a new kind of structure on the string worldsheet. The physical degrees of freedom of the string are the spacetime coordinates transverse to the worldsheet. These degrees of freedom behave like bosons living on the string worldsheet. Now, we introduce partners to some or all of these degrees of freedom. These are fermionic variables that pair up with bosonic variables. Remarkably, it turns out there are five different ways of doing this in a way consistent with quantum mechanics.

In each case, the first thing we discover is that instead of being consistent in only 26 spacetime dimensions, the superstring is only consistent in 10 spacetime dimensions. The second thing we discover is that there are no tachyons in the spectrum of the superstring. Lastly, although there are massive string states as before, there are still massless string states but now as well as representing bosons, there are fermions too. For historical reasons, these massless string states all contain gravity as well as a collection of other fields. The five different string theories are usually referred to as Type IIA string theory, type IIB string theory, $SO(32)$ string theory and two heterotic string theories, $SO(32)$ and $E_8 \otimes E_8$. The contain different spectra of massless particles. In each case, the fields are the same as those found in various ten-dimensional supergravity theories. Supergravity is an extension of general relativity that is supersymmetric. The two type II string theories are related to $N = 2$ pure supergravity and the other theories are related to $N = 1$ supergravity theories coupled to supersymmetric Yang-Mills theories with either the $SO(32)$ or $E_8 \otimes E_8$ gauge groups. Any of these theories has the potential to turn into realistic models of physics.

There are connections which enable one to translate any problem in any one of the string theories into any of the other string theories. These are known as duality symmetries. The complete picture of the five string theories and the eleven dimensional theory in which the connections between all of them are realized is known as M theory. Whilst quite a lot is known about M theory, it is still a work in progress.

M theory is our candidate theory of everything. It does however seem a long way from something we recognize. What would we want from a theory that describes all the phenomena we observe? Firstly, it must be a theory with four space time dimensions not ten or eleven. Secondly at low energies, it must contain the Standard Model as well as gravitation. One thing we can be pretty certain of, and that is at around the GUT scale of roughly 10^{16} GeV, our theory must be described by a four-dimensional theory that has supersymmetry. So assuming we start in eleven spacetime dimensions, we need to be able to get rid of seven of them. Suppose you look at a tree trunk from close up. What you see is the trunk with all kinds of wrinkles and defects in its bark. As you go further away, the trunk appears to get smaller and smaller until when you are very far away, you can barely see a structureless vertical line. At even greater distances, you see nothing. The lesson is that if you are very far away from something you font really notice it.

It is by this kind of method that we get rid of seven space time dimensions. We assume that they are so small, that on the scale on which we exist, they just don't get noticed. We imagine that these directions of space are on the unification scale or smaller. In a sense, this is derivable in both string and M-theory. The wrinkles on space in these directions do have a physical consequence however. The precise shape and size of these extra dimensions is reflected in low energy physics by the existence of light scalars or some new fields called moduli fields and axions that describe relations among the small dimensions. These wrapped up spatial directions need not be smooth. They can have spikes and these spikes turn out to be interesting. If the space is spiky it results in Yang-Mills fields appearing in our part of space together with their fermionic superpartners. the nature of the spikiness determines what kind of gauge group these Yang-Mills fields have. It could be that the gauge group is $SU(5)$, the gauge group expected in Grand Unified Theories. In this case the fermionic partners can be multiplets of quarks and leptons. By adjusting the spikiness, one can instead generate the low-energy gauge group of the Standard Model, namely $SU(3) \otimes SU(2) \otimes U(1)$. Thee fields will then be exactly those found in the Standard Model namely, colored quarks associated with the $SU(3)$, the $W^{\pm}$ and Z associated with the $SU(2)$ and a photon. The fermionic superpartners are then the quarks, the electron and the neutrinos. In this case, an electron is the result of spikiness in the hidden extra dimensions. One might wonder if there is somekind of limit in complexity of the gauge group. The most complicated form of spikiness results in the gauge group E_8. In a way this observation contains a miracle in that from a theory involving a single E_8 type of spike, one finds a theory with three and only three generations of quarks and leptons, exactly as appears to be the case.

There is however another picture that seems just as plausible, In the heterotic string, one finds exactly the same Yang-Mills fields and superpartners as a result of just oscillations of the string. So for heterotic string, the electron come about as oscillations of the string and not as a result of any particular kind of properties of space.

How is it possible that there are two different descriptions of the same thing in string theory? The answer is that there is a curious collection of symmetries which relate one picture in string or M theory to another. There is a translation between phenomena in the heterotic string and phenomena in M theory. It just so happens that what is described by spikiness of hidden parts of space in the M theory picture is just string excitation in the the heterotic string. It does not make sense to ask which one is right. They are both legitimate explanations of what is happening in nature. In this sense, the phenomenon is exactly like wave-particle duality in quantum mechanics. whether you decide to think of an electron or a wave is up to you. Its behavior is accurately described by the quantum mechanical theory however you care to describe it. M theory has its now dualities which allow one to describe the electron as one thing in the eleven dimensional picture and a different thing in the heterotic picture. However, if you ask questions about physics you can measure, you will get the same answers in either description.

What we have described here is a how one should think of an electron, or indeed any other elementary particle in string theory. In quantum field theory, these are all quantized field excitations. However in string theory, everything is much more geometrical. There appear to many equivalent ways of looking at the same particle indicating a huge hidden symmetry in string theory that has yet to be completely explored. This is a huge challenge as the mathematics of string theory is hard. Nonetheless, humanity has in the past risen to such challenges and reached our current understanding of the nature of the Universe. Even 50 years ago, it would have been impossible to predict how much progress has been made. We can only hope that our civilization can continue to encourage and support progress in this field whose aim to understand what we are ultimately made of.

I would like to thank the Michigan Center for Theoretical Physics for its hospitality whilst this written and Gordy Kane for many stimulating conversations.

2

The What and Why of Moduli

Joseph Conlon

*Rudolf Peierls Centre for Theoretical Physics,
1 Keble Road, Oxford, OX1 3NP, UK*
j.conlon1@physics.ox.ac.uk

I discuss moduli: what they are, and what they are not, where they come from and where they are useful, where they are physically irrelevant and where they are crucial.

1. What are Moduli?

The existence of moduli is one of the most generic, and therefore one of the most interesting, predictions of string compactifications.

The effective field theories of string compactifications can differ in many different ways. Different vacua of string theory can involve very different gauge groups and representation content for matter particles. For example, the exceptional groups appear in compactifications of the heterotic string or F-theory, but not in weakly coupled compactifications of branes in type IIA or type IIB theory. This landscape of possibilities creates an apparent problem for making connections to observational physics.

However one feature common to many different string compactifications is the generic existence of moduli. Furthermore, the types of moduli and the interactions present often take a very similar form between different string theories, and are largely independent of the detailed form of the gauge group. This makes the study of moduli one of the most important aspects of string phenomenology, as this sector is the most model-independent aspect of string compactifications.

What are moduli? The quickest definition is as light scalar particles with no gauge interactions and only gravitational-strength interactions. Let us give some examples of moduli. Possibly the most universal modulus is the dilaton modulus, which also exhibits many features that are often present in the effective action.

The dilaton modulus (conventionally denoted as S) is the modulus whose vev determines the value of the string coupling. As the string coupling is present in all string theories, the dilaton modulus is a universal feature of string compactifications. The real part of the vev gives the string coupling, possibly with a volume factor, while the imaginary part is axionic. In heterotic string theory,

$$S = e^{-2\phi}\mathcal{V} + ia_0. \tag{1}$$

Here $e^{-\phi} = g_s^{-1}$ and $\mathcal{V}$ is the compact six-dimensional volume written in units of the string length. In type IIB string theory

$$S = e^{-\phi} + ia_0, \tag{2}$$

where again $e^{-\phi} = g_s^{-1}$ sets the string coupling. In both cases a_0 is an axionic field.

The presence of the axionic part a_0 implies the presence of a axionic shift symmetry for the vev of S, $S \to S + i\epsilon$. This shift symmetry is a good symmetry of string perturbation theory in both the α' and g_s expansion. It is violated by effects non-perturbative in g_s (for example D-instantons in type IIB string theory). The fact that the symmetry is however a good symmetry of perturbative string theory implies one important feature: the perturbative action cannot depend on $\text{Im}(S)$, and is a function only of $S + \bar{S}$.

The latter is a surprisingly powerful constraint. As $S + \bar{S}$ is non-holomorphic, it cannot appear in the superpotential. The immediate consequence is then that the perturbative superpotential is independent of S: $W(S)$ only features non-perturbative contributions. Such contributions do indeed arise — for example from gaugino condensation. The shift symmetry $S \to S + i\epsilon$ also restricts the form of the Kähler potential to be $K = K(S + \bar{S})$ — again, up to corrections that are non-perturbative in e^{-S}.

Another example of a 'universal' modulus is the volume modulus, denoted by T. This mode corresponds to a homogeneous rescaling of the metric of the extra dimensions, $g_{i\bar{j}} \to R^2 g_{i\bar{j}}$. This modulus is called 'universal' as it is present whenever the extra dimensions are fundamentally geometric. This is not always true, but is certainly true in many cases. If so, the existence of the volume modulus is also independent of the detailed properties of the compactification manifold or the branes and fluxes that are used to define it. The volume modulus also has similar properties to the dilaton in terms of axionic symmetries and its imaginary component is axionic, thus generating a shift symmetry $T \rightarrow T + i\epsilon$ that remains good within perturbation theory.

Let us describe the microscopic origin of these shift symmetries. In heterotic string theory, the imaginary component of T arises from the worldsheet 2-form, $b = \int_{\Sigma_2} B_2$. This vanishes for all worldsheets that are topologically trivial, and so can only be non-zero for worldsheets that wrap cycles in the extra dimensions. Such worldsheets are known as worldsheet instantons. They are perturbative in g_s: they have the topology of the sphere. However they are non-perturbative in α', and generate terms that are suppressed by $e^{-2\pi T}$. It therefore follows that the action can only depend on b via effects that are non-perturbative in α'.

In type IIB compactifications with D3/D7 planes and O-planes, the Kähler moduli are the gauge kinetic functions for D7 branes wrapped on 4-cycles. In this case their real parts are

$$T_i = e^{-\phi} \int_{\Sigma_{4,i}} \sqrt{g} + i \int_{\Sigma_{4,i}} C_4. \tag{3}$$

The volume modulus corresponds to the linear combination in the direction of overall rescaling, $g_{i\bar{j}} \rightarrow \lambda^2 g_{i\bar{j}}$. In general there are many Kähler moduli — $h^{1,1}$, where $h^{1,1}$ is the appropriate Hodge number for the Calabi-Yau. $h^{1,1}$ is typically $\mathcal{O}(100)$, which is illustrative of the fact that many moduli can survive to the low energy effective theory. Each cycle has an associated axion, coming from reduction of an axionic form on that cycle. In IIB theory, this arises from reduction of C_4 on the 4-cycle, whereas in heterotic compactifications it arises from reduction of the NS-NS 2-form B_2 on an appropriate cycle.

There are also complex structure moduli. These parametrise the complex structure of the compactification manifold — colloquially, its shape. For a Calabi-Yau these are counted by $h^{2,1}$. As the name suggests, these are generally *complex*. Neither component has an axionic interpretation: both the real and imaginary components are on the same footing. Again, the multiplicity may be high - $h^{2,1}$ is again $\mathcal{O}(100)$ for a Calabi-Yau, and so very many moduli can survive into the low energy effective theory.

There are also other sources of moduli. These can arise for example from deformation of vector bundles on the Calabi-Yau: generally these bundles have continuous deformations that are consistent with the supersymmetry conditions, and these deformations manifest themselves as moduli in the 4-dimensional theory. Another source of moduli comes from the motion of branes: the position fields of D3 branes are uncharged under gauge interactions, and manifest themselves in the 4d theory as moduli.

While the above discussion has focused on the case of Calabi-Yau compactifications, it should be clear that it extends to any (approximately) supersymmetric compactification. Deformations of the extra-dimensional geometry — where 'geometry' is understood in the broadest possible sense — that preserve the supersymmetry of the comapctification will appear in the 4-dimensional theory as moduli. The same is also true for non-supersymmetric compactifications, although as discussed below it is harder there to preserve the sense of moduli as being approximately massless.

How should one then classify a particle as a modulus? The rough definition is as a scalar particle with gravitational couplings that is approximately massless. The last condition is the most vexing. In certain contexts — for example supersymmetric gauge theory — the word modulus is used to describe an exactly flat direction. In string compactifications, this would not be appropriate, as moduli must get non-zero masses as otherwise they would generally give rise to unobserved fifth forces. Furthermore, the dynamics of supersymmetry breaking, required for a realistic vacuum, will necessarily also generate a potential and thereby a mass for the moduli.

2. What are Moduli Not?

It is also important to say what moduli are *not*. Moduli are generally neither the only scalars present, nor the only gravitationally coupled particles, nor even the only gravitationally coupled scalars. What properties should a particle *not* have if it is to be called a modulus?

First of all, moduli are not fermions and moduli are not vectors. Moduli have to be scalar particles. The reason for this is simple and mundane. The interesting physics of moduli comes from the many possible vevs they can take. Scalars however are the only particle type that can take a vev without simultaneously breaking 4-dimensional Lorentz invariance. As Lorentz invariance is so well tested, this leaves scalars as the interesting case where we can consider potential large displacements in vevs without violating Lorentz invariance.

Another exclusion property follows from the requirement that moduli belong to the four dimensional effective field theory. Modes that are genuinely higher dimensional, such as Kaluza-Klein modes, or modes that are fundamentally stringy, such as excited string harmonics, should not be classified as moduli. This is despite the fact that some of these string or KK modes may transform as scalars under the four-dimensional Lorentz group.

The point is that these modes are never massless: there is no limit in which they are massless, they cannot be counted as 'light' and it makes no sense to include these modes within the four dimensional effective field theory without also including all other modes within the tower of string excited states.

For a good decoupling limit to exist, the mass spectrum is then required to look like

$$m_{moduli} \ll m_{KK} \leq M_s.$$

If we also have $m_{KK} \ll M_s$, then there is a further region in which the theory is described by an effective higher-dimensional supergravity theory.

How are moduli different from scalars found in conventional four dimensional theories? After all, one apparently fundamental scalar was recently found at CERN in the form of the Higgs. Supersymmetric extensions of the Standard Model also contain many scalar particles, one for each fermion of the Standard Model, and these particles are not regarded as moduli. What are the distinctive features of moduli that separate them from the scalars encountered in regular particle physics model building?

One difference is that moduli do not have a good concept of zero vev. Many 'regular' scalar fields often come with a well defined notion of zero vev, where the zero vev locus corresponds to a restored or enhanced symmetry, and a non-zero vev corresponds to a breaking of that symmetry. This is certainly true of the Higgs: at $\langle \phi_h \rangle = 0$, the $SU(2) \times U(1)$ symmetry is unbroken, whereas once $\langle \phi_h \rangle \neq 0$ this symmetry is broken down to $U(1)_Y$, and mass terms are generated in the Lagrangian for both the $W^{\pm}$ and Z bosons and the fermionic particles of the Standard Model. This point is also true for the scalars normally considered in field theory: there is a good notion of zero vev $\langle \phi \rangle = 0$, and thus a preferred locus in moduli space (this is manifestly true for any charged scalar, where $\langle \phi \rangle = 0$ is the only locus with unbroken gauge symmetry). In these cases one often also performs an expansion about zero vev, where non-renormalisable effects are suppressed by $\frac{v}{\Lambda_{UV}}$, where Λ_{UV} is the appropriate UV scale.

This is not true of moduli. There are certain moduli where the notion of 'zero vev' can indeed have a sensible meaning: one example would be a modulus that controls the blow up of a certain cycle away from a singularity. Here zero vev corresponds to the point where the cycle is collapsed at the singularity and has zero size, while a non-zero vev corresponds to the resolution of the cycle to finite volume, and the greater the vev the greater the size of the blown-up cycle. However, for many cases there is no such 'zero vev' point in moduli space. Examples are the string coupling or the overall volume modulus. These can take a continuum of values, and there is no preferred locus or expectation value that corresponds to zero vev. The metric on moduli space for these fields are

$$\frac{M_P^2}{(S+\bar{S})^2}\partial_\mu S \partial^\mu \bar{S} \qquad \text{and} \qquad \frac{3M_P^2}{(T+\bar{T})^2}\partial_\mu T \partial^\mu \bar{T}.$$

Denoting $\tau = \text{Re}(T)$ and $g_s^{-1} = \text{Re}(S)$, the canonically normalised fields are then

$$\Phi_s = \frac{M_P}{2}\ln g_s^{-1} \qquad \text{and} \quad \Phi_t = \frac{M_P\sqrt{3}}{2}\ln \tau_b.$$

Note that no single value is preferred - both the $g_s \to 0$ and the $g_s \to \infty$ limits are at an infinite distance in field space from any finite value of g_s. So although we might naively think that $g_s = 0$ should correspond to 'zero vev', we see that this is not so. We also see that what appear to be relatively small changes in g_s (e.g. from $g_s = 0.1$ to $g_s = 0.2$) are actually at Planckian separations in field space. Moduli therefore inhabit a space where there is no *a priori* special value, and where in terms of canonically normalised fields they can range over several Planckian distances in field space.

Of course, one can always choose the eventual stabilised minimum of the modulus potential and define this value as 'zero vev'. If we work close to the stabilised vacuum, we can then expand fluctuations of the modulus field about this value. However the point is that this value is one we have put in the end, once we know the potential. In earlier epochs of the universe — for example during the inflationary epoch — this value would not have been special in any way. In particular, it does not appear special from the start, in the way that for the Higgs potential the point of zero vev is always a special point in moduli space.

3. How Heavy are Moduli?

In exactly supersymmetric $\mathcal{N} = 2$ compactifications to Minkowski space, moduli are exactly massless. In this case the flatness of moduli space is protected to all orders by the extended supersymmetry. In $\mathcal{N} = 1$ compactifications, the supersymmetry is no longer sufficient to protect flat directions to all orders in perturbation theory. However it is still useful to think of moduli as approximately massless degrees of freedom, whose masses arise from small perturbative or non-perturbative effects.

Most work on string compactifications assumes that the supersymmetry breaking scale is low: there is a clear hierarchy $m_{3/2} \ll M_P$. This is motivated for both practical reasons (it is easier to control the computations if $m_{3/2} \ll M_P$) and phenomenological ones (supersymmetry may be relevant for addressing the weak hierarchy problem). In this context, the supersymmetry breaking scale is approximately zero, at leading order in an expansion in $\left(\frac{m_{3/2}}{M_P} \right)$. The dynamics of supersymmetry breaking typically generates moduli masses at this same order in smallness,

$$m_\phi \sim m_{3/2}. \tag{4}$$

We note that there are also often logarithmic enhancements to this expression,

$$m_\phi \sim m_{3/2} \ln \left(\frac{M_P}{m_{3/2}} \right). \tag{5}$$

These originate from non-perturbative dynamics and their competition with tree-level terms. These may be non-trivial as $\ln \left(10^{18} \text{GeV}/10^3 \text{GeV} \right) \sim 30$, but do not change the parametric picture.

If a low supersymmetry breaking scale — a small value for $m_{3/2}$ — arises in a controlled fashion, then this gives a natural separation of scales between the moduli masses and the UV scales. In the cases where $m_{3/2}$ is not far separated from the Planck scale, then the notion of moduli becomes harder to define as there is no longer a good separation between the scale of moduli masses and the scale of KK modes (for example).

It is worth commenting on scalings different from (4) or (5). One interesting example is the scaling of the volume modulus in the LARGE Volume Scenario,

$$m_\phi \sim m_{3/2} \left(\frac{m_{3/2}}{M_P} \right)^{1/2}. \tag{6}$$

In this case the contribution of $\mathcal{O}(m_{3/2})$ to the mass of the volume modulus is absent and $m_\phi \ll m_{3/2}$. This can be traced to the underlying no scale structure. The basic feature of no scale is the presence of a flat potential together with non-zero supersymmetry breaking, and thus for no scale models the mass of a modulus can be lighter than it 'ought' to be. Even lighter moduli can sometimes be found in fibered versions of the LARGE volume scenario, with a mass scaling

$$m_\psi \sim m_{3/2} \left(\frac{m_{3/2}}{M_P} \right)^{2/3}. \tag{7}$$

This again traces to the underlying no-scale structure present in these models.

A contrasting example is the case where moduli are much heavier than the gravitino mass. This however can generally only be accomplished with fine-tuning, so the gravitino mass is smaller than it naturally ought to be. For example, in many KKLT scenarios, the complex structure moduli are stabilised at a mass scale much greater than that of the gravitino mass,

$$m_U \gg m_{3/2}. \tag{8}$$

In this context, this arises as the gravitino mass is much smaller than it 'ought' to be. The gravitino mass is tuned small through a random cancellation of fluxes. The natural scale of the complex struture moduli is $m_U \sim m_s R^{-3}$, which is also the natural scale of the gravitino mass, as

$$m_{3/2} \sim e^{K/2} W \sim \frac{M_P}{R^6} \int G_3 \wedge \Omega, \tag{9}$$

and the last expression is semi-topological being independent of the volume of the space.

However a cancellation occurs in W ensuring a small value for $m_{3/2}$: as eq. (90 shows, the gravitino mass arises from the sum of many different terms. If there happens to be an accidental cancellation (the difference of two large numbers being small), the gravitino mass is much smaller than would be expected. Put another way: the magnitude of the potential is anomalous small compared to the curvature of the potential.

4. Why are Moduli Important?

Why are moduli so crucial for making contact with low-energy physics? The simplest reason for this is that, within string theory, the value of almost every observable quantity originates as the vev of a modulus. Let us give some examples:

(1) The strong (and weak, and electromagnetic) gauge couplings. The high-scale values of these couplings are set in string theory by the vev of the modulus that controls the gauge kinetic function: for gauge groups realised by wrapping branes on cycles, this would be the modulus that controls the size of this cycle. Moduli vevs are important not just for the high-scale gauge coupling. The low-scale gauge coupling at a scale

Λ_{low} is given by

$$\frac{1}{g^2}(\Lambda_{low}) = \frac{1}{g^2}(\Lambda_{UV}) + \frac{\beta}{16\pi^2} \ln\left(\frac{\Lambda_{UV}}{\Lambda_{low}}\right). \tag{10}$$

However in string theory Λ_{UV} is also a function of moduli. The UV scale will typically be the string scale or the compactification scale, and the numerical value of the string scale or compactification scale in relation to the 4-dimensional Planck mass $M_P = 2.4 \times 10^{18}$GeV is set by moduli vevs.

(2) The Yukawa couplings of the Standard Model. These parameters, which are dimensionless in the Standard Model, arise in string theory as the vevs of moduli. As an example, the superpotential would involve a term

$$Y_{ijk}(\Phi)H_u Q_L \bar{U}_D, \tag{11}$$

where Φ are moduli fields. The vev of Φ would therefore affect the numerical value of the Yukawa couplings. There may be other, additional structure in any model (for example a Froggatt-Nielsen structure). However the above effect will always be present.

(3) The scale of supersymmetry breaking and the structure of soft terms (in any string compactification, these are in principle observable quantities). This is set, in the first approximation, by $m_{3/2}$, which is in turn set by the moduli vevs. The structure of soft terms is determined by the couplings between the source of supersymmetry breaking and the Standard Model; these couplings are also determined by moduli vevs.

One of the most noticeable features about the world is the presence of hierarchies. There are many examples. They include the ratio of the weak scale to the Planck scale $M_W/M_P \sim 10^{-16}$, the ratio of the electron mass to the weak scale, $m_e/m_H \sim 10^{-6}$, the smallness of the cosmological constant compared to almost any other scale, $\Lambda_{cc}^4 \sim 10^{-120}M_P^4$. It is expected that these hierarchies have some physical origin: the exponential differences in scales that arise in several places is not simply a fluke, but rather a reflection of deeper underlying physics. As moduli vevs control many of these scales, we see that moduli physics is expected to play an important role in understanding hierarchies: we expect the vevs to be such as to generate these hierarchies.

Cosmology provides another arena for the importance of moduli when connecting string compactifications to observations. It is highly plausible that the early universe went through a period of inflationary expansion, during which the size of the universe grew exponentially. In this scenario, the observed density perturbations originated as the quantum fluctuations of the scalar inflaton field. During the inflationary epoch, the universe was in an approximate de Sitter solution, violated by a slowly rolling scalar field.

Both the existence and stability of an approximately de Sitter solution require control of all scalar fields in the potential. Inflation requires the absence of unstable directions and an approximately flat direction along which the inflaton can roll. Ensuring the absence of unstable directions requires control of the moduli potential: even a single unstabilised direction is sufficient to cause fast roll in that direction.

In this respect, it does not matter that moduli have only Planck-suppressed couplings to the visible sector. As is well known, the flatness of the inflationary potential is vulnerable against Planckian corrections to the Kähler potential. This is manifest for example in the supergravity eta problem. The structure of the supergravity scalar potential,

$$V = e^{K/M_P^2} \left(K^{i\bar{j}} D_i W D_{\bar{j}} W - \frac{3}{M_P^2} |W|^2 \right) \tag{12}$$

implies that a Planck suppressed correction to the kähler potential,

$$K \rightarrow K \left(1 + \frac{\Phi\bar{\Phi}}{M_P^2} \right) \tag{13}$$

leads, on expanding the exponential term e^{K/M_P^2}, to an $\mathcal{O}(1)$ correction to the η parameter,

$$\eta = M_P^2 \frac{V''}{V}. \tag{14}$$

Control of any inflationary model therefore requires control of any Planck-suppressed contributions to that model.

It follows that while in certain areas of physics (for example in computing strong interaction cross-sections) the existence of moduli can just be neglected without affecting the physics in any way, when constructing inflationary models this is no longer true. If the existence of moduli is ignored when building an inflationary model, restoring the moduli will just destroy the model: the moduli introduce unstable fast-roll directions and the original model does not survive in any sense.

This is particularly clear when considering the universal moduli such as the dilaton or the volume modulus. Whatever the original source of the inflationary potential, its scale should depend on the fundamental scale of the theory, $M_s \equiv \frac{1}{\sqrt{\alpha'}}$. This is physically clear: any potential ultimately has a microscopic origin, and the microscopic magnitude always depends on the fundamental scale of the theory. However, the fundamental scale is related to $M_P \equiv 2.4 \times 10^{18}$GeV through the moduli,

$$M_s = g_s \frac{M_P}{\sqrt{\mathcal{V}}}, \tag{15}$$

and therefore any inflationary potential automatically induces a potential for the dilaton and volume directions. As the string coupling and volume are exponential in these fields, the potential would by itself be an exponential runaway in these directions. It is thus essential to include the potential for these modes explicitly. Without doing this, any inflationary model is automatically destabilised on including the volume and dilaton moduli.

The above reflects the fact that inflation is UV-sensitive: the details and properties of inflationary models are sensitive to Planck-suppressed physics. Moduli are also a virtue in this respect: there are many string inflation models which include moduli as the inflaton field.

There is another place in cosmology where moduli play a crucial role. The fact that moduli have Planck-suppressed interactions means that when considering the *production* of an ensemble of particles, the existence of moduli can essentially be ignored: they couple so much more weakly than any other particle and so moduli are to all practical purposes not produced. This is why it is not worth considering moduli when computing QCD processes at the LHC: particles that can be produced by renormalisable interactions are produced by renormalisable interactions.

The converse however is true when considering the *decays* of an initial ensemble of particles. Here the late-time physics is dominated by the particles which have the *weakest* interactions and thus the longest lifetimes. This is particularly true in the expanding early universe, where radiation is diluted more rapidly than matter (a^{-4} versus a^{-3}). Now it is the particles that have renormalisable decay channels that are irrelevant: they decay rapidly and their decay products are diluted. It is the long-lived particles with only gravitational interactions that survive.

This is the situation that prevails after inflation. A large inflationary energy density will cause the production through the misalignment mechanism of many scalars, including moduli: the inflationary energy density gives a contribution to the potential, and so the fields are initially misaligned from their final zero-temperature minimum. The subsequent dynamics sees the relaxation of these fields to zero vev and their subsequent decay, giving the epoch of reheating. However this should be dominated by the particles with the *longest* lifetimes — and these are the moduli.

Moduli therefore play a crucial role in understanding the physics of reheating — viewed from a different angle, this goes by the name of the cosmological moduli problem. Moduli live a long time, and if they are too light $m_\Phi \lesssim 30\text{TeV}$ they decay after the time by which nucleosynthesis must have occurred.

5. Summary

Moduli are the light, weakly coupled scalar degrees of freedom that arise almost ubiquitously in string compactifications. Their existence is one of the most generic predictions of string compactifications, as they arise in all limits of the theory. They also play a crucial role in connecting the theory to observations: in string theory the values of low-energy parameters are set by the vevs of moduli, and the dynamics of moduli is crucial for generating the hierarchies and scales that we see in nature.

3

Perspective on the Weakly Coupled Heterotic String

Mary K. Gaillard

*Department of Physics, University of California and
Theoretical Physics Group, 50A-5101,
Lawrence Berkeley National Laboratory,
Berkeley, CA 94720, USA*

Since the first "string revolution" of 1984, the weakly coupled $E_8 \otimes E_8$ heterotic string theory has been a promising candidate for the underlying theory of the Standard Model. The particle spectrum and the issue of dilaton stabilization are reviewed. Specific models for hidden sector condensation and supersymmetry breaking are described and their phenomenological and cosmological implications are discussed. The importance of T-duality is emphasized. Theoretical challenges to finding a satisfactory vacuum, as well as constraints from LHC data are addressed.

1. Prologue

String theory was first introduced into particle physics as a candidate theory of strong interactions, in its bosonic version, that required 26 spacetime dimensions. The supersymmetric version requires only 10 space-time dimensions, but still many more spatial dimensions than the three that we observe. Moreover, string theory requires a massless spin-two particle, and no such animal was to be found among the light hadrons. However this made string theory a prime candidate for a quantum theory of gravity which requires a massless graviton with spin two. Supersymmetry is essential for the consistency of a superstring theory of gravity, but not sufficient. Any superstring theory that might describe the universe we observe has to be a ten dimensional supersymmetric theory of gravity coupled to gauge fields. A stumbling block to this approach was the presence of "hexagon anomalies", the ten dimensional analogue of the triangle anomalies which, if present in gauge theories, would break gauge invariance and render them unrenormalizable. The hexagon anomalies break Lorentz invariance, as well as gauge invariance. The "first string revolution" occurred in 1984 when Mike Green and John Schwarz showed [1] that there were two gauge groups, with the same dimension (496) and the same rank (16), that were anomaly-free, namely $SO(32)$, the group of rotations in 32 Euclidean dimensions, and the direct product of the maximal exceptional group E_8 with itself, or $E_8 \otimes E_8$. An $O(32)$ string theory had already been constructed, and, not long after the Green-Schwarz result, the $E_8 \otimes E_8$ "heterotic" string was found by Gross, Harvey, Martinec and Rohm [2]. This theory appeared promising for phenomenology, especially when the six extra spacial dimensions were compactified on a Calabi-Yau manifold, first achieved by Candelas, Horowitz, Strominger and Witten [3].

Eventually it was established that there were five consistent theories, a situation that disturbed some theorists until the advent of the "second string revolution", when it was discovered that these five theories were related to one another by dualities: S-duality, the inversion of the fine structure constant α at the string scale, or T-duality: the inversion of a radius of compactification when one of the nine spatial dimensions dimensions is curled up into circle of radius R. It was further discovered, thanks to these dualities, that all five string theories can be obtained from an eleven dimensional theory of membranes known as M-theory, as illustrated in Figure 1. If the extra eleventh dimension of M-theory is curled up into a circle, one gets the type IIA string theory in ten dimensions. Curling up one of the radii of the IIA theory into a circle gives a nine dimensional theory that can also be obtained by a similar operation on the type IIB ten dimensional string theory, or by compactifying two dimensions of M-theory on a torus (combining the two circles). On the other hand, restricting the eleventh dimension of M-theory to a finite distance on a line gives a theory with two ten dimensional worlds, each with an E_8 gauge theory. If the length of this line is considerably larger than the radii of the hidden six dimensional sectors of the two four dimensional worlds, one of which we live in, we get the Hořava-Witten (HW) scenario [6]. If instead the length of the eleventh dimension is shrunk to zero, we get back the string theory in ten dimensions found by Gross *et al.* Compactifying one dimension of this theory on a circle gives a nine dimensional theory that can also be reached by compactifying one of the two $O(32)$ theories on a circle, or directly from M-theory by compactifying two dimensions on a cylinder (a line and a circle). The two $O(32)$ theories (type I and heterotic) are related to one other by S-duality, and the two nine dimensional theories are related to one another T-duality. Another limit of M-theory gives supergravity in eleven dimensions; when

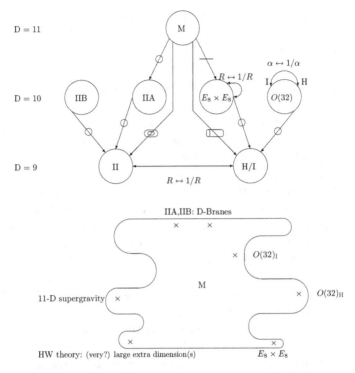

Fig. 1.   M-theory according to John Schwarz [4] (top) and Mike Green. [5].

seven of these are reduced to a point, one gets $N = 8$ supergravity in four dimensions.

Many theorists were heartened by the discovery that the string theories could be unified into a single theory. But there was still the problem of finding the right vacuum: how did we end up in the world we live in? Not only are there five different string theories, as well as many other points in the M-theory landscape of Bousso and Polchinski [7], but each string theory has many different possible vacua, depending on the size and shape of the hidden space. In particular, there has been a lot of activity in trying to count the number of type IIB vacua; the number is very large—so large that it can in practice be thought of as infinite. At some point the theoretical community seemed to divide into two camps. One camp takes a probabilistic and anthropic approach: we should be living at a point in the landscape with the features that are the most probable within the subset of vacua that are capable of supporting some kind of observers.

The other group tries to find a specific vacuum that reproduces what we observe, and postpones worrying about how we got here. This camp focuses largely (but not exclusively) on the $E_8 \otimes E_8$ heterotic string or its nearby neighbor, the HW theory. These efforts will be the main focus of this perspective.

2. Compactifications of the $E_8 \otimes E_8$ heterotic string

The zero-slope, or infinite string tension, limit of the theory is ten dimensional supergravity coupled to a supersymmetric Yang-Mills theory with an $E_8 \otimes E_8$ gauge group. To make contact with the real world, six of these ten dimensions must be compact and here they are generally assumed to be of order of the (reduced) Planck length $(m_P)^{-1} \sim 10^{-32}$cm. The spectrum of particles that appears in the effective low energy theory below the scale of compactification, as well as the residual symmetry of that theory, depends on the way in which these six extra dimensions curl up.

For example, if the topology of the extra dimensions were three two-tori, each of which has a flat geometry, the 8-component spinor operator of $N = 1$ supergravity in ten dimensions would appear as the four two-component operators of $N = 4$ supergravity in four dimensions, giving a uniquely defined theory with matter in the adjoint of the gauge group, which does not resemble the world in which we live. On the other hand, compactification on a six-sphere would leave no supersymmetry in four dimensions, because there is no spin component that is invariant under the $SO(6)$ holonomy group (the group of transformations under parallel transport) of that geometry. A world with no low energy supersymmetry may be the one we live in, but supersymmetry not too far above the scale of electroweak symmetry breaking would go a long way towards addressing the notorious gauge hierarchy problem: the mass gap of about sixteen orders of magnitude between the electroweak scale–a quarter of a TeV–and the Planck scale of 2×10^{15}TeV.

The attractive property of a Calabi-Yau (CY) manifold is that it leaves only one two-component spinor invariant under the holonomy group, which is an $SU(3)$ subgroup of the maximal $SU(4) \cong SO(6)$ holonomy group of a six dimensional compact space. This breaks $N = 4$ supersymmetry to $N = 1$ in four dimensions. This is the only phenomenologically viable supersymmetric theory at low energies, because it is the only one that admits complex representations of the gauge group that are needed to describe quarks and leptons.

Curvature in the compact 6-space implies that some components of the graviton acquire vacuum values (*vev*'s); these are related by supersymmetry to fermions. Nonvanishing fermion vacuum values would break four dimensional Lorentz invariance; to prevent this relations must be imposed among the vacuum values of the bose fields of ten dimensional super-gravity that include, besides the graviton g_{MN} and the Yang-Mills fields A_M, a scalar, the dilaton ϕ, and an antisymmetric tensor b_{MN}, where $M, N = 0, \ldots, 9$. Candelas *et al.* chose the simplest solution, with $\langle \phi \rangle = \langle b \rangle = 0$, and nonvanishing *vev*'s for some of the Yang-Mill fields with spin components along compact directions. For this solution, the affine connection of general coordinate transformations on the compact space, described by three complex dimensions, is identified with the gauge con-nection of an $SU(3)$ subgroup of one of the E_8's: $E_8 \supset E_6 \otimes SU(3)$, resulting in $E_6 \otimes E_8$ as the gauge group in four dimensions.

Since the early 1980's, E_6 has been considered the largest group that is a phenomenologically viable candidate for a Grand Unified Theory (GUT) of the Standard Model (SM). Hence E_6 is identified as the gauge group of the "observable sector", and the additional E_8 is attributed to a "hid-den sector", that interacts with the former only with gravitational strength couplings. Orbifolds, which are flat spaces except for points of infinite curvature, are more easily studied than CY manifolds, and orbifold com-pactifications that closely mimic CY compactification, and that yield real-istic spectra with just three generations of quarks and leptons, have been found [8–10]. In this case the surviving gauge group is

$$E_6 \otimes \mathcal{G}_o \otimes E_8, \qquad \mathcal{G}_o \in SU(3).$$

The low energy effective field theory is determined by the massless spec-trum, *i.e.*, the spectrum of states with masses very small compared with the string tension and compactification scale. Massless bosons have zero triality under an $SU(3)$ which is the diagonal of the $SU(3)$ holonomy group and the (broken) $SU(3)$ subgroup of one E_8. The ten-vectors A_M appear in four dimensions as four-vectors A_μ, $\mu = M = 0, 1, \ldots, 3$, and as scalars A_m, $m = M - 3 = 1, \ldots, 6$. Under the decomposition $E_8 \supset E_6 \otimes SU(3)$, the E_8 adjoint contains the adjoints of E_6 and $SU(3)$, and the represen-tation $(\mathbf{27}, \mathbf{3}) + (\overline{\mathbf{27}}, \overline{\mathbf{3}})$. Thus the massless spectrum includes gauge fields in the adjoint representation of $E_6 \otimes \mathcal{G}_o \otimes E_8$ with zero triality under both $SU(3)$'s, and scalar fields in $\mathbf{27} + \overline{\mathbf{27}}$ of E_6, with triality ± 1 under both $SU(3)$'s, together with their fermionic superpartners. The number of $\mathbf{27}$ and $\overline{\mathbf{27}}$ chiral supermultiplets that are massless depends on the topology of the compact manifold. The important point for phenomenology is the decomposition under $E_6 \to SO(10) \to SU(5)$:

$$(\mathbf{27})_{E_6} = (\mathbf{16} + \mathbf{10} + \mathbf{1})_{SO(10)} = (\{\overline{\mathbf{5}} + \mathbf{10} + \mathbf{1}\} + \{\mathbf{5} + \overline{\mathbf{5}}\} + \mathbf{1})_{SU(5)}. \quad (1)$$

A $\overline{5} + 10 + 1$ contains one generation of quarks and leptons of the SM, a right-handed neutrino and their scalar superpartners; a $5 + \overline{5}$ contains the two Higgs doublets needed in the supersymmetric extension of the SM and their fermion superpartners, as well as color-triplet supermultiplets. While all the states of the SM and its minimal supersymmetric extension are present, there are no scalar particles in the adjoint representation of the gauge group.[a] In conventional models for grand unification, these (or other large representations) are needed to break the GUT group to the SM. (A counter example is flipped $SU(5)$ [11], in which the GUT symmetry can be broken by a 10 of $SU(5)$, and which can also be obtained [12] from the heterotic string.) In string theory, gauge symmetry breaking can also be achieved by the Hosotani or "Wilson line", mechanism in which gauge flux is trapped around "holes" or "tubes" in the compact manifold, in a manner reminiscent of the Aharonov-Bohm effect. The vacuum value of the trapped flux $< \int d\ell^m A_m >$ has the same effect as an adjoint Higgs, without the difficulties of constructing a potential for large Higgs representations that actually reproduces the observed vacuum. When this effect is included, the gauge group in four dimensions is

$$\mathcal{G}_{obs} \otimes \mathcal{G}_{hid}, \quad \mathcal{G}_{obs} = \mathcal{G}_{SM} \otimes \mathcal{G}' \otimes \mathcal{G}_o, \quad \mathcal{G}_{SM} \otimes \mathcal{G}' \in E_6, \quad \mathcal{G}_o \in SU(3),$$
$$\mathcal{G}_{hid} \in E_8, \quad \mathcal{G}_{SM} = SU(3)_c \otimes SU(2)_L \otimes U(1)_w. \tag{2}$$

There are many other four dimensional vacua of the heterotic string, in addition to those described above. The attractiveness of the above picture is that the requirement of $N = 1$ SUSY naturally results in a phenomenologically viable gauge group and particle spectrum, and the gauge symmetry can be broken to a product group embedding the SM without introducing large Higgs representations.

3. Gauge coupling unification

In the weakly coupled heterotic string, the scale of unification—namely the string scale:

$$m_s = g_s m_{Pl}, \qquad m_{Pl} = (8\pi G_N)^{-\frac{1}{2}} \approx 2 \times 10^{18} \text{GeV}, \tag{3}$$

is determined [13–15] by the value of the gauge coupling constants g_a at that scale:

$$g_a(m_s) = g_s = \left\langle (\text{Res})^{-\frac{1}{2}} \right\rangle, \tag{4}$$

---

[a]We do not consider higher affine level $k = n > 1$ for non-Abelian gauge groups.

where the dilaton Res is the real part of the scalar component of the dilaton chiral supermultiplet S:

$$\langle s \rangle = \langle S| \rangle = g_s^{-2} - i\theta/8\pi^2. \tag{5}$$

More precisely, in the $\overline{MS}$ scheme one gets [16]

$$g_a(\mu) = g_s, \qquad \mu = g_s m_{Pl}/\sqrt{2e}, \tag{6}$$

which, for $g_s^2 = \frac{1}{2}$ is almost two orders of magnitude larger than the value $\mu \approx 10^{16}$ GeV found by extrapolating the measured couplings in the Minimal Supersymmetric Standard Model (MSSM) [17]. It was argued that this result favored the Hořava-Witten model, where the presence of a relatively large eleventh dimension, $r_{11} \sim 10^{-16}$ GeV^{-1} lowers the unification scale to the inverse radius where gravitational interactions become important in that scenario.

On the other hand there are other effects that can modify the renormalization group evolution in the weakly coupled heterotic string. In orbifold compactifications, these include threshold corrections [18] that depend on the Kähler moduli T^I whose *vev*'s determine the size and shape of the compact dimensions. However these are absent in the Z_3 and Z_7 orbifolds that have been found promising for phenomenology, and threshold effects are small in the weak coupling limit $\langle t^I \rangle \sim 1$, where t^I is the scalar component of T^I.

In the context of the weakly coupled heterotic string, the most natural resolution to this discrepancy in scales is the presence of vector-like pairs of chiral multiplets that do not form complete $SU(5)$ representations [16, 19], and that get masses at some intermediate scale between the Planck and electroweak scales. Such additional states are commonplace in orbifold compactifications of the heterotic string.

4. Supersymmetry breaking

If supersymmetry provides a correct description of nature, we live in a world of broken supersymmetry, because we have not observed the supersymmetric partners of the known particles. There are two ways to break supersymmetry while preserving its ability to provide a technical solution to the "gauge hierarchy" problem. The first is spontaneous symmetry breaking. The second is to introduce soft terms that explicitly break supersymmetry; these include scalar masses and mixing, gaugino masses, and trilinear (A-terms) and bilinar (B-terms) holomorphic functions of scalars. This route leads to a large number of arbitrary parameters that have to be introduced into the theory. The more attractive route, namely spontaneous symmetry breaking, has been shown to be impossible in supersymmetric extensions of the Standard Model (SM); it inevitability leads to the existence of at least one quark scalar superpartner that is lighter than all the quarks [20]. For this reason it was proposed that there might be a "hidden" world in which supersymmetry is broken, and that the information of this is transmitted to our world only by gravitational strength interactions.

The $E_8 \otimes E_8$ heterotic string provides this hidden world with the second E_8 (or whatever it becomes after Hosotani breaking of the E_8 symmetry). Supersysmmetry can be spontaneously broken in this hidden world by strong interaction effects of the gauge couplings in that world. This scenario was initially plagued by the "runaway dilaton problem": the dilaton coupling to the Yang-Mills sector[b]

$$\mathcal{L}_{YM} = \frac{1}{8} \int d^4\theta \frac{E}{R} S W_a^\alpha W_\alpha^a + h.c., \tag{7}$$

where W_α^a is the Yang-Mills superfield strength, induces [22] a superpotential for the dilaton of the form

$$W(S) \propto e^{-S/b_c}, \tag{8}$$

where b_c determines the coefficient of the one-loop beta-function of the condensing gauge group:

$$\mu \frac{\partial g_c}{\partial \mu} = -\frac{3}{2} b_c g_c^3. \tag{9}$$

The result (8) follows from the R-symmetry and scale invariance of this sector, which is most easily seen in flat-space supersymmetry, with the Yang-Mills action

$$S_{YM}^{susy} = \frac{1}{4} \int d^4x \int d^2\theta S W_a^\alpha W_\alpha^a + h.c. = \frac{1}{4} \int d^4x \frac{d^2}{d\theta^2} S W_a^\alpha W_\alpha^a + h.c. \tag{10}$$

At the classical level this action is invariant under the transformations:

$$x \to e^a x, \quad \theta \to e^{(a-i\alpha)/2}\theta, \quad dx \to e^a dx, \quad d\theta \to e^{-(a-i\alpha)/2}\theta$$

$$W_\alpha(x,\theta) \to e^{-3(a+i\alpha)/2} W_\alpha(x,\theta), \qquad S(x,\theta) \to S(x,\theta). \tag{11}$$

The classical invariance (11) is broken by the chiral and conformal anomalies; invariance is maintained at the quantum level by modifying the last

---

[b]We use the Kähler superspace formalism of ref. [21]. E is the superdeterminant of the supervielbein, and R is an auxiliary field of the supergravity multiplet.

transformation property in (11)

$$S(x, \theta) \rightarrow S(x, \theta) + 3b_c(a + i\alpha), \tag{12}$$

such that the variation of the Lagrangian due to anomalies is exactly canceled by the variation in S. Then the superpotential action

$$S_{\text{superpot}}^{\text{susy}} = \int d^4x \int d^2\theta W(S) + h.c. = \int d^4x \frac{d^2}{d\theta^2} W(S) + h.c. \tag{13}$$

is also invariant. Writing explicitly

$$W(S) = \Lambda^3 e^{-S/b_c}, \tag{14}$$

with Λ of the order of the string scale or the Planck scale, one recognizes the (cubed) condensation scale as determined by the renormalization group equations (RGE). The same result can be obtained [23] by introducing a chiral field for the gaugino condensate (interpreted as the lightest bound state in the strongly coupled sector) and constructing [24] its superpotential by anomaly matching. The above arguments go through in the same way for supergravity [25, 26]; in this case local supersymmetry assures [14, 26, 27] that the higher order corrections to the beta function are encoded in the result.

The "runaway dilaton" problem was that the resulting potential for the dilaton $s = S|$, $V(s) \propto e^{-2s/b_c}$, has its minimum for vanishing gauge coupling $\langle s \rangle \rightarrow \infty$, vanishing gaugino condensate and no supersymmetry breaking. One class of solutions to this problem used the so-called "racetrack models", which involved large gauge groups with very similar beta-functions. These models tend to give a negative vacuum energy, and are not easily accommodated in the heterotic string, where the condensing group is a subgroup of E_8.

In any case the above picture is incomplete; the superpotential for the dilaton does not respect T-duality, which is known [28] to be an invariance of the heterotic string to all orders in perturbation theory. T-duality transformations effect a Kähler transformation:

$$K \rightarrow K + F(T) + \bar{F}(\bar{T}), \qquad W \rightarrow e^{-F}W, \qquad W_\alpha \rightarrow e^{-\frac{i}{2}\text{Im}F}W_\alpha, \tag{15}$$

with an appropriate redefinition of θ, as in (11), such that the classical supergravity terms (7) and

$$\mathcal{L}_{\text{superpot}} = \frac{1}{2} \int d^4\theta \frac{E}{R} e^{K/2} W + h.c., \tag{16}$$

are invariant.[c] Here F is a holomorphic function of the "Kähler moduli" T; for example in orbifolds with three (folded up) tori, there are three "diagonal moduli" T^I. The real parts of their scalar components $t^I = T^I|$ determine the radii of compactification of each torus in string units (3):

$$R^I = \langle (\mathrm{Re} t^I) \rangle m_s^{-1}.$$ (17)

The group of T-duality transformations includes an $SL(2, \mathbf{Z})$ group under which the diagonal Kähler moduli transform as

$$T^I \to T'^I = \frac{aT^I - ib}{icT^I + d}, \qquad a, b, c, d \in \mathbf{Z}, \qquad ad - bc = 1.$$ (18)

In this case

$$F(T) = \sum_I F^I(T^I), \qquad F^I = icT^I + d.$$ (19)

In the effective field theory, invariance under T-duality is broken at the quantum level by the chiral and conformal anomalies. Specifically, under (18), the Yang Mills Lagrangian in (16) shifts by the amount

$$\delta \mathcal{L}_{\mathrm{YM}} = -\frac{1}{8} \int d^4\theta \frac{E}{R} \sum_a W_a^\alpha W_\alpha^a \frac{1}{8\pi^2} \sum_I \left[C_a - \sum_i C_a^i \left(1 - q_i^I \right) \right] + h.c.,$$ (20)

where C_a and C_a^i are adjoint and matter quadratic Casimirs, and q_i^I is the "modular weight" of the matter chiral superfield Φ^i; under (18)

$$\Phi^i \to e^{-\sum_I q_i^I F^I} \Phi.$$ (21)

This anomaly signals that the theory is incomplete: a four-dimensional version [29] of the Green-Schwarz (GS) counterterm is needed to cancel the field theory anomalies.

The classical Kähler potential for the dilaton chiral superfield S,

$$K(S, \bar{S}) = -\ln(S + \bar{S}),$$ (22)

is invariant under T-duality if S is invariant. However, if the Kähler potential (22) is modified to read

$$K(S, \bar{S}) = -\ln(S + \bar{S} + bV_G),$$ (23)

where the function $V_G(T, \bar{T}, |\Phi|^2)$ of chiral superfields transforms under (15) as

$$V_G \to V_G + F(T) + \bar{F}(\bar{T}),$$ (24)

---

[c] $\int d^4\theta E/R$ in (7) and (16) transforms like $\int d^2\theta$ in (10) and (13).

invariance of (23) requires that S transform as

$$S \to S - bF. \tag{25}$$

Then the anomaly is canceled *provided*

$$b = \frac{1}{8\pi^2} \left[C_a - \sum_i C_a^i \left(1 - q_i^I\right) \right], \quad \forall \ a, I. \tag{26}$$

This universal anomaly condition is indeed satisfied in Z_3 and Z_7 orbifolds with no moduli-dependent threshold corrections. In other orbifolds the anomaly is canceled by a combination of the "Green-Schwarz" term in (23) and threshold corrections of the form

$$\mathcal{L}_{\text{thresh}} = -\frac{1}{8} \int d^4\theta \frac{E}{R} \sum_a W_a^\alpha W_\alpha^a \sum_I \frac{b_a^I}{8\pi^2} \eta^2(T^I),$$

$$b_a^I = 8\pi^2 b - C_a + \sum_i C_a^i \left(1 - q_i^I\right), \tag{27}$$

where the Dedekind η-function transforms under (18) as

$$\eta(T^I) \to e^{\frac{1}{2}F^I} \eta(T^I). \tag{28}$$

The dependence of the GS function V_G on the diagonal moduli T^I is determined by matching [14] the field theory and string theory coupling of the Yang-Mills fields to $\mathrm{Im}t^I$, giving

$$V_G = g(T, \bar{T}) + O(|\Phi|^2), \qquad g = \sum_I g^I, \qquad g^I = -\ln(T^I + \bar{T}^{\bar{I}}), \tag{29}$$

which is the same as the dependence of the Kähler potential on these moduli. This Kähler potential leads to a "no-scale" structure [30] of the potential V if the superpotential W is independent of T:

$$V(T, S) = e^K \left(\sum_I D_I W K^{I\bar{I}} D_{\bar{I}} \bar{W} + D_S W K^{S\bar{S}} D_{\bar{S}} \bar{W} - 3|W| \right)^2$$

$$= D_S W K^{S\bar{S}} D_{\bar{S}} \bar{W} = V(S), \tag{30}$$

where

$$K^{n\bar{m}} e^{K/2} D_{\bar{m}} \bar{W} = K^{n\bar{m}} e^{K/2} \left(\bar{W}_{\bar{m}} + K_{\bar{m}} \bar{W} \right) = -F^n \tag{31}$$

is the auxiliary field for the chiral supermultiplet Φ^n.

The earliest studies invoked a second contribution to the superpotential for S in order stabilize the potential at nonvanishing coupling and to cancel its contribution to the vacuum energy. This led to T mediated supersymmetry breaking:

$$\langle F^S \rangle = 0, \qquad \langle F^T \rangle = \left\langle 2\mathrm{Re}t^I e^{K/2} \bar{W} \right\rangle \neq 0, \tag{32}$$

which is considered problematic for flavor changing neutral currents, because the Kähler moduli couplings to matter are not universal. Another problem is that the Kähler moduli *vev*'s are not determined, and the vacuum is degenerate. (This could be a discrete degeneracy [31] if the condensate superpotential is canceled by quantized flux [22], *i.e.* the *vev* of the three form of 10-d gravity: $\langle H_{lmn} \rangle \neq 0$, but this breaks supersymmetry at a very high scale.)

When it was realized that modular invariance had to be restored to the effective potential for S, this was first done "by hand" [26, 32, 33], that is, by simply inserting a factor $\eta^6(T)$, so that the condensate–induced superpotential $W_c(S) \propto \eta^6(T)e^{-S/b_c}$ transforms as in (15). The T-dependence in the superpotential destroys the no-scale property of the potential and generally gives a negative vacuum energy, but it does stabilize the moduli, for example at the oft-quoted value [32] $\langle t \rangle = 1.23$ for a single Kähler modulus under the assumptions (22) and (32).

This approach ignored the GS term, which could be incorporated by using (23) and (25), and taking

$$W_c(S) = \Lambda^3 e^{-S/b_c} \eta^{2b/b_c - 1}. \tag{33}$$

However the presence of the GS term in the Kähler potential for the dilaton by itself destroys the no-scale structure of the potential, and destabilizes it with a run-away direction towards strong coupling [34].

In all these cases the no-scale structure, with a positive semi-definite potential, can be restored by reinterpreting [35, 36] the wave function renormalization of the strongly coupled gauge fields as a correction to the Kähler potential of the composite field, rather than to its superpotential, in conformity with nonrenormalization theorems of supersymmetry. However in this case we still have moduli mediated supergravity, with the above mentioned problems, and it is difficult to break supersymmetry (except at a very high scale) when the GS term is included [36].

All the above potentials for the dilaton can be obtained by introducing gaugino and matter [37, 38] condensates as ordinary chiral superfields, but this is actually a questionable [36] procedure. The composite superfield $W_a^\alpha W_\alpha^a$ is a chiral field that satisfies a constraint; in supergravity this reads

$$\left(\mathcal{D}^\alpha \mathcal{D}_\alpha - 24\bar{R}\right) W_a^\alpha W_\alpha^a - \left(\mathcal{D}_{\dot{\alpha}} \mathcal{D}^{\dot{\alpha}} - 24R\right) W_{\dot{\alpha}}^a W_a^{\dot{\alpha}} = \text{total derivative}, \tag{34}$$

and one would expect the interpolating field for the gaugino condensate to satisfy the same constraint. This can be achieved by taking the condensate chiral superfield U to be the chiral projection of a vector superfield V [39, 40]:

$$U = - \left(\mathcal{D}_{\dot{\alpha}} \mathcal{D}^{\dot{\alpha}} - 8R \right) V, \tag{35}$$

which automatically satisfies (34), and, as it turns out, is the natural formulation [41] in the context of the heterotic string. Supergravity in ten dimensions contains a dilaton ϕ and gauge invariant two-form potential b_{MN}. Their remnants in four dimensions include the dilaton Res and a two-form $b_{\mu\nu}$ whose three-form field strength $h_{\mu\nu\rho}$ provides the axionic companion $a = \text{Im}s$ of the dilaton through the duality transformation

$$\epsilon_{\mu\nu\rho\sigma} \partial^{\sigma} a \frac{\partial^2 K}{\partial (\text{Re}s)^2} = h_{\mu\nu\rho}. \tag{36}$$

It follows from supersymmetry that the chiral supermultipet is dual to a linear supermultiplet [42], whose components are a scalar ℓ, a two-form and a majorana spinor (with no auxiliary field). The coupling of the chiral superfield S to the Yang-Mills superfield strength implies that it is dual to a *modified* linear superfield, defined by

$$\left(\bar{\mathcal{D}}^2 - 8R \right) L = - \sum_a W_a^{\alpha} W_{\alpha}^a. \tag{37}$$

When one of the superfield strength bilinears on the RHS forms a condensate, we replace that bilinear with its interpolating chiral supermultiplet U, and (34) is automatically satisfied. In other words we identify the vector superfield V in (35) with the modified linear superfield L.

It is not intuitively obvious how to generate a potential for an antisymmetric tensor (Kalb-Raymond field) $b_{\mu\nu}$ or for a linear supermultiplet L. The former problem was solved in 1981 by Aurilia and Takahashi [43]. Following their approach, the latter problem has been solved for global supersymmetry [39, 40] and for local supersymmetry, first using gauge fixed superconformal supergravity [40], and then in the Kähler superspace of [21] with the GS term incorporated [36].

In fact, the 4d Green-Schwarz term is more elegantly introduced in the linear multiplet formulation [44]; it appears [45] as a constant of integration in the superfield duality transformation:

$$S + \bar{S} + V_G(T, \Phi) = - \int \frac{dL}{L} \frac{\partial K}{\partial L}. \tag{38}$$

It has been argued [39] that the two formulations are equivalent in the presence of nonperturbative effects such as gaugino condensation. Although there are some indications [36, 46] that these formulations are not exactly equivalent, they have been shown to give the same results for condensate-induced potentials [47]. The construction of an effective potential for the dilaton is more straightforward in the linear formulation, but the results can easily be translated [48] to the more familiar language of the chiral formulation. However the linear formulation has the advantage that that there is no mixing of the dilaton ℓ with the Kähler moduli t^I, and the salient features of the model are therefore more transparent.

5. Kähler stabilization of the dilaton

Because the GS term destabilizes the dilaton in the direction of strong coupling, nonperturbative corrections to the dilation potential, either field theoretic [49] or string theoretic [50] must be included. Since the latter, proportional to $e^{-a/g}$, are less suppressed for weakish coupling, $\alpha^{-1} = 4\pi\langle\mathrm{Res}\rangle > 1$, than the former,[d] proportional to e^{-a/g^2}, they are expected to dominate. It was argued [49, 52] that these effects could stabilize the dilaton potential, and explicit realizations of this scenario were subsequently constructed, first for a single gaugino condensate [34, 53]. In this case, when T-duality is imposed [34] by including the GS term, the moduli-dependence of the anomalous quantum correction is canceled, and Kähler moduli are not stabilized. The general (and more realistic) case of several gaugino condensates as well as matter condensates was considered in [38]. When matter condensates and/or threshold corrections (27) are included the moduli are stabilized at self-dual points, $\langle t^I \rangle = 1$ or $e^{i\frac{\pi}{6}}$, with vanishing F-terms $\langle F^I \rangle = 0$. As a consequence, supersymmetry breaking is dilaton-dominated, with universal couplings to matter.

The effective potential for the condensates is constructed by anomaly matching. Once the heavy condensates have been integrated out, and the Kähler moduli fixed at their *vev*'s, the potential for the dilaton takes the

---

[d]These are in fact implicitly included in the approach of [34, 38]; it has been shown [51] that those constructions, using static condensates, give the same result as obtained after integrating out dynamical condensates, the potential source of the field theoretic corrections.

form

$$V(\ell) = \sum_a \frac{|u_a|^2}{16\ell} \left[k'(\ell) (1 - b_a\ell)^2 - \ell b_a^2 \right], \quad |u_a(\ell)|^2 = c_a e^{k(\ell) - 2s(\ell)/b_a}.$$

(39)

Here $k(\ell)$ is the Kähler potential for the dilaton, $u_a = U_a|$, and

$$b_a = \frac{1}{8\pi^2} \left(C_a - \frac{1}{3} \sum_i C_a^i \right)$$

(40)

is related to the one-loop β-function of the group $\mathcal{G}_a$ as in (9). The constant c_a depends on the various Casimirs of the group $\mathcal{G}_a$, and contains as a factor the *vev* of the modular (T-duality) invariant term

$$\prod_i \left[(t^I + \bar{t}^I) |\eta^4(t^I)| \right] e^{\frac{b}{b_a} - 1}.$$

(41)

Evaluated at self-dual points, the terms in square brackets are approximately unity, The value of $|u_a(\ell)|$ is dominated by the standard RGE factor $e^{-1/g_a^2 b_a}$ with now $g_a^{-2} = \langle s(\ell) \rangle$, $\forall\, a$ at the string scale. The function $s(\ell)$ satisfies the differential equation

$$k'(\ell) = -2\ell s'(\ell),$$

(42)

which is just the lowest component of the derivative of (38) if we identify $s(\ell) = \text{Re}s + \frac{1}{2}V_G(t, \phi)$.

Unless there are condensing gauge groups with nearly–but not quite–degenerate values of b_a (which is difficult to achieve in this context), the potential is dominated by the contribution from the group(s) with the largest value of b_a. The phenomenology[e] is the same as for a single condensing gauge group with one important exception. If there is just one condensate (or more than one with the same beta-function coefficient) larger than the QCD condensate, there is an unbroken R-symmetry [49]; as a consequence the axion partner of the dilaton is massless above the scale of QCD condensation, and is a candidate for the Peccei-Quinn axion [55].

To determine [56] the spectrum of the remaining particles in this scenario, it is sufficient to consider a single (or dominant) condensate u with beta-function b_c, and drop the sum over a in (39). If we impose the conditions that the vacuum energy is vanishingly small and that $g^{-2}(m_s) = \langle s(\ell) \rangle$ is close to the measured value of about 2, we can make a two or three parameter fit to the nonperturbative corrections to $k(\ell)$ of

---

[e] For a detailed review of the phenomenology and cosmology of this class of models see Ref. [54].

the form [56], for example, $k(\ell) = \ln(\ell) + \left(a_1 + a_2/\sqrt{\ell}\right)\exp(-b/\sqrt{\ell})$ with $a_i, b \sim 1$; other parameterizations have been considered in [57]. The phenomenology is insensitive to this choice.

The requirement of (nearly) vanishing vacuum energy requires that

$$\frac{k'(\ell_0)}{\ell_0} = \frac{3b_c^2}{(1 - \ell_0 b_c)^2} \approx 3b_c^2, \qquad \ell_0 = \langle\ell\rangle, \tag{43}$$

since generally $b_c \ll 1$, and in the weak coupling regime we expect $\ell_0 \sim 1$. The dilaton kinetic energy term is given by

$$\mathcal{L}_{KE}(\ell) = \frac{1}{4}\frac{\partial^2 k(s)}{\partial s^2}\partial^\mu s \partial_\mu s = \frac{k'(\ell)}{4\ell}\partial^\mu \ell \partial_\mu \ell = \frac{k'(\ell_0)}{4\ell_0}\partial^\mu \hat{\ell}\partial_\mu \hat{\ell}\left[1 + O(\hat{\ell})\right], \tag{44}$$

where here $s = s(\ell)$ and $\hat{\ell} = \ell - \ell_0$. Since (43) implies that $k'(\ell_0)/\ell_0$ is very small, the dilaton mass m_d, proportional to $\ell_0/k'(\ell_0)$ is quite large; if $V''(\ell) \sim 1$ (in reduced Planck mass units)

$$m_d \sim \frac{1}{b_c^2}m_{\frac{3}{2}} \gg m_{\frac{3}{2}}, \qquad m_{\frac{3}{2}} = \frac{b_c}{4}|u_c|. \tag{45}$$

The Kähler moduli have smaller masses:

$$m_t \approx \frac{2\pi}{3}\left(\frac{b}{b_c} - 1\right)m_{\frac{3}{2}}, \tag{46}$$

but can still be considerably larger than the gravitino mass $m_{\frac{3}{2}}$ if $b \gg b_c$, thus alleviating the cosmological "moduli problem" [58]. The matter scalar masses m_i depend on their couplings to the GS term (29):

$$m_{\frac{3}{2}} \leq m_i \leq m_t. \tag{47}$$

The lower bound is satisfied if V_G is independent of Φ^i, and the upper bound holds if the $O(\Phi^2)$ term is just the completion of the Kähler potential $K(T + \bar{T}, |\Phi|^2)$. An analysis of the viability of electroweak symmetry breaking favors the former [59] for the MSSM scalars, at least for the stop and the Higgs. On the other hand the condition (43) suppresses the gaugino masses:

$$m_{\frac{1}{2}}^a(\Lambda_c) \approx g_a^2(\Lambda_c)\frac{s'(\ell_0)}{4}\bar{u}_c = -g_a^2(\Lambda_c)\frac{k'(\ell_0)}{8\ell}\bar{u}_c, \qquad |m_{\frac{1}{2}}| \approx \frac{3}{2}g_a^2(\Lambda_c)b_c^2|m_{\frac{3}{2}}|. \tag{48}$$

For the scalars, the running of the wave function from the string scale to Λ_c is canceled by the running of bilinear couplings of scalars to the condensate, which are proportional to the Kähler metric. So they are degenerate at the condensation scale. However the gaugino wave function renormalization runs as usual between these scales, and the ratios among gaugino masses are the same as in the minimal supergravity (mSUGRA— or cMSSM, for constrained MSSM) model [60], up to additional radiative corrections [61]. In fact, because the gauginos are much lighter than the scalars, these can be important; in some regions of parameter space the gaugino spectrum resembles that of an anomaly mediated scenario [62, 63]. Radiative corrections are especially important if the upper bound in (47) is satisfied for some scalars.

Possibilities for generating the observed energy density of dark matter [64] and a successful inflationary epoch [65] have also been explored in this class of models with promising results. However, these scenarios will have to be revisited [66] in light of results from the Large Hadron Collider (LHC).

6. The virtues of T-duality

T-duality assures that the self-dual points, $t = 1$, $e^{i\frac{\pi}{6}}$ are extrema of the potential. When the dilaton is stabilized by corrections to the Kähler potential, with cancellation of the modular anomaly, these are always minima, with vanishing F-terms for the Kähler moduli. This assures that supersymmetry breaking is dilaton dominated; the universal coupling of the dilaton to squarks in turn assures that no flavor changing neutral current (FCNC) effects are introduced at tree level in the effective theory, provided matter couplings to the GS term are flavor diagonal (or absent). However it has been argued [67] that loop corrections can introduce large FCNC effects because they involve the Kähler Ricci tensor $R_{i\bar{m}}$, that itself is a contraction of the Kähler Riemann tensor which involves a sum over *all* the scalars of the effective 4-d theory:

$$\mathcal{L}_{1-\text{loop}} \ni -\frac{\Lambda^2}{16\pi^2} R_{i\bar{m}} F^i \bar{F}^{\bar{m}} = -\frac{\Lambda^2}{16\pi^2} R^j_{ij\bar{m}} F^i \bar{F}^{\bar{m}}, \tag{49}$$

with the auxiliary fields F^i given (31). The large number of terms in the sum over j could potentially compensate for the loop suppression factor. There are several ways in which these terms can be suppressed in the present context [68]. The tree level Lagrangian contains the term

$$\mathcal{L}_{\text{tree}} \ni -K_{i\bar{m}} F^i \bar{F}^{\bar{m}}. \tag{50}$$

The *vev* of the Kähler metric $\langle K_{i\bar{m}} \rangle$ has to be flavor diagonal to a high degree of accuracy to avoid kinetic flavor mixing, and the strong suppression of FCNC at tree level requires that the *vev* of MSSM scalar derivatives of the whole expression in (50) be flavor diagonal. If conservation of flavor symmetry in the Kähler metric is due to an isometry of the Kähler geometry, the symmetry of the Kähler potential is also a symmetry of the Ricci tensor. For example, the Kähler potential for the "untwisted" sector of orbifold compactifications with just three diagonal Kähler moduli:

$$K_{\text{untw}} = \sum_{I=1}^{3} K^I = -\sum_{I} \ln\left(T^I + \bar{T}^I \sum_{a=1}^{N_I} |\Phi^{Ia}_{\text{untw}}|^2 \right), \qquad (51)$$

has, in addition to the $SL(2, \mathbf{Z}) \cong SL(1,1)$ symmetry defined by (18) and (21), a much larger symmetry, namely $\prod_{I=1}^{3} SU(N_I + 1, 1)$, which assures that the Kähler Ricci tensor is proportion to the Kähler metric for each untwisted sector I:

$$R^I_{i\bar{m}} = (N_n + 2)K^I_{i\bar{m}}. \qquad (52)$$

If the full Kähler metric possessed a similar isometry, there would be no significant FCNC contributions from loop corrections. More generally, if supersymmetry is dilaton dominated [48]:

$$\left\langle F^{i \neq S} \right\rangle = 0, \qquad \left\langle F^S \right\rangle \approx -\frac{k'(\ell_0)}{\ell_0}\bar{u} \approx -3b_c m_{\frac{3}{2}}, \qquad (53)$$

no flavor violation is introduced at one loop. Regularization of the quadratic divergences with a constant momentum cut-off Λ does not preserve supersymmetry. Supersymmetric Pauli Villars regularization introduces additional terms in such a way that the entire quadratically divergent contribution amounts to [69, 70] a redefinition of the vierbein e^a_μ and a renormalization of the Kähler metric: $K(\Phi^a, T^I, S) \to K^R(\Phi^a, T^I, S)$, and the masses for MSSM scalars ϕ^a are given by (in the absence of D-term contributions, and assuming for concreteness the lower bound in (47))

$$(m^2)^i_j = \delta^i_j m^2_{\frac{3}{2}} - \left\langle (R_R)^i_{jk\bar{m}} F^k_R \bar{F}^{\bar{m}}_R \right\rangle \qquad (54)$$

with the subscript R denoting the renormalized Riemann tensor and auxiliary fields. $R^i_{js\bar{s}}$ vanishes at tree level and is proportional to δ^i_j at one loop. Even under more general assumptions, it seems difficult to generate large quadratically divergent contributions in the heterotic string context; their detailed expression depends on parameters in the Pauli-Villars sector that are not completely determined by the cancellation of quadratic and linear divergences—and that reflect the underlying UV physics that make the theory finite [68].

In addition, T-duality suppresses [71] higher dimension operators that could break [49] the R-symmetry that keeps the "universal" axion—that is, the axion partner of the dilaton—massless down to the scale of QCD condensation. An analysis [55] using $b_c = .038$, the preferred value at the time for the correct abundance of dark matter [64], found that this axion could be the QCD–or Peccei-Quinn–axion, provided the group of T-duality transformations is larger than the minimal one defined by (18). As it happens, that value of b_c is very close to a critical value $8\pi^2 b_c = N_c = 3$ (for low energy QCD) where the axion mass vanishes, and the residual R-symmetry does not include a transformation on the quarks, so in that limit there is no solution to the QCD CP violation problem. In addition the axion coupling is quite high in this scenario, and its viability requires a very small misalignment angle in the early universe [72]. However the preferred value of b_c has to be reconsidered taking into account LHC data [66]; a considerably larger value of b_c could alleviate these problems.

Finally, T-duality might provide a rationale for R-parity, or an even stronger symmetry[f] [73] that may be required to suppress the dimension four superpotential operator $(U^c)^2 D^c E^c$, which leads to dimension five operators in the effective Lagrangian that, if suppressed only by a factor m_P^{-1} or m_s^{-1}, is in conflict [74] with limits on proton decay. The transformations (21) and (28) actually have phase factors on the RHS with constant phases [75] δ that depend on the parameters a, b, c, d in (18). These are commonly dropped in the literature, because they do not appear in the potential, and can be reabsorbed [76] into the definition of the chiral superfields Φ^I. However, it is precisely the phases in (28) that forbid operators of higher order in superfield strength bilinears from appearing in the Lagrangian, thereby protecting the mass of the universal axion.

When these phases are kept explicit one gets selection rules for superpotential couplings that are identical to string selection rules [77], but now expressed in terms of the symmetries of the effective field theory. The group of modular invariance is broken by the *vev*'s of the moduli and the gaugino condensate, but a subgroup that leaves the self-dual points and the gaugino condensate invariant, remains a symmetry of the theory. This may be further broken by the large *vev*'s of scalars, for example, when an anomalous $U(1)$ symmetry is broken. The resulting discreet symmetry is a combination of T-duality and $U(1)$ transformations that leaves the scalars with nonvanishing *vev*'s invariant, under which chiral superfields Φ transform with phase factors $e^{i\gamma/\pi}$, where γ is an arbitrary rational number. (For example $\gamma = n/33$ in the extensively studied model of Ref. [9].) This

---

[f]See also Section 6.1 of [54].

allows for the possibility of a symmetry stronger than R-parity that also forbids the dangerous dimension four operators in the superpotential [73].

7. Challenges

Orbifolds with Wilson lines, generally have an anomalous $U(1)$, often denoted by $U(1)_X$, that is not an anomaly of the underlying string theory, and must be canceled by an additional GS term [78]. This is accomplished by the substitution in (23) or (38) $V_G \rightarrow V_G - \delta_X V_X$, where V_X is the $U(1)_X$ vector superfield, and

$$\delta_X = -\frac{1}{48\pi}Q_X = -\frac{1}{2\pi}\sum_{a\neq X} C_a^i q_i^X = -\frac{1}{6\pi}Q_X^3. \tag{55}$$

The matrix $Q_X = \mathrm{diag}(q_1^X,\ldots,q_N^X)$ is the $U(1)_X$ charge operator. The conditions in (55), required for full anomaly cancellation, hold in all orbifold compactifications. The presence of $\delta_X V_X$ in the GS term leads to a D-term, $D_X \neq 0$, which is canceled by the vev's of $U(1)_X$-charged scalars. Since these are generally charged under additional $U(1)$'s, full cancellation of of the D-term requires some number m of $U(1)$'s being broken by $n \geq m$ scalar vev's. This is in fact a welcome feature since it removes many matter fields from the effective theory below the $U(1)_X$-breaking scale and allows for an asymptotically free gauge group that can condense and break supersymmetry. This additional GS term can also stabilize the dilaton and provide a flat, inflationary potential during the early universe [65]

For "minimal" models with $n = m$, the phenomenology[g] [57] of gaugino condensation with Kähler stabilization of the dilaton does not change substantially from that discussed above $except$ for MSSM scalar masses, that now get an additional contribution

$$\Delta m_i^2 \approx \frac{\zeta_i}{b_c^2 \ell^2} m_{\frac{3}{2}}^2, \qquad \zeta_i = \sum_{a,A} Q_a^A q_i^a, \tag{56}$$

where the $m \times m$ matrix Q_a^A is the inverse of the matrix Q_A^a whose elements are the $U(1)_a$ charges q_A^a of the m scalars ϕ^A that get large vev's at the $U(1)$-breaking scale. If $\zeta \sim 1$ (56) will dominate over (47). It is not positive definite, and can destabilize the MSSM vacuum. It can also generate FCNC effects unless MSSM states with the same SM quantum numbers also have the same $U(1)_a$ charges (which could easily be the case in Z_n orbifolds). Another issue is the large degeneracy of the vacuum which can

---

[g]See also Section 5 of [54].

leave massless "D-moduli" [79]. In minimal models all the complex scalars that vanish in the vacuum get vev's as in (47), but their fermionic partners remain massless—a cosmological disaster [58, 80]—unless they get masses through Yukawa couplings. Models with $n > m$ are even more problematic for both cosmology and phenomenology. The challenge for string phenomenologists is to find orbifold compactifications with $\zeta \ll 1$, as well as a mechanism for sufficient suppression of the proton lifetime. Models with R-parity conservation have been found [81] in Z_6 orbifold compactifications, but as discussed in the previous section, a stronger symmetry is needed to suppress dimension-four superpotential terms.

Another obvious challenge is finding a compactification that produces, not just the SM particle content, but the pattern of quark and lepton fermion masses, and in particular explaining the very small neutrino masses. For example, it was found [82] that a simple seesaw mechanism does not arise in the well-studied standard model-like Z_3 orbifold compactifications.

Finally, the effective supergravity theory described above is probably still incomplete. In addition to the anomalous terms containing Yang-Mills superfield strength bilinears, there are many others. In particular, every logarithmic divergence has an associated conformal anomaly. The full chiral [83, 84] and conformal [84] anomalies in supergravity have been calculated. A subset of the latter combine with chiral anomalies to form "F-term" anomalies like (20); others have no chiral counterpart and form "D-term" anomalies. Determining the conformal anomalies requires regulating the theory; using a supersymmetric Pauli-Villars regulation [85] the anomalous coefficients of the Yang-Mills bilinears, as well as operators associated with space-time curvature and the Kähler connection, are canceled by the GS mechanism. There are other operators, that depend on the modular weights and $U(1)_X$ charges of the matter fields, whose cancellation appears to require constraints on unknown twisted matter Kähler potential terms, and on the detailed choice of Pauli-Villars regulators. It has been shown [86] that the GS term completely cancels the anomalies in Z_3 and Z_7 orbifold compactifications (without Wilson lines) at the string level. Achieving this at the effective field theory level could provide information relevant for the suppression of FCNC and for loop corrections to soft supersymmetry breaking terms [48].

The ultimate challenge for any theory is confrontation with experiment. Signatures for Kähler stabilized heterotic models with anomaly cancellation were studied in detail [87] before the startup of the LHC. The absence of supersymmetry signals at the LHC and the discovery of an SM-like Higgs with a mass considerably larger than the MSSM tree level prediction $m_h <$ m_Z for the lightest Higgs scalar constrain supersymmetry in general and the class of models discussed above in particular. An analysis [66] of Kähler stabilized, modular invariant models without D-term couplings to MSSM scalar masses, and with no MSSM couplings in the GS term, led to the following constraints on these models: (1) scalar superpartners are too heavy to be accessible to the LHC and their contribution to deviations from SM prediction are negligible, (2) gluino masses are no larger tha 2900 GeV, and more likely (for smaller values of $b_c \leq 0.1$ as expected in these models) less than 2100–and LHC accessible, (3) the Higgs mass can be no larger than 127 GeV and has SM-like couplings, (4) the ratio of the two Higgs vev's $\tan\beta > 40$ is large, implying a small μ parameter, (5) there will be a collection of neutralinos and charginos, much lighter than gluinos, but about twice as heavy as those in a similarly constrained mSugra model, with small mass splittings, (6) the LSP (lightest supersymmetric particle, which is stable and can provide dark matter) will be mostly Higgsino, and (7) evidence for dark matter should show up in the Large Underground Xenon (LUX) experiment within two years. However, there remains a significant region of parameter space still to be explored, including variations on the above assumptions.

Acknowledgments

I wish to thank my many collaborators, especially Pierre Binétruy, Joel Giedt, Brent Nelson, Tom Taylor and Yi-Yen Wu, from whom I have learned a lot. This work was supported in part by the Director, Office of Energy Research, Office of High Energy and Nuclear Physics, Division of High Energy Physics of the U.S. Department of Energy under Contract DE-AC03-76SF00098 and in part by the National Science Foundation under grant PHY-0098840.

References

[1] M. B. Green and J. H. Schwarz, *Phys. Lett. B* **149**, 117 (1984).
[2] D. J. Gross, J. A. Harvey, E. J. Martinec and R. Rohm, *Phys. Rev. Lett.* **54**, 502 (1985).
[3] P. Candelas, G. T. Horowitz, A. Strominger and E. Witten, *Nucl. Phys. B* **258**, 46 (1985).
[4] J.H. Schwarz, *Nucl. Phys. Proc. Suppl. B* **55**, 1 (1997).
[5] M. Green, a seminar.

[6] P. Hořava and E. Witten, *Nucl. Phys. B* **460**, 506 (1996) and B **475**, 94 (1996).

[7] R. Bousso and J. Polchinski, *Sci. Am.* **291**, 60 (2004).

[8] L.E. Ibàñez, H.-P. Nilles and F. Quevedo, *Phys. Lett. B* **187**, 25 (1987); A. Font, L. Ibàñez, D. Lust and F. Quevedo, *Phys. Lett. B* **245**, 401 (1990).

[9] A. Font, L. E. Ibàñez, F. Quevedo and A. Sierra, *Nucl. Phys. B* **331**, 421 (1990).

[10] O. Lebedev, H. -P. Nilles, S. Raby, S. Ramos-Sanchez, M. Ratz, P. K. S. Vaudrevange and A. Wingerter, *Phys. Rev. Lett.* **98**, 181602 (2007)

[11] I. Antoniadis, J. R. Ellis, J. S. Hagelin and D. V. Nanopoulos, *Phys. Lett. B* **194**, 231 (1987).

[12] B. A. Campbell, J. R. Ellis, J. S. Hagelin, D. V. Nanopoulos and R. Ticciati, *Phys. Lett. B* **198**, 200 (1987).

[13] V. S. Kaplunovsky, *Nucl. Phys. B* **307**, 145 (1988) [Erratum-ibid. B **382**, 436 (1992)].

[14] M. K. Gaillard, T. R. Taylor, *Nucl. Phys. B* **381**, 577 (1992);

[15] P. Mayr and S. Stieberger, *Nucl. Phys. B* **412**, 502 (1994); V. Kaplunovsky and J. Louis, *ibid.* **422**, 57 (1994).

[16] M. K. Gaillard and R. Xiu, *Phys. Lett. B* **296**, 71 (1992).

[17] U. Amaldi, W. de Boer, P. H. Frampton, H. Furstenau and J. T. Liu, *Phys. Lett. B* **281**, 374 (1992).

[18] L.J. Dixon, V.S. Kaplunovsky and J. Louis, *Nucl. Phys. B* **355**, 649 (1991); I. Antoniadis, K.S. Narain and T.R. Taylor, *Phys. Lett. B* **267**, 37 (1991).

[19] R. Xiu, *Phys. Rev. D* **49**, 6656 (1994) L. E. Ibanez, *Phys. Lett. B* **318**, 73 (1993); S. P. Martin and P. Ramond, *Phys. Rev. D* **51** (1995) 6515-6523; K. R. Dienes, A. E. Faraggi and J. March-Russell, *Nucl. Phys. B* **467**, 44 (1996); K. R. Dienes, *Phys. Rep.* **287** (1997) 447-525; J. Giedt, *Mod. Phys. Lett. A* **18**, 1625 (2003).

[20] S. Dimopoulos and H. Georgi, *Nucl. Phys. B* **193**, 150 (1981).

[21] P. Binétruy, G. Girardi, R. Grimm, *Phys. Rept.* **343**, 255 (2001).

[22] M. Dine, R. Rohm, N. Seiberg and E. Witten, *Phys. Lett. B* **156**, 55 (1985).

[23] T. R. Taylor, *Phys. Lett. B* **164** (1985) 43.

[24] G. Veneziano and S. Yankielowicz, *Phys. Lett. B* **113**, 231 (1982).

[25] P. Binétruy and M. K. Gaillard, *Phys. Lett. B* **232**, 83 (1989).

[26] S. Ferrara, N. Magnoli, T. R. Taylor, G. Veneziano, *Phys. Lett. B* **245**, 409 (1990).

[27] P. Binétruy, M. K. Gaillard, *Nucl. Phys. B* **358**, 121 (1991).

[28] E. Alvarez and M. A. R. Osorio, *Phys. Rev. D* **40**, 1150 (1989); A. Giveon, N. Malkin and E. Rabinovici, *Phys. Lett. B* **220**, 551 (1989).

[29] G.L. Cardoso and B.A. Ovrut, *Nucl. Phys. B* **369**, 315 (1993); J.-P. Derendinger, S. Ferrara, C. Kounnas and F. Zwirner, *Nucl. Phys. B* **372**, 145 (1992).

[30] J. R. Ellis, A. B. Lahanas, D. V. Nanopoulos and K. Tamvakis, *Phys. Lett. B* **134**, 429 (1984).

[31] P. Binétruy, S. Dawson, M. K. Gaillard and I. Hinchliffe, *Phys. Rev. D* **37**, 2633 (1988).

[32] A. Font, L. E. Ibanez, D. Lust and F. Quevedo, *Phys. Lett. B* **249**, 35 (1990).
[33] H. P. Nilles and M. Olechowski, *Phys. Lett. B* **248**, 268 (1990); M. Cvetic, A. Font, L. E. Ibanez, D. Lust and F. Quevedo, *Nucl. Phys. B* **361**, 194 (1991).
[34] P. Binétruy, M. K. Gaillard and Y.-Y. Wu, *Nucl. Phys. B* **481**, 109 (1996).
[35] P. Binétruy and M. K. Gaillard, *Phys. Lett. B* **253**, 119 (1991).
[36] P. Binétruy and M. K. Gaillard, *Phys. Lett. B* **365**, 87 (1996).
[37] T. R. Taylor, G. Veneziano and S. Yankielowicz, *Nucl. Phys. B* **218**, 493 (1983).
[38] P. Binétruy, M. K. Gaillard and Y.-Y. Wu, *Nucl. Phys. B* **493**, 27 (1997).
[39] C. P. Burgess, J. -P. Derendinger, F. Quevedo and M. Quiros, *Phys. Lett. B* **348**, 428 (1995).
[40] P. Binétruy, M. K. Gaillard and T. R. Taylor, *Nucl. Phys. B* **455**, 97 (1995)
[41] S. J. Gates, Jr., *Nucl. Phys. B* **184**, 381 (1981).
[42] W. Siegel, *Phys. Lett. B* **85**, 333 (1979); U. Lindstrom and M. Rocek, *Nucl. Phys. B* **222**, 285 (1983).
[43] A. Aurilia and Y. Takahashi, *Prog. Theor. Phys.* **66**, 693 (1981); A. Aurilia, Y. Takahashi and P. K. Townsend, *Phys. Lett. B* **95**, 265 (1980).
[44] S. Ferrara and M. Villasante, *Phys. Lett. B* **186**, 85 (1987).
[45] P. Binétruy, G. Girardi and R. Grimm, *Phys. Lett. B* **265**, 111 (1991); P. Adamietz, P. Binétruy, G. Girardi and R. Grimm, *Nucl. Phys. B* **401**, 257 (1993).
[46] J. de Boer and K. Skenderis, *Nucl. Phys. B* **481**, 129 (1996); S. J. Gates, private communication.
[47] T. Barreiro, B. de Carlos and E. J. Copeland, *Phys. Rev. D* **57**, 7354 (1998); J. Giedt and B. D. Nelson, *JHEP* **0405**, 069 (2004).
[48] P. Binétruy, M.K. Gaillard and B. Nelson, *Phys. Lett. B* **412**, 288 (1997).
[49] T. Banks and M. Dine, *Phys. Rev. D* **50**, 7454 (1994).
[50] S.H. Shenker, in *Random Surfaces and Quantum Gravity*, Eds. O. Alvarez, E. Marinari, and P. Windey, NATO ASI Series B262 (Plenum, NY, 1990); E. Silverstein, *Phys. Lett. B* **396**, 91 (1997).
[51] Y.-Y. Wu, *Supersymmetry breaking in superstring theory by gaugino condensation and its phenomenology*, hep-th/9706040.
[52] C. P. Burgess, J. P. Derendinger, F. Quevedo and M. Quiros, *Annals Phys.* **250**, 193 (1996).
[53] J. A. Casas, *Phys. Lett. B* **384**, 103 (1996).
[54] M. K. Gaillard and B. D. Nelson, *Int. J. Mod. Phys. A* **22**, 1451 (2007).
[55] M. K. Gaillard and B. Kain, *Nucl. Phys. B* **734**, 116 (2006).
[56] P. Binétruy, M. K. Gaillard and Y.-Y. Wu, *Phys. Lett. B* **412**, 288 (1997).
[57] M. K. Gaillard, J. Giedt and A. L. Mints, *Nucl. Phys. B* **700**, 205 (2004) [Erratum-ibid. **713**, 607 (2005)].
[58] G.D. Coughlan, W. Fischler, E.W. Kolb, S. Raby and G.G. Ross, *Phys. Lett. B* **131**, (1983) 59.
[59] M.K. Gaillard and B. Nelson, *Nucl. Phys. B* **571**, 3 (2000).
[60] A. H. Chamseddine, R. L. Arnowitt and P. Nath, *Phys. Rev. Lett.* **49**, 970 (1982).

[61] M. K. Gaillard, B. D. Nelson and Y-Y. Wu, *Phys. Lett. B* **459**, 549 (1999).

[62] L. Randall and R. Sundrum, *Nucl. Phys. B* **557**, 557 (1999).

[63] G. Giudice, M. Luty, H. Murayama and R. Rattazzi, *JHEP* **9812**, 027 (1998).

[64] A. Birkedal-Hansen and B. D. Nelson, *Phys. Rev. D* **64**, 015008 (2001); *ibid.* **67**, 095006 (2003).

[65] M. K. Gaillard, D. H. Lyth and H. Murayama, *Phys. Rev. D* **58**, 123505 (1998); M. J. Cai and M. K. Gaillard, *ibid.* **62**, 047901 (2000); T. Barreiro, B. de Carlos and E. J. Copeland, *ibid.* **58**, 083513 (1998); D. Skinner, *ibid.* **67**, 103506 (2003); B. Kain, *Nucl. Phys. B* **800**, 270 (2008).

[66] B. L. Kaufman, B. D. Nelson and M. K. Gaillard, *Phys. Rev. D* **88**, 025003 (2013).

[67] K. Choi, J. S. Lee and C. Muñoz, *Phys. Rev. Lett.* **80**, 3686 (1998); K. Choi, H. B. Kim and C. Muñoz, *ibid* **57**, 7521 (1998).

[68] M. K. Gaillard and B. D. Nelson, *Nucl. Phys. B* **751**, 75 (2006).

[69] P. Binétruy and M. K. Gaillard, *Phys. Lett. B* **220**, 68 (1989).

[70] M. K. Gaillard, *Phys. Lett. B* **342**, 125 (1995).

[71] D. Butter and M. K. Gaillard, *Phys. Lett. B* **612**, 304 (2005).

[72] P. Fox, A. Pierce and S. D. Thomas, *Probing a QCD string axion with precision cosmological measurements*, hep-th/0409059.

[73] M. K. Gaillard, *Phys. Rev. Lett.* **94**, 141601 (2005).

[74] R. Harnik, D. T. Larson, H. Murayama and M. Thormeier, *Nucl. Phys. B* **706**, 372 (2005).

[75] A. N. Schellekens and N. P. Warner, *Nucl. Phys. B* **287**, 317 (1987); S. Ferrara, D. Lust, A. D. Shapere and S. Theisen, *Phys. Lett. B* **225**, 363 (1989).

[76] S. Ferrara, .D. Lust and S. Theisen, *Phys. Lett. B* **233**, 147 (1989).

[77] S. Hamidi and C. Vafa, *Nucl. Phys. B* **279**, 465 (1987).

[78] M. Dine, N. Seiberg and E. Witten, *Nucl. Phys. B* **289**, 589 (1987); J. Attick, L. Dixon and A. Sen, *Nucl. Phys. B* **292**, 109 (1987); M. Dine, I. Ichinose and N. Seiberg, *Nucl. Phys. B* **293**, 253 (1988).

[79] F. Buccella, J.P. Derendinger, S. Ferrara and C.A. Savoy, *Phys. Lett. B* **115**:375 (1982); M.K. Gaillard and J. Giedt, *ibid.* **479**, 308 (2000).

[80] A. S. Goncharov, A. D. Linde and M. I. Vysotsky, *Phys. Lett. B* **147**, 279 (1984); J. R. Ellis, D. V. Nanopoulos and M. Quiros, *ibid.* **174**, 176 (1986).

[81] O. Lebedev, H. P. Nilles, S. Raby, S. Ramos-Sanchez, M. Ratz, P. K. S. Vaudrevange and A. Wingerter, *Phys. Rev. D* **77**, 046013 (2008).

[82] J. Giedt, G. L. Kane, P. Langacker and B. D. Nelson, *Phys. Rev. D* **71**, 115013 (2005)

[83] D. Z. Freedman and B. Kors, *JHEP* **0611**, 067 (2006).

[84] D. Butter and M. K. Gaillard, *Phys. Lett. B* **679**, 519 (2009); M. K. Gaillard, *Pramana* **78**, 875 (2012).

[85] M. K. Gaillard, *Phys. Rev. D* **58**, 105027 (1998); *ibid.* **61**, 084028 (2000).

[86] C. A. Scrucca and M. Serone, *JHEP* **0102**, 019 (2001).

[87] G. L. Kane, J. D. Lykken, S. Mrenna, B. D. Nelson, L.-T. Wang and T. T. Wang, *Phys. Rev. D* **67**, 045008 (2003).

4

Geography of Fields in Extra Dimensions:
String Theory Lessons for Particle Physics

Hans Peter Nilles[a] and Patrick K. S. Vaudrevange[b]

[a] Bethe Center for Theoretical Physics
and
Physikalisches Institut der Universität Bonn,
Nussallee 12, 53115 Bonn, Germany

[b] Excellence Cluster Universe, Technische Universität München,
Boltzmannstr. 2, D-85748, Garching, Germany
and
Arnold Sommerfeld Center for Theoretical Physics, LMU,
Theresienstraße 37, 80333 München, Germany

String theoretical ideas might be relevant for particle physics model building. Ideally one would hope to find a unified theory of all fundamental interactions. There are only few consistent string theories in $D = 10$ or 11 space-time dimensions, but a huge landscape in $D = 4$. We have to explore this landscape to identify models that describe the known phenomena of particle physics. Properties of compactified six spatial dimensions are crucial in that respect. We postulate some useful rules to investigate this landscape and construct realistic models. We identify common properties of the successful models and formulate lessons for further model building.

1. Introduction

One of the main goals of string theory is the inclusion of the Standard Model (SM) of particle physics in an ultraviolet complete and consistent theory of quantum gravity. The hope is a unified theory of all fundamental interactions: gravity as well as strong and electroweak interactions within the $SU(3) \times SU(2) \times U(1)$ SM. Recent support for the validity of the particle physics Standard Model is the 2012 discovery of the "so-called" Higgs boson.

How does this fit into known string theory? Ideally one would have hoped to derive the Standard Model from string theory itself, but up to now such a program has not (yet) been successful. It does not seem that the SM appears as a prediction of string theory. In view of that we have to ask the question whether the SM can be embedded in string theory. If this is possible we could then scan the successful models and check specific properties that might originate from the nature of the underlying string theory.

Known superstring theories are formulated in $D = 10$ space time dimensions (or $D = 11$ for M theory) while the SM describes physics in $D = 4$. The connection between $D = 10$ and $D = 4$ requires the compactification of six spatial dimensions. The rather few superstring theories in $D = 10$ give rise to a plethora of theories in $D = 4$ with a wide spectrum of properties. The search for the SM and thus the field of so-called "String Phenomenology" boils down to a question of exploring this compactification process in detail.

But how should we proceed? As the top-down approach is not successful we should therefore analyse in detail the properties of the SM and then use a bottom-up approach to identify those regions of the "string landscape" where the SM is most likely to reside. This will provide us with a set of "rules" for $D = 4$ model constructions of string theory towards the constructions of models that resemble the SM.

The application of these rules will lead us to "fertile patches" of the string landscape with many explicit candidate models. Given these models we can then try to identify those properties of the models that make them successful. They teach us some lessons towards selecting the string theory in $D = 10$ as well as details of the process of compactification.

In the present paper we shall describe this approach to "string phenomenology". In Section 2 we shall start with "five golden rules" as they have been formulated some time ago [1]. These rules have been derived in a bottom-up approach exploiting the particular properties of quark- and lepton representations in the SM. They lead to some kind of (grand) unified picture favouring SU(5) and SO(10) symmetries in the ultraviolet. However, these rules are not to be understood as strict rules for string model building. You might violate them and still obtain some reasonable models. But, as we shall see, life is more easy if one follows these rules.

In Section 3 we shall start explicit model building along these lines. We will select one of those string theories that allow for an easy incorporation of the rules within explicit solvable compactifications. This leads us to orbifold compactifications of the heterotic $E_8 \times E_8$ string theory [2, 3] as an example. We shall consider this example in detail and comment on generalizations and alternatives later. The search for realistic models starts with the analysis of the so-called $\mathbb{Z}_6$-II orbifold [4–11]. We define the search strategy in detail and present the results known as the "MiniLandscape" [8, 11], a fertile patch of the string landscape for realistic model building. We analyse the several hundred models of the MiniLandscape towards specific properties, as e.g. the location of fields in extra-dimensional space. The emerging picture leads to a notion of "Local Grand Unification", where some of the more problematic aspects of grand unification (GUT) can be avoided. We identify common properties of the successful models and formulate "lessons" from the MiniLandscape that should join the "rules" for realistic model building.

Section 4 will be devoted to the construction of new, explicit MSSM-like models using all $\mathbb{Z}_N$ and certain $\mathbb{Z}_N \times \mathbb{Z}_M$ orbifold geometries resulting in approximately 12000 orbifold models. Then, in Section 5 we shall see how the lessons of the MiniLandscape will be tested in this more general "OrbifoldLandscape".

In Section 6 we shall discuss alternatives to orbifold compactifications, as well as model building outside the heterotic $E_8 \times E_8$ string. The aim is a unified picture of rules and lessons for successful string model building. Section 7 will be devoted to conclusions and outlook.

2. Five Golden Rules

Let us start with a review of the "Five golden rules for superstring phenomenology", which can be seen as phenomenologically motivated guidelines to successful string model building [1]. The rules can be summarized as follows: we need

(1) spinors of SO(10) for SM matter
(2) incomplete GUT multiplets for the Higgs pair
(3) repetition of families from geometrical properties of the compactification space
(4) $\mathcal{N} = 1$ supersymmetry
(5) R-parity and other discrete symmetries

Let us explain the motivation for these rules in some detail in the following.

2.1. *Rule I: Spinors of* SO(10) *for SM matter*

It is a remarkable fact that the spinor **16** of SO(10) is the unique irreducible representation that can incorporate exactly one complete generation of quarks and leptons, including the right-handed neutrino. Thereby, it can explain the absence of gauge-anomalies in the Standard Model for each generation separately. Furthermore, it offers a simple explanation for the observed ratios of the electric charges of all elementary particles. In addition, there is a strong theoretical motivation for Grand Unified Theories like SO(10) from gauge coupling unification at the GUT scale $M_{\mathrm{GUT}} \approx 3 \times 10^{16}$ GeV. Hence, the first golden rule for superstring phenomenology suggests to construct string models in such a way that at least some generations of quarks and leptons reside at a location in compact space, where they are subject to a larger gauge group, like SO(10). Hence, these generations come as complete representations of that larger group, e.g. as **16** of SO(10).

The heterotic string offers this possibility through the natural presence of the exceptional Lie group E_8, which includes an SO(10) subgroup and its spinor representation. Furthermore, using orbifold compactification the four-dimensional Standard Model gauge group can be enhanced to a local GUT, i.e to a GUT group like SO(10) which is realized locally at an orbifold singularity in extra dimensions. In addition, there are matter fields (originating from the so-called twisted sectors of the orbifold) localised at these special points in extra dimensions and hence they appear as complete multiplets of the local GUT group, for example as **16**-plets of SO(10).

On the other hand, the spinor of SO(10) is absent in (perturbative) type II string theories, which can be seen as a drawback of these theories. Often this drawback manifests itself in an unwanted suppression of the top quark Yukawa coupling. On the other hand, F-theory (and M-theory) can cure this through the non-perturbative construction of exceptional Lie groups like e.g. E_6. When two seven-branes with SO(10) gauge group intersect in the extra dimensions, a local GUT can appear at the intersection. There, the gauge group can be enhanced to a local E_6 and a spinor of SO(10) can appear as matter representation.

2.2. *Rule II: Incomplete GUT multiplets for the Higgs pair*

Beside complete spinor representations of SO(10) for quarks and leptons, the (supersymmetric extension of the) Standard Model needs split, i.e. incomplete SO(10) multiplets for the gauge bosons and the Higgs(–pair). Their unwanted components inside a full GUT multiplet would induce rapid proton decay and hence need to be ultra-heavy. In the case of the Higgs doublet, this problem is called the doublet–triplet splitting problem, because for the smallest GUT SU(5) a Higgs field would reside in a five-dimensional representation of SU(5), which includes beside the Higgs doublet an unwanted Higgs triplet of SU(3). This problem might determine the localisations of the Higgs pair and of the gauge bosons in the compactification space: they need to reside at a place in extra dimensions where they feel the breaking of the higher-dimensional GUT to the 4D SM gauge group. Hence, incomplete GUT multiplets, e.g. for the Higgs, can appear. This is the content of the second golden rule.

In this way local GUTs exhibit grand unified gauge symmetries only at some special "local" surroundings in extra dimensions, while in 4D the GUT group seems to be broken down to the Standard Model gauge group. This allows us to profit from some of the nice properties of GUTs (like complete representations for matter as described in the first golden rule), while avoiding the problematic properties (like doublet–triplet splitting).

In the case of the heterotic string on orbifolds the so-called untwisted sector (i.e. the 10D bulk) can naturally provide such split SO(10) multiplets for the gauge bosons and the Higgs. In particular, when the orbifold twist acts as a $\mathbb{Z}_2$ in one of the three complex extra dimensions, one can obtain an untwisted Higgs pair that is vector-like with respect to the full (i.e. observable and hidden) gauge group. Combined with an (approximate) R-symmetry this can yield a solution to the μ-problem of the MSSM. Furthermore, as all charged bulk fields originate from the 10D $E_8 \times E_8$ vector multiplet this scenario naturally yields gauge–Higgs–unification.

Finally, an untwisted Higgs pair in the framework of heterotic orbifolds can relate the top quark Yukawa coupling to the gauge coupling and hence give a nice explanation for the large difference between the masses of the third generation compared to the first and second one. In order to achieve this, the top quark needs to originate either from the bulk (as it is often the case in the MiniLandscape [8] of $\mathbb{Z}_6$-II orbifolds) or from an appropriate fixed torus, i.e. a complex codimension one singularity in the extra dimensions.

2.3. *Rule III: Repetition of families*

The triple repetition of quarks and leptons as three generations with the same gauge interactions but different masses is a curiosity within the Standard Model and asks for a deeper understanding. One approach from a bottom-up perspective is to engineer a so-called flavour symmetry: one introduces a (non-Abelian) symmetry group, discrete or gauge, and unifies the three generations of quarks and leptons in, for example, a single three-dimensional representation of that flavour group. However, as the Yukawa interactions violate the flavour symmetry, it must be broken spontaneously by the vacuum expectation value of some Standard Model singlet, the so-called flavon. This might explain the mass ratios and mixing patterns of quarks and leptons.

The third golden rule for superstring phenomenology asks for the origin of such a flavour symmetry. The rule suggests to choose the compactification space such that some of its geometrical properties lead to a repetition of families and hence yields a discrete flavour symmetry. In this case, the repetition of the family structure comes from topological properties of the compact manifold. Within the framework of type II string theories, the number of families can be related to intersection numbers of D-branes in extra dimensions, while for the heterotic string it can be due to a degeneracy between orbifold singularities. In the latter case, one can easily obtain non-Abelian flavour groups which originate from the discrete symmetry transformations that interchange the degenerate orbifold singularities, combined with a stringy selection rule that is related to the orbifold space group [12]. In any case the number of families will be given by geometrical and topological properties of the compact six-dimensional manifold.

2.4. *Rule IV:* $\mathcal{N} = 1$ *supersymmetry*

Superstring theories are naturally equipped with $\mathcal{N} = 1$ or 2 supersymmetry in 10D. However, generically all supersymmetries are broken by the compactification to 4D. The fourth golden rule suggests to choose a "nongeneric" compactification space such that $\mathcal{N} = 1$ survives in 4D. Examples for such special spaces are Calabi–Yaus, orbifolds and orientifolds. Motivation for this is a solution of the so-called "hierarchy problem" between the weak scale (a TeV) and the string (Planck) scale. Supersymmetry can stabilize this large hierarchy. Since such a supersymmetry appears naturally in string theory, we assume that $\mathcal{N} = 1$ supersymmetry will survive down to the TeV-scale.

2.5. *Rule V: R-parity and other discrete symmetries*

Apart from the gauge symmetries of string theory, we need more symmetries to describe particle physics phenomena of the supersymmetric Standard Model. These could provide the desired textures of Yukawa couplings, explain the absence of flavour changing neutral currents, help to avoid too fast proton decay, provide a stable particle for cold dark matter and solve the so-called μ-problem. We know that (continuous) global symmetries might not be compatible with gravitational interactions. Hence, local discrete symmetries might play this role in string theory.

One of these symmetries is the well-known matter parity of the minimal supersymmetric extension of the Standard Model (MSSM): it forbids proton decay via dim. 4 operators and leads to a stable neutral WIMP candidate. Other discrete gauge symmetries are required to explain the flavour structure of quark/lepton masses and mixings.

3. The MiniLandscape

As we have seen in our review in Section 2, the five golden rules [1] naturally ask for exceptional Lie groups. SO(10), although it is not an exceptional group, fits very well in the chain of exceptional groups $E_8 \rightarrow E_7 \rightarrow E_6 \rightarrow SO(10) \rightarrow SU(5) \rightarrow SM$. Therefore, the $E_8 \times E_8$ heterotic string is the prime candidate and we choose it as our starting point. Alternatives to obtain E_8 in string theory are M- and F-theory, where such gauge groups can appear in non-perturbative constructions.

The implementation of the rules in string theory started with the consideration of orbifold compactifications of the $E_8 \times E_8$ heterotic string. This lead to the so-called "heterotic brane world" [14] where toy examples have

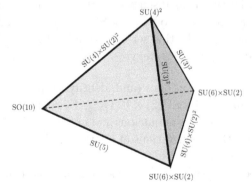

Fig. 1. Gauge group topography from Ref. [13]. At different fixed points (corners of the tetrahedron), E_8 gets broken to different subgroups (U(1) factors are suppressed). At the edges we display the intersection of the two local gauge groups realised at the corners. The 4D gauge group is the standard model gauge group.

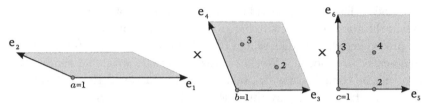

Fig. 2. The six-dimensional torus $(e_1, \ldots, e_6)$ of the $\mathbb{Z}_6$-II orbifold. In the θ-, ω-twisted sector the second, third torus is left invariant, respectively, while in the $\theta\omega$-sector there are fixed points (labelled by a, b, c).

been constructed in the framework of the $\mathbb{Z}_2 \times \mathbb{Z}_2$ orbifold. There, the explicit "geographical" properties of fields in extra dimensions have been presented and the local GUTs at the orbifold fixed points were analysed, see e.g. Fig. 1.

3.1. *Exploring the $\mathbb{Z}_6$-II orbifold*

A first systematic attempt at realistic model constructions [8, 11] was based on the $\mathbb{Z}_6$-II orbifold [4] of the $E_8 \times E_8$ heterotic string. This orbifold considers a six-torus defined by the six-dimensional lattice of $G_2 \times SU(3) \times SO(4)$ modded out by two twists, each acting in four of the six extra dimensions: θ of order 2 ($\theta^2 = 1$) and ω of order 3 ($\omega^3 = 1$), see Fig. 2.

In Ref. [8] the embedding of the twists into the $E_8 \times E_8$ gauge group was chosen in such a way that at an intermediate step there are local $SO(10)$ GUTs with localised **16**-plets for quarks and leptons. This choice can be motivated by rule I, as discussed in the previous section. Further breakdown of the gauge group to $SU(3) \times SU(2) \times U(1)$ is induced by two orbifold Wilson lines [15]. In this set-up, a scan for realistic models was performed using the following strategy:

- choose appropriate Wilson lines (and identify inequivalent models)
- SM gauge group $SU(3) \times SU(2) \times U(1)_Y \subset E_8$ times a hidden sector
- Hypercharge $U(1)_Y$ is non-anomalous and in $SU(5)$ GUT normalisation
- (net) number of three generations of quarks and leptons
- at least one Higgs pair
- exotics are vector-like w.r.t. the SM gauge group and can be decoupled

Using the above criteria, the computer assisted search led to a total of some 200 and 300 MSSM-like models in Refs. [8] and [11], respectively. The models typically have additional vector-like exotics as well as unbroken

U(1) gauge symmetries, one of which is anomalous. This anomaly induces an Fayet–Iliopoulos-term (FI-term), hence a breakdown of the additional U(1)'s and thus allows for a decoupling of the vector-like exotics. Explicit examples are given in Ref. [9] as benchmark models.

All fields of the models can be attributed to certain sectors with specific geometrical properties. In the present case there is an untwisted sector with fields in 10D (bulk), as well as twisted sectors where fields are localised at certain points (or two-tori) in the six-dimensional compactified space. The $\theta\omega$ twisted sector (Fig. 2) has fixed points and thus yields fields localised at these points in extra dimensions that can only propagate in our four-dimensional space–time. The θ and ω twisted sectors, in contrast, have fixed two-tori in extra dimensions. Fields in these sectors are confined to six space–time dimensions. Many properties of the models depend on these "geographic" properties of the fields in extra dimensions. For example, Yukawa couplings between matter and Higgs fields and in particular their coupling strengths are determined by the "overlap" of the fields in extra dimensions.

3.2. *Lessons from the* $\mathbb{Z}_6$-*II MiniLandscape*

Given this large sample of realistic models, we can now analyse their properties and look for similarities and regularities. Which geometrical and geographical properties in extra dimensions are important for realistic models?

By construction, all the models have observable sector gauge group $SU(3) \times SU(2) \times U(1)$ and possibly some hidden sector gauge group relevant for supersymmetry breakdown. There is a net number of three generations of quarks and leptons and at least one pair of Higgs doublets H_u and H_d. The Higgs-triplets are removed and the doublet–triplet splitting problem is solved. A first question concerns a possible "μ-term": $\mu H_u H_d$ and we shall start our analysis with the Higgs-system, following the discussion of Ref. [16].

3.2.1. *Lesson 1: Higgs-doublets from the bulk*

The Higgs-system is vector-like and a μ-term $\mu H_u H_d$ is potentially allowed. As this is a term in the superpotential we would like to understand why μ is small compared to the GUT-scale: This is the so-called μ-problem. To avoid this problem one could invoke a symmetry that forbids the term. However, we know that μ has to be non-zero. Hence, the symmetry has to be broken and this might reintroduce the μ-problem again. In string theory the problem is often amplified since typically we find several (say N) Higgs doublet pairs. In the procedure to remove the vector-like exotics (as described above) we have to make $N - 1$ pairs heavy while keeping one light. In fact, in many cases the small μ-parameter is the result of a specific fine-tuning in such a way to remove all doublet pairs except for one. We do not consider this as a satisfactory solution. Fortunately, the models of the MiniLandscape are generically not of this kind.

Many MiniLandscape models provide one Higgs pair that resists all attempts to remove it. This is related to a discrete R-symmetry [9] that can protect the μ-parameter in the following way: In some cases the discrete R-symmetry is enlarged to an approximate $U(1)_R$ [17, 18].[a] Therefore, a μ-parameter is generated at a higher order M in the superpotential W, where the approximate $U(1)_R$ is broken to its exact discrete subgroup. This yields a suppression $\mu \sim \langle W \rangle \sim \epsilon^M$, where $\epsilon < 1$ is set by the FI parameter.

The crucial observation for this mechanism to work is the localisation of the Higgs pair H_u and H_d in agreement with our second golden rule: both reside in the 10D bulk originating from gauge fields in extra dimensions. Furthermore, the Higgs pair is vector-like with respect to all symmetries, gauge and discrete. This is related to the $\mathbb{Z}_2$ orbifold action in one of the two-tori. Hence, each term in the superpotential $f(\Phi_i) \subset W$ also couples to the Higgs pair, i.e. $f(\Phi_i)H_uH_d \subset W$. As SUSY breakdown requires a non-vanishing VEV of the superpotential the μ-term is related to the gravitino mass, i.e. $\mu = f(\langle\Phi_i\rangle) = \langle W \rangle \sim \epsilon^M \sim m_{3/2}$. This is a reminiscent of a field theoretical mechanism first discussed in Ref. [19].

3.2.2. Lesson 2: Top-quark from the bulk

Among all quarks and leptons the top-quark is very special: its large mass requires a large top-quark Yukawa coupling. Many MiniLandscape models address this naturally via the localisation of the top-quark in extra dimensions: both (t,b) and $\bar{t}$ reside in the 10D bulk, along with the Higgs pair. Hence, we have gauge-Yukawa unification and the trilinear Yukawa coupling of the top is given by the gauge coupling.

Typically the top-quark is the only matter field with trilinear Yukawa coupling. The location of the other fields of the third family is strongly model-dependent, but in general they are distributed over various sectors: the third family could be called a "patchwork family".

---

[a]In addition, $U(1)_R$ symmetries can explain vanishing vacuum energy in SUSY vacua.

3.2.3. *Lesson 3: Flavour symmetry for the first two families*

The first two families are found to be located at fixed points in extra dimensions (Fig. 2). As such they live at points of enhanced symmetries, both gauge and discrete.

The discrete symmetry is the reason for their suppressed Yukawa couplings. In the $\mathbb{Z}_6$-II example shown in the figure two families live at adjacent fixed points in the third extra-dimensional torus: one family is located at $a = b = c = 1$, the other at $a = b = 1$ and $c = 3$ (see Fig. 2). Technically, this is a consequence of a vanishing Wilson line in the e_6 direction. This leads to a D_4 flavour symmetry [4, 12, 20]. The two localised families form a doublet, while the third family transforms in a one-dimensional representation of D_4. This set-up forbids sizeable flavour changing neutral currents and thus relieves the so-called "flavour problem". Furthermore, the geometric reason for small Yukawa couplings of the first and second family is their minimal overlap with the bulk Higgs fields. This leads to Yukawa couplings of higher order and a hierarchical generation of masses based on the Froggatt–Nielsen mechanism [21], where the FI-term provides a small parameter ϵ that controls the pattern of masses.

In addition, the first two families live at points of enhanced gauge symmetries and therefore build complete representations of the local grand unified gauge group, e.g. as **16**-plets of SO(10). Hence, they enjoy the successful properties of "Local Grand Unification" outlined in the first golden rule.

3.2.4. *Lesson 4: The pattern of SUSY breakdown*

The question of supersymmetry breakdown is a complicated process and we shall try to extract some general lessons that are rather model-independent. Specifically we would consider gaugino condensation in the hidden sector [22–25] realized explicitly in the MiniLandscape [26], see also Section 5.4.

A reasonable value for the gravitino mass can be obtained if the dilaton is fixed at a realistic value $1/g^2(M_{\mathrm{GUT}}) = \mathrm{Re}S \approx 2$. Thus, the discussion needs the study of moduli stabilization, which, fortunately, we do not have to analyse here. In fact we can rely on some specific pattern of supersymmetry breaking which seems to be common in various string theories, first observed in the framework of Type IIB theory [27–33] and later confirmed in the heterotic case [34, 35]: so-called "mirage mediation". Its source is a suppression of the tree level contribution in modulus mediation (in particular for gaugino masses and A-parameters). The suppression factor is given by the logarithm of the "hierarchy" $\log(M_{\text{Planck}}/m_{3/2})$, which numerically is of the order $4\pi^2$. Non-leading terms suppressed by loop factors can now compete with the tree-level contribution. In its simplest form the loop corrections are given by the corresponding β-functions, leading to "anomaly mediation" if the tree level contribution is absent. Without going into detail, let us just summarise the main properties of mirage mediation:

- gaugino masses and A-parameters are suppressed compared to the gravitino mass by the factor $\log(M_{\text{Planck}}/m_{3/2})$
- we obtain a compressed pattern of gaugino masses (as the $SU(3)$ β-function is negative while those of $SU(2)$ and $U(1)$ are positive)
- soft scalar masses m_0 are more model-dependent. In general we would expect them to be as large as $m_{3/2}$ [29].

The models of the MiniLandscape inherit this generic picture. But they also teach us something new on the soft scalar masses, which results in lesson 4. The scalars reside in various localisations in the extra dimensions that feel SUSY in different ways: First, the untwisted sector is obtained from simple torus compactification of the 10D theory leading to extended $\mathcal{N} = 4$ supersymmetry in $D = 4$. Hence, soft terms of bulk fields are protected (at least at tree level) and broken by loop corrections when they communicate to sectors with less SUSY. Next, scalars localised on fixed tori feel a remnant $\mathcal{N} = 2$ SUSY and might be protected as well. Finally, fields localised at fixed points feel only $\mathcal{N} = 1$ SUSY and are not further protected [36, 37]. Therefore, we expect soft terms $m_0 \sim m_{3/2}$ for the localised first two families, while other (bulk) scalar fields, in particular the Higgs bosons and the stop, feel a protection from extended SUSY. Consequently, their soft masses are suppressed compared to $m_{3/2}$ (by a loop factor of order $1/4\pi^2$). This constitutes lesson 4 of the MiniLandscape.

4. The OrbifoldLandscape

The 10D heterotic string compactified on a six-dimensional toroidal orbifolds provides an easy and calculable framework for string phenomenology [2, 3]. A toroidal orbifold is constructed by a six-dimensional torus divided out by some of its discrete isometries, the so-called point group. For simplicity we assume this discrete symmetry to be Abelian. Combined with the condition on $\mathcal{N} = 1$ supersymmetry in 4D one is left with certain $\mathbb{Z}_N$ and $\mathbb{Z}_N \times \mathbb{Z}_M$ groups, in total 17 different choices. For each choice, there are in general several inequivalent possibilities, e.g. related to the underlying six-torus. Recently, these possibilities have been classified using methods from crystallography, resulting in 138 inequivalent orbifold geometries with Abelian point group [38].

The orbifolder [39] is a powerful computer program to analyse these Abelian orbifold compactifications of the heterotic string. The program includes a routine to automatically generate a huge set of consistent (i.e. modular invariant and hence anomaly-free) orbifold models and to identify those that are phenomenologically interesting, e.g. that are MSSM-like.

A crucial step in this routine is the identification of inequivalent orbifold models in order to avoid an overcounting: even though the string theory input parameters of two models (i.e. so-called shifts and orbifold Wilson lines) might look different, the models can be equivalent and share, for example, the same massless spectrum and couplings. The current version (1.2) of the orbifolder uses simply the massless spectrum in terms of the representations under the full non-Abelian gauge group in order to identify inequivalent models. However, there are typically five to ten $U(1)$ factors and the corresponding charges are neglected for this comparison of spectra, because they are highly dependent on the choice of $U(1)$ basis. As pointed out by Groot Nibbelink and Loukas [40] one can easily improve this by using in addition to the non-Abelian representations also the $U(1)_Y$ hypercharge as it is uniquely defined for a given MSSM model. We included this criterion into the orbifolder. However, it turns out that using this refined comparison method the number of inequivalent MSSM-like orbifold models increases only by 3%.

4.1. *Search in the "OrbifoldLandscape"*

Using the improved version of the orbifolder we performed a scan in the landscape of all $\mathbb{Z}_N$ and certain $\mathbb{Z}_N \times \mathbb{Z}_M$ heterotic orbifold geometries for MSSM-like models, where our basic requirements for a model to be MSSM-like are:

- SM gauge group $SU(3) \times SU(2) \times U(1)_Y \subset E_8$ times a hidden sector
- Hypercharge $U(1)_Y$ is non-anomalous and in $SU(5)$ GUT normalisation
- (net) number of three generations of quarks and leptons
- at least one Higgs pair
- all exotics must be vector-like with respect to the SM gauge group

We identified approximately 12000 MSSM-like orbifold models that suit the above criteria. Given the large number of promising models we call them the "OrbifoldLandscape". A summary of the results can be found in the appendix in Tabs. A.1 and A.2. Furthermore, the orbifolder input files needed to load these models into the program can be found at [41]. The scan did not reveal any MSSM-like models from orbifold geometries with $\mathbb{Z}_3$, $\mathbb{Z}_7$ and $\mathbb{Z}_2 \times \mathbb{Z}_6$-II point group. This is most likely related to the condition of SU(5) GUT normalisation for hypercharge.

Note that this search for MSSM-like orbifold models is by far not complete. For example, we only used the standard $\mathbb{Z}_N \times \mathbb{Z}_M$ orbifold geometries (i.e. those with label (1-1) following the nomenclature of Ref. [38]). In addition, our search was performed in a huge, but still finite parameter set of shifts and Wilson lines. Finally, the routine to identify inequivalent orbifold models can surely be improved further. Hence, presumably only a small fraction of the full heterotic orbifold Landscape has been analysed here.

4.2. *Comparison to the literature*

Let us compare our findings to the literature. The $\mathbb{Z}_6$-II (1-1) orbifold has been studied intensively in the past, see e.g. [4, 6, 7, 10]. Also the MiniLandscape [8, 11] was performed using this orbifold geometry, see Section 3.1. In the first paper [8] local SO(10) and E_6 GUTs were used as a search strategy and thus one was restricted to four out of 61 possible shifts, resulting in 223 MSSM-like models. In the second paper [11] this restriction was lifted, resulting in almost 300 MSSM-like models. They are all included in our set of 348 MSSM-like models from $\mathbb{Z}_6$-II (1-1), see Tab. A.1 in the appendix.

Similar to $\mathbb{Z}_6$-II, the $\mathbb{Z}_2 \times \mathbb{Z}_4$ orbifold geometry has been conjectured to be very promising for MSSM model-building [42]. Here, we can confirm this conjecture: we found 3632 MSSM-like models from $\mathbb{Z}_2 \times \mathbb{Z}_4$ (1-1) — the largest set of models in our scan. Also from a geometrical point of view, the $\mathbb{Z}_2 \times \mathbb{Z}_4$ orbifold is very rich: there are in total 41 different orbifold geometries with $\mathbb{Z}_2 \times \mathbb{Z}_4$ point group, i.e. based on different six-tori and roto-translations [38]. We considered only the standard choice here, labelled (1-1). Hence, one can expect a huge landscape of MSSM-like models to be discovered from $\mathbb{Z}_2 \times \mathbb{Z}_4$.

Recently, Groot Nibbelink and Loukas performed a model scan in all $\mathbb{Z}_8$-I and $\mathbb{Z}_8$-II geometries [40]. They also used a local GUT search strategy (based on SU(5) and SO(10) local GUTs) and hence started with 120 and 108 inequivalent shifts for $\mathbb{Z}_8$-I and $\mathbb{Z}_8$-II, respectively. Their scan resulted in 753 MSSM-like models. Without imposing the local GUT strategy our search revealed in total 1713 MSSM-like models from $\mathbb{Z}_8$, see Tab. A.1.

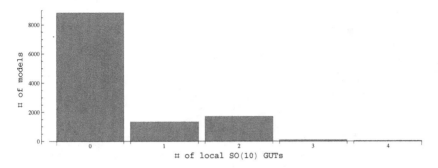

Fig. 3. Number of MSSM-like orbifold models vs. number of local SO(10) GUTs with **16**-plets for matter.

Further orbifold MSSM-like models have been constructed using the $\mathbb{Z}_{12}$-I orbifold geometry [43, 44]. This orbifold seems also to be very promising as we identified 750 MSSM-like models in this case, see Tab. A.1. Finally, we confirm the analysis of Ref. [45] for the orbifold geometries $\mathbb{Z}_6$-I and $\mathbb{Z}_N$ with $N = 3, 4, 7$ and standard lattice (1-1).

In the next section we will apply the strategies described by the "Five golden rules of superstring phenomenology" to our OrbifoldLandscape and search for common properties of our 12000 MSSM-like orbifold models. Thereby, we will see how many MSSM-like models would have been found following the "Five golden rules" strictly and how many would have been lost. Hence, we will estimate the prosperity of the "Five golden rules".

5. Five Golden Rules in the OrbifoldLandscape

In the following we focus on the golden rules I–IV. A detailed analysis of rule V is very model-dependent and will thus not be discussed here.

5.1. *Rule I: Spinor of* SO(10) *for SM matter*

As discussed in Section 2.1 at least some generations of quarks and leptons might originate from spinors of SO(10) sitting at points in extra dimensions with local SO(10) GUT.[b] Hence, we perform a statistic on the number of such localisations in our 12000 MSSM-like orbifold models. The results are summarised in Tab. A.2 and displayed in Fig. 3.

It turns out that 25% of our models have at least one local SO(10) GUT. Furthermore, we find that some orbifolds seem to forbid local SO(10) GUTs

---

[b]See also [5, 14] and for an overview on local GUTs Ref. [10].

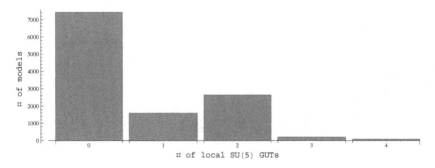

Fig. 4.   Number of MSSM-like orbifold models vs. number of local SU(5) GUTs with **10**-plets for matter.

with **16**-plets (for example $\mathbb{Z}_6$-I [10]). On the other hand, the MSSM-like models from $\mathbb{Z}_6$-II and $\mathbb{Z}_8$-I (1-1) and (2-1) prefer zero or two localised **16**-plets of SO(10). Three local **16**-plets are very uncommon, they mostly appear in $\mathbb{Z}_2 \times \mathbb{Z}_4$.

Note that the number of local GUTs can be greater than three even though the model has a (net) number of three generations of quarks and leptons. Obviously, an anti-generation of quarks and leptons is needed in such a case. The maximal number we found in our scan is four local SO(10) GUTs with **16**-plets for matter in the cases of $\mathbb{Z}_2 \times \mathbb{Z}_2$ and $\mathbb{Z}_2 \times \mathbb{Z}_4$ orbifold geometries.

5.1.1. *Other local GUTs*

In addition, we analyse our 12000 models for local SU(5) GUTs with local matter in **10**-plets. The results are summarised in Table A.2 and displayed in Fig. 4. We find this case to be very common: almost 40% of our MSSM-like models have at least one local **10**-plet of a local SU(5) GUT.

Next, we also look for local E_6 GUTs with **27**-plets. We find only a few cases, most of them appear in $\mathbb{Z}_N \times \mathbb{Z}_M$ orbifold geometries, see Table A.2.

Finally, we scan our models for localised SM generations (i.e. localised left-handed quark-doublets) transforming in a complete multiplet of any local GUT group that unifies the SM gauge group. Again, our results are listed in Table A.2 and visualised in Fig. 5. We find most of our models, i.e. 70%, have at least one local GUT with a localised SM generation.

In summary, the first golden rule, which demands for local GUTs in extra dimensions in order to obtain complete GUT multiplets for matter, is very successful: most of our 12000 MSSM-like models share this property

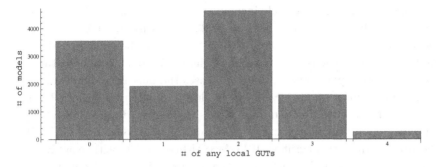

Fig. 5. Number of MSSM-like orbifold models vs. number of local GUTs with local GUT multiplets for SM matter.

automatically, it was not imposed by hand in our search.

5.2. *Rule II: Incomplete GUT multiplets for the Higgs pair*

Since the Higgs doublets reside in incomplete GUT multiplets, they might be localised at some region of the orbifold where the higher-dimensional GUT is broken to the 4D Standard Model gauge group. This scenario yields a natural solution to the doublet–triplet splitting problem. The untwisted sector (i.e. bulk) would be a prime candidate for such a localisation, but there can be further possibilities. The numbers of such GUT breaking localisations are summarised in Table A.1 and displayed in Fig. 6.

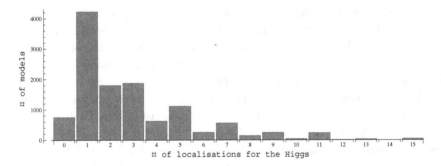

Fig. 6. Number of orbifold models vs. number of localisations with broken local GUT such that only a Higgs doublet but not the triplet survives. The 10D bulk is the most common localisation of this kind.

We see that GUT breaking localisations are very common among our MSSM-like models. Only a very few models do not contain any GUT breaking localisations that yield incomplete GUT multiplets for at least one Higgs. On the other hand, there are 4223 cases with one GUT breaking localisation — in most cases (4097 out of 4223) this is the bulk. In addition, there are many models that have more than one possibility for naturally split Higgs multiplets, but in almost all cases the bulk is among them.

Note that most of our MSSM-like models have additional exotic Higgs-like pairs, mostly two to six additional ones. In contrast to the MSSM Higgs pair they often originate from complete multiplets of some local GUT. On the other hand, we identified 1011 MSSM-like models with exactly one Higgs pair. Cases with exactly one Higgs pair, originating from the bulk might be especially interesting.

In summary, the second golden rule, which explains incomplete GUT multiplets for the Higgs using GUT breaking localisations in extra dimensions, is very successful — as in the case of the first golden rule, most of our 12000 MSSM-like models follow this rule automatically.

5.3. *Rule III: Repetition of families*

The Standard Model contains three generations of quarks and leptons with a peculiar pattern of masses and mixings. This might be related to a (discrete) flavour symmetry.[c]

From the orbifold perspective discrete flavour symmetries naturally arise from the symmetries of the orbifold geometry [12, 20]. However, certain background fields (i.e. orbifold Wilson lines [15]) can break these symmetries. The maximal number of orbifold Wilson lines is six corresponding to the six directions of the compactified space. The orbifold-rotation, however, in general identifies some of those directions. Hence, the corresponding Wilson lines have to be equal. For example, the $\mathbb{Z}_3$ orbifold allows for maximally three independent Wilson lines.

In general, one can say that the more Wilson lines vanish the larger is the discrete flavour symmetry. On the other hand, non-vanishing Wilson lines are generically needed in order to obtain the Standard Model gauge group and to reduce the number of generations to three. Hence, it is interesting to perform a statistic on the number of vanishing Wilson lines for our 12000 MSSM-like orbifold models, see Tab. A.1 in the appendix and Fig. 7.

There are orbifold geometries, like $\mathbb{Z}_4$, $\mathbb{Z}_6$-I and $\mathbb{Z}_{12}$-I, apparently de-

---

[c]A gauged flavour symmetry like SU(2) or SU(3) is also possible. Some of the models in our OrbifoldLandscape realise this possibility, but we do not analyse these cases in detail here.

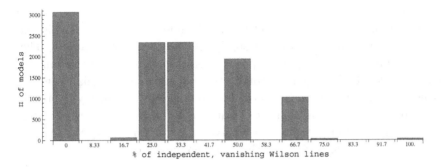

Fig. 7. Number of MSSM-like orbifold models vs. the percentage of independent Wilson lines that are vanishing (e.g. the $\mathbb{Z}_2 \times \mathbb{Z}_2$ orbifold allows for six independent Wilson lines. If one is vanishing, the percentage is 16.7%). Higher percentages generically correspond to larger flavour symmetries.

manding for all possible orbifold Wilson lines to be non-trivial in order to yield the MSSM, see Tab. A.1. These MSSM-like models are expected to have no discrete, non-Abelian flavour symmetries. On the other hand, there are several orbifold geometries that seem to require at least one vanishing Wilson line in order to reproduce the MSSM with its three generations, for example $\mathbb{Z}_6$-II, $\mathbb{Z}_2 \times \mathbb{Z}_2$, $\mathbb{Z}_2 \times \mathbb{Z}_4$, $\mathbb{Z}_3 \times \mathbb{Z}_3$ and $\mathbb{Z}_4 \times \mathbb{Z}_4$. In general, the case of vanishing Wilson lines is very common: we see that in 75% of our MSSM-like orbifold models at least one allowed orbifold Wilson line is zero. In these cases non-Abelian flavour symmetries are expected. For example, most of the MSSM-like models from $\mathbb{Z}_6$-II (1-1) have a D_4 flavour symmetry with the first two generations transforming as a doublet and the third one as a singlet [4, 12, 20].

In summary, the third golden rule, which explains the origin of three generations of quarks and leptons by geometrical properties of the compactification space, is generically satisfied for our 12000 MSSM-like orbifold models.

5.4. *Rule IV: $\mathcal{N} = 1$ supersymmetry*

By construction, i.e. by choosing the appropriate orbifold geometries, our 12000 MSSM-like orbifold models preserve $\mathcal{N} = 1$ supersymmetry in four dimensions. This is expected to be broken by non-perturbative effects, i.e. by hidden sector gaugino condensation [22–25]. Here, we follow the discussion of [26] where low energy supersymmetry breaking in the Mini-Landscape of $\mathbb{Z}_6$-II orbifolds was analysed. See also [46, 47] for a related discussion.

In detail, our MSSM-like models typically possess a non-Abelian hidden sector gauge group with little or no charged matter representations. The corresponding gauge coupling depends via the one-loop β-functions on the energy scale. If the coupling becomes strong at some (intermediate) energy scale Λ the respective gauginos condensate and supersymmetry is broken spontaneously by a non-vanishing dilaton F-term. Assuming that SUSY breaking is communicated to the observable sector via gravity the scale of soft SUSY breaking is given by the gravitino mass, i.e.

$$m_{3/2} \sim \frac{\Lambda^3}{M_{\text{Plank}}^2}\,, \tag{1}$$

where M_{Plank} denotes the Planck mass and the scale of gaugino condensation Λ is given by

$$\Lambda \sim M_{\text{GUT}} \exp\left(-\frac{1}{2\beta}\frac{1}{g^2(M_{\text{GUT}})}\right)\,. \tag{2}$$

For every MSSM-like orbifold model we compute the β-function of the largest hidden sector gauge group under the assumption that any non-trivial hidden matter representation of this gauge group can be decoupled in a supersymmetric way. Furthermore, we assume dilaton stabilization at a realistic value $1/g^2(M_{\text{GUT}}) = \text{Re}S \approx 2$. Hence, we obtain the scale Λ of gaugino condensation. Our results are displayed in Fig. 8. For an intermediate scale $\Lambda \sim 10^{13}\text{GeV}$ one obtains a gravitino mass in the TeV range, which is of phenomenological interest.

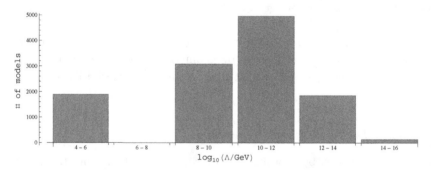

Fig. 8.   Number of MSSM-like orbifold models vs. scale of gaugino condensation for the largest hidden sector gauge group.

The models in the OrbifoldLandscape seem to prefer low energy SUSY breaking. This result is strongly related to the heterotic orbifold construction: the $E_8 \times E_8$ gauge group in 10D is broken by orbifold shift and Wilson lines, which are highly constrained by string theory (i.e. modular invariance). Therefore, both E_8 factors get broken and not only the observable one. It turns out that the unbroken gauge group from the hidden E_8 has roughly the correct size to yield gaugino condensation at an intermediate scale and hence low energy SUSY breaking.

Note that our analysis is just a rough estimate as various effects have been neglected, for example the decoupling of hidden matter, the identification of the gaugino condensation and (string) threshold corrections. These effects can in principle affect the scale of SUSY breaking even by 2-3 orders of magnitude.

6. The General Landscape

With these considerations we have only scratched the surface of the parameter space of potentially realistic models. In addition, we have used "five golden rules" as a prejudice for model selection and it has to be seen whether this is really justified.

For general model building in the framework of (perturbative) string theory we have the following theories at our disposal:

- type I string with gauge group $SO(32)$
- heterotic $SO(32)$
- heterotic $E_8 \times E_8$
- type IIA and IIB orientifolds
- intersecting branes with gauge group $U(N)^M$

As we explained in detail, our rule I points towards exceptional groups and hence towards the $E_8 \times E_8$ heterotic string. On the other hand, type II orientifolds typically provide gauge groups of type $SO(M)$ or $U(N)$ and products thereof. Although we have $SO(2N)$ gauge groups in these schemes, matter fields do not come as spinors of $SO(2N)$, but originate from adjoint representations. In the intersecting brane models based on $U(N)^M$ gauge groups matter transforms in bifundamental representations of $U(N) \times U(L)$ (originating from the adjoint of $U(N + L)$). While this works nicely for the standard model representations, it appears to be difficult to describe a grand unified picture with e.g. gauge group $SU(5)$. Trying to obtain a GUT yields a gauge group at least as large as $U(5)$ and one has problems with a perturbative top-quark Yukawa coupling. One possible way out is the construction of string models without the prejudice for GUTs, see e.g. [48].

A comprehensive review on these intersecting brane model constructions can be found in the book of Ibáñez and Uranga [49] or other reviews [50]. These models have a very appealing geometric interpretation, see e.g. [51]: Fields are located on branes of various dimensions. Thus, physical properties of the models can be inferred from the localisation of the brane-fields in the extra dimensions and by the overlap of their wave functions, similar to the heterotic MiniLandscape. This nice geometrical set-up leads to attempts to construct so-called "local models". Here, one assumes that all particle physics properties of the model are specified by some local properties at some specific point or sub-space of the compactified dimensions and that the "bulk" properties can be decoupled. However, the embedding of the local model into an ultraviolet complete and consistent string model is an assumption and its validity remains an open question.

Further schemes include "non-perturbative" string constructions:

- M-theory in $D = 11$
- heterotic M-theory $E_8 \times E_8$
- F-theory

These non-perturbative constructions are conjectured theories that generalize string theories or known supergravity field theories in higher dimensions. The low energy limit of M-theory is 11-dimensional supergravity. Heterotic M-theory is based on a $D = 11$ theory bounded by two $D = 10$ branes with gauge group E_8 on each boundary and F-theory is a generalization of type IIB theory, where certain symmetries can be understood geometrically. This non-perturbative construction allows for singularities in extra dimensions that lead to non-trivial gauge groups according to the so-called A-D-E classification. Groups of the A-type ($SU(N)$) and D-type ($SO(2N)$) can also be obtained in the perturbative constructions with D-branes and orientifold branes, while exceptional gauge groups can only appear through the presence of E-type singularities. This allows for spinors of $SO(10)$ and can produce a non-trivial top-quark Yukawa coupling within an $SU(5)$ grand unified theory. In that sense, F-theory can be understood as an attempt to incorporate rule I within type IIB theory. Unfortunately, it is difficult to control the full non-perturbative theory and the search for realistic models is often based on local model building. Many questions are still open but there is enough room for optimism that promising models can be embedded in a consistent ultraviolet completion.

A general problem of string phenomenology is the difficulty to perform the explicit calculations needed to check the validity of the model.

This is certainly true for the non-perturbative models, where we have (at best) some effective supergravity description. But also in the perturbative constructions we have to face this problem. We have to use simplified compactification schemes to be able to do the necessary calculations — we need a certain level of "Berechenbarkeit". In our discussion we used the flat orbifold compactification that allows the use of conformal field theory methods. In principle, this enables us to do all the necessary calculations to check the models in detail. In the $\mathbb{Z}_6$-II MiniLandscape this has been elaborated to a large extend. For the more general orbifold landscape, this still has to be done. Other constructions with full conformal field theory control are the free fermionic constructions [52] and the "tensoring" of conformal field theory building blocks: so-called Gepner models [53]. They share "Berechenbarkeit" with the flat orbifold models, but the geometric structure of compactified space is less transparent.

We have to hope that these simplified compactifications (or approximations) lead us to realistic models. In the generic situation one needs smooth manifolds, e.g. Calabi-Yau spaces, and some specific models have been constructed [54, 55]. However, these more generic compactifications require more sophisticated methods for computations that are only partially available, for example in order to determine Yukawa couplings. More recently a simplification based on the embedding of line bundles has allowed the constructions of many models [56, 57]. Still the calculational options are limited. It would be interesting to get a better geometric understanding of the compact manifold. At the moment the "determination" of couplings is based on a supergravity approximation using U(1) symmetries. These symmetries are exact in this approximation at the "stability wall" but are expected to be broken to discrete symmetries in the full theory. This is in concord with rule V asking for the origin of discrete symmetries. Furthermore, this question has recently been analysed intensively within the various string constructions [58–62].

7. Summary

We have seen that there is still a long way to go in the search for realistic particle physics models from string theory. There are many possible roads but we are limited by our calculational techniques. Thus, in the near future we are still forced to make choices. Here, we have chosen to follow "five golden rules" outlined in Section 2, which are mainly motivated by the quest for a unified picture of particle physics interactions. This strategy seems to require an underlying structure provided by exceptional groups pointing towards the $E_8 \times E_8$ heterotic string and F-theory.

Even given these rules, there are stumbling blocks because of the complexity of the compact manifolds. We cannot resolve these problems in full generality: we have to use simplified compactification schemes or approximations. We have to hope that nature has chosen a theory that is somewhat close to these simplified schemes. Of course, any method to go beyond this simplified assumptions should be seriously considered. However, there is some hope that this assumption might be justified: The orbifold models studied in this work have enhanced (discrete) symmetries that could be the origin of symmetries of the standard model, especially with respect to the flavour structure and symmetries relevant for proton stability as well as the absence of other rare processes. Generically, these symmetries are slightly broken as we go away from the orbifold-point. This gives rise to some hierarchical structures, for example for the ratio of quark masses in the spirit of Froggatt and Nielsen [21].

The analysis of the MiniLandscape can be seen as an attempt to study these questions in detail. Based on the availability of conformal field theory techniques we can go pretty far in the analysis of explicit models. A detailed analysis of the "OrbifoldLandscape" has not been performed yet, but should be possible along the same lines. In Section 4 we started this enterprise of model building by constructing 12000 MSSM-like models. In a next step, the detailed properties of promising models have to be worked out. Especially the framework of the $\mathbb{Z}_2 \times \mathbb{Z}_4$ [42] should provide new insight into the properties of realistic models and might teach us further key properties shared by successful models.

One key property that we have learned is the geography of fields in the extra dimensions. The localisation of matter fields and the gauge group profiles in extra dimensions are essential for the properties of the low energy model. This is the first message of the heterotic orbifold construction and shared by the "braneworld" constructions in type II string theory and F-theory. Further lessons are:

- The Higgs pair is a bulk field. This allows for a convincing solution of the μ-problem using a (discrete) R-symmetry and yields doublet–triplet splitting.
- A sizeable value of the top-quark Yukawa coupling requires a sufficient overlap with the Higgs fields in extra dimensions. Thus, the top-quark should extend to the bulk as well.

- The matter fields of the first and second generation should be localised in a region of the extra-dimensional space where they are subject to an enhanced gauge symmetry, like SO(10). This local GUT forces them to appear as complete representations, e.g. as spinors of SO(10). Furthermore, the geometrical structure can manifest itself in a discrete flavour symmetry.

- The quest for low energy supersymmetry is the guiding principle in string model building. Still, it has to be seen whether this is realised in nature. At the moment no sign of supersymmetry has been found at the LHC, although the value of the Higgs mass is consistent with SUSY. The analysis of the models of the MiniLandscape and the location of the fields suggests a certain structure where even some remnants of extended supersymmetry (for fields in the bulk) seem to be at work. This picture of "heterotic supersymmetry" [36, 37] can hopefully be tested experimentally in the not too far future.

Acknowledgments

This work was partially supported by the SFB–Transregio TR33 "The Dark Universe" (Deutsche Forschungsgemeinschaft) and the DFG cluster of excellence "Origin and Structure of the Universe" (www.universe-cluster.de).

Appendix A. Summary of the OrbifoldLandscape

Table A.1. Statistics on MSSM-like models (using the search criteria listed in Section 4.1) obtained from a random scan in all $\mathbb{Z}_N$ and certain $\mathbb{Z}_N \times \mathbb{Z}_M$ heterotic orbifold geometries. The first column labels the geometry following the nomenclature from [38]. The second column gives the number of inequivalent MSSM-like models found in our scan. Next, we give the maximal number of independent Wilson lines (WLs) possible for the respective orbifold geometry and in the fourth column we count the number of MSSM-like models with a certain number (i.e. 0,1,2,3,4) of vanishing Wilson lines, see Section 5.3. In the fifth column we count the number of locations with broken local GUT such that Higgs-doublets without triplets appear, see Section 5.2. Finally, in the last column we give the number of models without $U(1)_{anom}$, i.e. without FI term.

orbifold		# MSSM	max. # of indep. WLs	# models with indep. vanishing WLs					# models with locations for split Higgs							# MSSM without $U(1)_{anom}$
				0	1	2	3	≥4	0	1	2	3	4	5	≥6	
$\mathbb{Z}_3$	(1,1)	0	3	0	0	0	0	0	0	0	0	0	0	0	0	0
$\mathbb{Z}_4$	(1,1)	0	4	0	0	0	0	0	0	0	0	0	0	0	0	0
	(2,1)	128	3	128	0	0	0	0	6	107	12	3	0	0	0	0
	(3,1)	25	2	25	0	0	0	0	0	25	0	0	0	0	0	0
$\mathbb{Z}_6$-I	(1,1)	31	1	31	0	0	0	0	0	31	0	0	0	0	0	0
	(2,1)	31	1	31	0	0	0	0	0	31	0	0	0	0	0	0
$\mathbb{Z}_6$-II	(1,1)	348	3	13	335	0	0	0	20	167	111	34	8	2	6	1
	(2,1)	338	3	10	328	0	0	0	19	162	107	33	9	2	6	2
	(3,1)	350	3	18	332	0	0	0	17	172	112	41	7	1	0	2
	(4,1)	334	2	39	295	0	0	0	17	161	113	32	11	0	0	3
$\mathbb{Z}_7$	(1,1)	0	1	0	0	0	0	0	0	0	0	0	0	0	0	0
$\mathbb{Z}_8$-I	(1,1)	263	2	221	42	0	0	0	0	128	85	50	0	0	0	7
	(2,1)	164	2	123	41	0	0	0	0	76	53	35	0	0	0	5
	(3,1)	387	1	387	0	0	0	0	27	150	175	32	3	0	0	27
$\mathbb{Z}_8$-II	(1,1)	638	3	212	404	22	0	0	12	257	165	123	16	50	15	7
	(2,1)	260	2	92	168	0	0	0	15	108	84	34	2	12	5	3
$\mathbb{Z}_{12}$-I	(1,1)	365	1	365	0	0	0	0	5	259	55	42	4	0	0	8
	(2,1)	385	1	385	0	0	0	0	7	271	63	44	0	0	0	9
$\mathbb{Z}_{12}$-II	(1,1)	211	2	135	76	0	0	0	9	40	107	31	12	4	8	3
$\mathbb{Z}_2 \times \mathbb{Z}_2$	(1,1)	101	6	0	59	42	0	0	79	0	10	3	8	0	1	0
$\mathbb{Z}_2 \times \mathbb{Z}_4$	(1,1)	3632	4	67	2336	1199	30	0	393	1194	160	690	83	449	663	10
$\mathbb{Z}_2 \times \mathbb{Z}_6$-I	(1,1)	445	2	332	113	0	0	0	54	118	105	79	27	13	49	5
$\mathbb{Z}_2 \times \mathbb{Z}_6$-II	(1,1)	0	0	0	0	0	0	0	0	0	0	0	0	0	0	0
$\mathbb{Z}_3 \times \mathbb{Z}_3$	(1,1)	445	3	1	369	75	0	0	27	212	1	15	102	0	88	9
$\mathbb{Z}_3 \times \mathbb{Z}_6$	(1,1)	465	1	441	24	0	0	0	4	39	64	82	88	110	78	0
$\mathbb{Z}_4 \times \mathbb{Z}_4$	(1,1)	1466	3	11	529	921	5	0	28	441	49	195	81	323	349	1
$\mathbb{Z}_6 \times \mathbb{Z}_6$	(1,1)	1128	0	1128	0	0	0	0	9	74	165	271	161	148	300	0
total		11940							748	4223	1796	1869	622	1114	1568	102

Table A.2. Statistics on MSSM-like models (using the search criteria listed in Section 4.1) obtained from a random scan in all $\mathbb{Z}_N$ and certain $\mathbb{Z}_N \times \mathbb{Z}_M$ heterotic orbifold geometries. The first column labels the geometry following the nomenclature from [38]. The next four columns display the number of MSSM-like models with 0,1,2,3 and (up to) 4 local GUTs of specified gauge group with corresponding local matter: local SO(10) GUTs with local 16-plets, local E_6 GUTs with local 27-plets, local SU(5) GUTs with local 10-plets and, finally, any local GUTs that unify SU(3) × SU(2) × U(1)$_Y$ in a single gauge group with corresponding local matter representations containing left-handed quark doublets.

orbifold	# models with local SO(10) GUTs					# models with local E_6 GUTs			# models with local SU(5) GUTs					# models with local GUTs				
	0	1	2	3	4	0	1	2	0	1	2	3	4	0	1	2	3	4
$\mathbb{Z}_3$ (1,1)	0	0	0	0	0	0	0	0	0	0	0	0	0	0	0	0	0	0
$\mathbb{Z}_4$ (1,1)	0	0	0	0	0	0	0	0	0	0	0	0	0	0	0	0	0	0
$\mathbb{Z}_4$ (2,1)	78	50	0	0	0	50	78	0	128	0	0	0	0	0	128	0	0	0
$\mathbb{Z}_4$ (3,1)	5	20	0	0	0	20	5	0	25	0	0	0	0	0	25	0	0	0
$\mathbb{Z}_6$-I (1,1)	31	0	0	0	0	31	0	0	31	0	0	0	0	31	0	0	0	0
$\mathbb{Z}_6$-I (2,1)	31	0	0	0	0	31	0	0	31	0	0	0	0	31	0	0	0	0
$\mathbb{Z}_6$-II (1,1)	155	2	187	4	0	332	6	10	203	12	133	0	0	2	3	293	4	46
$\mathbb{Z}_6$-II (2,1)	148	1	186	3	0	323	5	10	204	6	128	0	0	2	5	324	4	3
$\mathbb{Z}_6$-II (3,1)	164	1	185	0	0	328	11	11	202	12	136	0	0	2	11	293	9	35
$\mathbb{Z}_6$-II (4,1)	158	3	173	0	0	299	23	12	195	18	121	0	0	0	14	315	5	0
$\mathbb{Z}_7$ (1,1)	0	0	0	0	0	0	0	0	0	0	0	0	0	0	0	0	0	0
$\mathbb{Z}_8$-I (1,1)	143	0	120	0	0	263	0	0	226	37	0	0	0	106	31	120	6	0
$\mathbb{Z}_8$-I (2,1)	92	0	72	0	0	164	0	0	147	17	0	0	0	75	15	74	0	0
$\mathbb{Z}_8$-I (3,1)	164	140	83	0	0	346	32	9	336	29	22	0	0	105	117	133	32	0
$\mathbb{Z}_8$-II (1,1)	428	77	133	0	0	638	0	0	276	155	207	0	0	79	194	355	10	0
$\mathbb{Z}_8$-II (2,1)	180	29	51	0	0	260	0	0	89	52	114	5	0	28	29	185	18	0
$\mathbb{Z}_{12}$-I (1,1)	365	0	0	0	0	259	0	106	365	0	0	0	0	250	0	115	0	0
$\mathbb{Z}_{12}$-I (2,1)	385	0	0	0	0	269	0	116	385	0	0	0	0	269	0	116	0	0
$\mathbb{Z}_{12}$-II (1,1)	110	69	32	0	0	177	31	3	86	78	47	0	0	0	80	131	0	0
$\mathbb{Z}_2 \times \mathbb{Z}_2$ (1,1)	72	6	12	1	10	66	33	2	75	0	11	0	15	3	18	8	30	42
$\mathbb{Z}_2 \times \mathbb{Z}_4$ (1,1)	2948	300	297	68	19	3181	358	93	2831	71	707	7	16	1918	70	670	911	63
$\mathbb{Z}_2 \times \mathbb{Z}_6$-I (1,1)	312	124	9	0	0	252	63	130	245	126	71	3	0	40	66	193	119	27
$\mathbb{Z}_2 \times \mathbb{Z}_6$-II (1,1)	0	0	0	0	0	0	0	0	0	0	0	0	0	0	0	0	0	0
$\mathbb{Z}_3 \times \mathbb{Z}_3$ (1,1)	444	1	0	0	0	445	0	0	289	3	2	151	0	246	2	3	194	0
$\mathbb{Z}_3 \times \mathbb{Z}_6$ (1,1)	396	33	36	0	0	463	2	0	77	294	42	12	40	3	291	116	15	40
$\mathbb{Z}_4 \times \mathbb{Z}_4$ (1,1)	1246	116	94	10	0	1293	173	0	703	31	709	13	10	353	205	674	224	10
$\mathbb{Z}_6 \times \mathbb{Z}_6$ (1,1)	761	349	18	0	0	1122	6	0	274	656	191	7	0	0	609	511	8	0
total	8816	1321	1688	86	29	10612	826	502	7423	1597	2641	198	81	3543	1913	4629	1589	266

References

[1] H. P. Nilles, Five golden rules for superstring phenomenology (2004). hep-th/0410160.

[2] L. J. Dixon, J. A. Harvey, C. Vafa, and E. Witten, Strings on Orbifolds, Nucl.Phys. **B261**, 678–686 (1985). doi: 10.1016/0550-3213(85)90593-0.

[3] L. J. Dixon, J. A. Harvey, C. Vafa, and E. Witten, Strings on Orbifolds. 2., Nucl.Phys. **B274**, 285–314 (1986). doi: 10.1016/0550-3213(86)90287-7.

[4] T. Kobayashi, S. Raby, and R.-J. Zhang, Searching for realistic 4d string models with a Pati-Salam symmetry: Orbifold grand unified theories from heterotic string compactification on a Z_6 orbifold, Nucl.Phys. **B704**, 3–55 (2005). doi: 10.1016/j.nuclphysb.2004.10.035.

[5] W. Buchmüller, K. Hamaguchi, O. Lebedev, and M. Ratz, Dual models of gauge unification in various dimensions, Nucl.Phys. **B712**, 139–156 (2005). doi: 10.1016/j.nuclphysb.2005.01.038.

[6] W. Buchmüller, K. Hamaguchi, O. Lebedev, and M. Ratz, Supersymmetric standard model from the heterotic string, Phys.Rev.Lett. **96**, 121602 (2006). doi: 10.1103/PhysRevLett.96.121602.

[7] W. Buchmüller, K. Hamaguchi, O. Lebedev, and M. Ratz, Supersymmetric Standard Model from the Heterotic String (II), Nucl.Phys. **B785**, 149–209 (2007). doi: 10.1016/j.nuclphysb.2007.06.028.

[8] O. Lebedev, H. P. Nilles, S. Raby, S. Ramos-Sánchez, M. Ratz, et al., A Mini-landscape of exact MSSM spectra in heterotic orbifolds, Phys.Lett. **B645**, 88–94 (2007). doi: 10.1016/j.physletb.2006.12.012.

[9] O. Lebedev, H. P. Nilles, S. Raby, S. Ramos-Sánchez, M. Ratz, et al., The Heterotic Road to the MSSM with R parity, Phys.Rev. **D77**, 046013 (2008). doi: 10.1103/PhysRevD.77.046013.

[10] M. Ratz, Notes on Local Grand Unification (2007). arXiv:0711.1582 [hep-ph].

[11] O. Lebedev, H. P. Nilles, S. Ramos-Sánchez, M. Ratz, and P. K. Vaudrevange, Heterotic MiniLandscape. (II). Completing the search for MSSM vacua in a Z_6 orbifold, Phys.Lett. **B668**, 331–335 (2008). doi: 10.1016/j.physletb.2008.08.054.

[12] T. Kobayashi, H. P. Nilles, F. Plöger, S. Raby, and M. Ratz, Stringy origin of non-Abelian discrete flavor symmetries, Nucl.Phys. **B768**, 135–156 (2007). doi: 10.1016/j.nuclphysb.2007.01.018.

[13] H. P. Nilles, S. Ramos-Sánchez, M. Ratz, and P. K. Vaudrevange, From strings to the MSSM, Eur.Phys.J. **C59**, 249–267 (2009). doi: 10.1140/epjc/s10052-008-0740-1.

[14] S. Förste, H. P. Nilles, P. K. Vaudrevange, and A. Wingerter, Heterotic brane world, Phys.Rev. **D70**, 106008 (2004). doi: 10.1103/PhysRevD.70.106008.

[15] L. E. Ibáñez, H. P. Nilles, and F. Quevedo, Orbifolds and Wilson Lines, Phys.Lett. **B187**, 25–32 (1987). doi: 10.1016/0370-2693(87)90066-9.

[16] H. P. Nilles, The strings connection: MSSM-like models from strings, Eur.Phys.J. **C74** (2014). doi: DOI10.1140/epjc/s10052-013-2712-3.

[17] R. Kappl, H. P. Nilles, S. Ramos-Sánchez, M. Ratz, K. Schmidt-Hoberg,

et al., Large hierarchies from approximate R symmetries, *Phys.Rev.Lett.* **102**, 121602 (2009). doi: 10.1103/PhysRevLett.102.121602.

[18] F. Brümmer, R. Kappl, M. Ratz, and K. Schmidt-Hoberg, Approximate R-symmetries and the mu term, *JHEP.* **1004**, 006 (2010). doi: 10.1007/JHEP04(2010)006.

[19] J. Casas and C. Muñoz, A Natural solution to the mu problem, *Phys.Lett.* **B306**, 288–294 (1993). doi: 10.1016/0370-2693(93)90081-R.

[20] H. P. Nilles, M. Ratz, and P. K. Vaudrevange, Origin of Family Symmetries, *Fortsch.Phys.* **61**, 493–506 (2013). doi: 10.1002/prop.201200120.

[21] C. Froggatt and H. B. Nielsen, Hierarchy of Quark Masses, Cabibbo Angles and CP Violation, *Nucl.Phys.* **B147**, 277 (1979). doi: 10.1016/0550-3213(79)90316-X.

[22] H. P. Nilles, Dynamically Broken Supergravity and the Hierarchy Problem, *Phys.Lett.* **B115**, 193 (1982). doi: 10.1016/0370-2693(82)90642-6.

[23] S. Ferrara, L. Girardello, and H. P. Nilles, Breakdown of Local Supersymmetry Through Gauge Fermion Condensates, *Phys.Lett.* **B125**, 457 (1983). doi: 10.1016/0370-2693(83)91325-4.

[24] J. Derendinger, L. E. Ibáñez, and H. P. Nilles, On the Low-Energy d = 4, N=1 Supergravity Theory Extracted from the d = 10, N=1 Superstring, *Phys.Lett.* **B155**, 65 (1985). doi: 10.1016/0370-2693(85)91033-0.

[25] M. Dine, R. Rohm, N. Seiberg, and E. Witten, Gluino Condensation in Superstring Models, *Phys.Lett.* **B156**, 55 (1985). doi: 10.1016/0370-2693(85)91354-1.

[26] O. Lebedev, H.-P. Nilles, S. Raby, S. Ramos-Sánchez, M. Ratz, et al., Low Energy Supersymmetry from the Heterotic Landscape, *Phys.Rev.Lett.* **98**, 181602 (2007). doi: 10.1103/PhysRevLett.98.181602.

[27] K. Choi, A. Falkowski, H. P. Nilles, M. Olechowski, and S. Pokorski, Stability of flux compactifications and the pattern of supersymmetry breaking, *JHEP.* **0411**, 076 (2004). doi: 10.1088/1126-6708/2004/11/076.

[28] K. Choi, A. Falkowski, H. P. Nilles, and M. Olechowski, Soft supersymmetry breaking in KKLT flux compactification, *Nucl.Phys.* **B718**, 113–133 (2005). doi: 10.1016/j.nuclphysb.2005.04.032.

[29] O. Lebedev, H. P. Nilles, and M. Ratz, De Sitter vacua from matter superpotentials, *Phys.Lett.* **B636**, 126–131 (2006). doi: 10.1016/j.physletb.2006.03.046.

[30] V. Nilles and J. Plank, *CCR.* **42**, 736–744 .

[31] O. Lebedev, H. P. Nilles, and M. Ratz, A Note on fine-tuning in mirage mediation. pp. 211–221 (2005). hep-ph/0511320.

[32] O. Loaiza-Brito, J. Martin, H. P. Nilles, and M. Ratz, Log(M(Pl) / m(3/2)), *AIP Conf.Proc.* **805**, 198–204 (2006). doi: 10.1063/1.2149698.

[33] O. Lebedev, V. Löwen, Y. Mambrini, H. P. Nilles, and M. Ratz, Metastable Vacua in Flux Compactifications and Their Phenomenology, *JHEP.* **0702**, 063 (2007). doi: 10.1088/1126-6708/2007/02/063.

[34] V. Löwen and H. P. Nilles, Mirage Pattern from the Heterotic String, *Phys.Rev.* **D77**, 106007 (2008). doi: 10.1103/PhysRevD.77.106007.

[35] V. Löwen and H. P. Nilles, Crosschecks for Unification at the LHC,

Nucl.Phys. **B827**, 337–358 (2010). doi: 10.1016/j.nuclphysb.2009.11.007.
arXiv:0907.4983 [hep-ph].

[36] S. Krippendorf, H. P. Nilles, M. Ratz, and M. W. Winkler, The heterotic
string yields natural supersymmetry, *Phys.Lett.* **B712**, 87–92 (2012). doi:
10.1016/j.physletb.2012.04.043.

[37] M. Badziak, S. Krippendorf, H. P. Nilles, and M. W. Winkler, The heterotic
MiniLandscape and the 126 GeV Higgs boson, *JHEP.* **1303**, 094 (2013). doi:
10.1007/JHEP03(2013)094.

[38] M. Fischer, M. Ratz, J. Torrado, and P. K. Vaudrevange, Classification
of symmetric toroidal orbifolds, *JHEP.* **1301**, 084 (2013). doi: 10.1007/
JHEP01(2013)084.

[39] H. P. Nilles, S. Ramos-Sánchez, P. K. Vaudrevange, and A. Wingerter, The
Orbifolder: A Tool to study the Low Energy Effective Theory of Heterotic
Orbifolds, *Comput.Phys.Commun.* **183**, 1363–1380 (2012). doi: 10.1016/j.
cpc.2012.01.026.

[40] S. Groot Nibbelink and O. Loukas, MSSM-like models on Z_8 toroidal orbifolds, *JHEP.* **1312**, 044 (2013). doi: 10.1007/JHEP12(2013)044.

[41] H. P. Nilles and P. K. Vaudrevange. The OrbifoldLandscape: Input files for the orbifolder (2014). http://www.th.physik.uni-bonn.de/
nilles/OrbifoldLandscape/.

[42] D. K. M. Pena, H. P. Nilles, and P.-K. Oehlmann, A Zip-code for
Quarks, Leptons and Higgs Bosons, *JHEP.* **1212**, 024 (2012). doi: 10.1007/
JHEP12(2012)024.

[43] J. E. Kim and B. Kyae, Flipped SU(5) from Z_{12}-I orbifold with Wilson line,
Nucl.Phys. **B770**, 47–82 (2007). doi: 10.1016/j.nuclphysb.2007.02.008.

[44] J. E. Kim, J.-H. Kim, and B. Kyae, Superstring standard model from Z_{12}-I
orbifold compactification with and without exotics, and effective R-parity,
JHEP. **0706**, 034 (2007). doi: 10.1088/1126-6708/2007/06/034.

[45] S. Ramos-Sánchez, Towards Low Energy Physics from the Heterotic
String, *Fortsch.Phys.* **10**, 907–1036 (2009). doi: 10.1002/prop.200900073.
Ph.D.Thesis (Advisor: H.P. Nilles).

[46] K. R. Dienes, Statistics on the heterotic landscape: Gauge groups and cosmological constants of four-dimensional heterotic strings, *Phys.Rev.* **D73**,
106010 (2006). doi: 10.1103/PhysRevD.73.106010.

[47] K. R. Dienes, M. Lennek, D. Senechal, and V. Wasnik, Supersymmetry versus Gauge Symmetry on the Heterotic Landscape, *Phys.Rev.* **D75**, 126005
(2007). doi: 10.1103/PhysRevD.75.126005.

[48] B. Gato-Rivera and A. Schellekens, GUTs without guts (2014).
arXiv:1401.1782 [hep-ph].

[49] L. E. Ibáñez and A. M. Uranga, String theory and particle physics: An
introduction to string phenomenology (2012). Cambridge, UK: Univ. Pr.
(2012) 673 p.

[50] R. Blumenhagen, B. Körs, D. Lüst, and S. Stieberger, Four-
dimensional String Compactifications with D-Branes, Orientifolds and
Fluxes, *Phys.Rept.* **445**, 1–193 (2007). doi: 10.1016/j.physrep.2007.04.003.

[51] G. Honecker, M. Ripka, and W. Staessens, The Importance of Being Rigid:

D6-Brane Model Building on T^6/Z_2xZ_6' with Discrete Torsion, *Nucl.Phys.* **B868**, 156–222 (2013). doi: 10.1016/j.nuclphysb.2012.11.011.

[52] A. E. Faraggi, A New standard - like model in the four-dimensional free fermionic string formulation, *Phys.Lett.* **B278**, 131–139 (1992). doi: 10.1016/0370-2693(92)90723-H.

[53] T. Dijkstra, L. Huiszoon, and A. Schellekens, Supersymmetric standard model spectra from RCFT orientifolds, *Nucl.Phys.* **B710**, 3–57 (2005). doi: 10.1016/j.nuclphysb.2004.12.032.

[54] R. Donagi, B. A. Ovrut, T. Pantev, and D. Waldram, Standard models from heterotic M theory, *Adv.Theor.Math.Phys.* **5**, 93–137 (2002).

[55] V. Braun, Y.-H. He, B. A. Ovrut, and T. Pantev, A Heterotic standard model, *Phys.Lett.* **B618**, 252–258 (2005). doi: 10.1016/j.physletb.2005.05.007.

[56] L. B. Anderson, J. Gray, A. Lukas, and E. Palti, Two Hundred Heterotic Standard Models on Smooth Calabi-Yau Threefolds, *Phys.Rev.* **D84**, 106005 (2011). doi: 10.1103/PhysRevD.84.106005.

[57] L. B. Anderson, J. Gray, A. Lukas, and E. Palti, Heterotic Line Bundle Standard Models, *JHEP.* **1206**, 113 (2012). doi: 10.1007/JHEP06(2012)113.

[58] M. Berasaluce-Gonzalez, L. E. Ibáñez, P. Soler, and A. M. Uranga, Discrete gauge symmetries in D-brane models, *JHEP.* **1112**, 113 (2011). doi: 10.1007/JHEP12(2011)113.

[59] L. Ibáñez, A. Schellekens, and A. Uranga, Discrete Gauge Symmetries in Discrete MSSM-like Orientifolds, *Nucl.Phys.* **B865**, 509–540 (2012). doi: 10.1016/j.nuclphysb.2012.08.008.

[60] M. Berasaluce-Gonzalez, P. Camara, F. Marchesano, D. Regalado, and A. Uranga, Non-Abelian discrete gauge symmetries in 4d string models, *JHEP.* **1209**, 059 (2012). doi: 10.1007/JHEP09(2012)059.

[61] P. Anastasopoulos, M. Cvetič, R. Richter, and P. K. Vaudrevange, String Constraints on Discrete Symmetries in MSSM Type II Quivers, *JHEP.* **1303**, 011 (2013). doi: 10.1007/JHEP03(2013)011.

[62] G. Honecker and W. Staessens, To Tilt or Not To Tilt: Discrete Gauge Symmetries in Global Intersecting D-Brane Models, *JHEP.* **1310**, 146 (2013). doi: 10.1007/JHEP10(2013)146.

5

The String Landscape: A Personal Perspective

Keith R. Dienes

Department of Physics, University of Arizona, Tucson, AZ 85721 USA
dienes@email.arizona.edu

In keeping with the "Perspectives" theme of this volume, this Chapter provides a personal perspective on the string landscape. Along the way, the perspectives of many other physicists are discussed as well. No attempt is made to provide a thorough and balanced review of the field, and indeed there is a slight emphasis on my own contributions to this field, as these contributions have been critical to forming my perspective. This Chapter is adapted from a Colloquium which I have delivered at a number of institutions worldwide, and I have attempted to retain the informal and non-technical spirit and style of this Colloquium presentation as much as possible.

1. Some background

This Chapter grew out of a Colloquium which I have given at a number of physics departments across the U.S. This colloquium was called "Probing the String Landscape: Implications, Applications, and Altercations", and my goal in this Colloquium was to introduce the string landscape in a non-technical way, to discuss the implications of the existence of the landscape, to present several applications of the landscape, and to give the audience some sense of the nature of the controversies it has spawned. Needless to say, such a Colloquium cannot avoid revealing the personal perspectives of the speaker, and this Colloquium was no exception. Indeed, the Colloquium was more in the style of social science rather than hard science, as I was more interested in conveying both the history of and current opinion about the string landscape than details concerning the landscape itself. Nevertheless, I also presented a considerable amount of my own work in this Colloquium, partly as illustration and also partly as a way of motivating my own ultimate opinions on this subject.

For this Chapter, I have opted to present what is essentially a written transcript of this Colloquium. Some parts of the Colloquium have been deleted, but others have been embellished. In keeping with its overall social-science flavor, the original Colloquium also liberally quoted a number of prominent physicists. I have opted to retain these quotes, even if (as in some cases) those very same physicists have themselves contributed Chapters to this volume. I apologize to those physicists for the extra exposure I am hereby giving to their words.

2. Motivation: We live in exciting times

I'll begin with a summary of the situation from one prominent physicist[a]:

> What we've discovered in the last several years is that string theory has an incredible diversity — a tremendous number of solutions — and allows different kinds of environments. A lot of the practitioners of this kind of mathematical theory have been in a state of denial about it. They didn't want to recognize it. They want to believe the universe is an elegant universe — and it's not so elegant. It's different over here. It's that over here. It's a Rube Goldberg machine over here. And this has created a sort of sense of denial about the facts about the theory. The theory is going to win, and physicists who are trying to deny what's going on are going to lose. — *Lenny Susskind, Felix Bloch Professor of Theoretical Physics, Stanford University.*

Any subject that can evince a reaction such as this is clearly an exciting one!

Indeed, as we shall see, discussions of the landscape invariably lead to a plethora of questions which are not the sorts of questions which normally appear in a physics Colloquium. These include, for example, questions such as:

---

[a]The majority of quotes in this Chapter are taken from http://edge.org/ conversation/the-landscape and/or http://edge.org/documents/archive/edge145. html, which may be taken to be the source of all quotes whose origins are otherwise uncredited.

- Can the ground state of a theory be completely irrelevant to that theory? Are we deluding ourselves by always focusing on the vacuum?
- What is the boundary between explanation and observation?
- How can we judge when a theory is "natural"? How can we judge whether one theory is more "natural" than another?
- What does it mean for a theory to be predictive? Falsifiable?
- What kinds of numbers should a fundamental theory of physics be capable of predicting? The mass of the electron? The radius of the Earth's orbit around the Sun? The price of tea in China?
- What tools are we allowed to use in formulating a scientific theory? The results of experiments? Theoretical expectations? *Ourselves*?
- Are we, once again, destined to be the center of the universe?
- To what extent can one talk meaningfully about alternative universes?
- Are all possible universes created equal? How many alternate universes might exist? What properties would they have that distinguish them from our own?
- If the nature of fundamental reality is universe-dependent, what becomes of the traditional methods of learning about the natural world which presuppose the uniqueness of a correct answer? Is this really science?

Of course, many of these questions don't sound like physics. Rather, they sound like something far more pernicious: the *philosophy* of physics. The *sociology* of physics. Indeed, one might suspect that any discussion of these questions represents nothing more than the the first steps on the slippery path to social science!

The point, however, is this: We are currently in the throes of a potentially huge paradigm shift in physics. My goal in this Chapter is therefore to explain what this is, and where it came from. Rather than present a finished story with a tidy outcome, I'll instead simply try to convey the sense of excitement and frustration that many in the string community are currently facing.

3. Introducing the string landscape

Most discussions of fundamental physics begin with the standard "tour" through relevant energy and length scales. They start with human-sized

things, proceed to atoms, then nuclei, and finally arrive at the Standard
Model of quarks and leptons. Finally, they push onwards into uncharted
BSM ("Beyond the Standard Model") territory and discuss (currently) hy-
pothetical notions such supersymmetry, GUT's, or strings.

3.1. (Not) Setting the stage

In this writeup, I'll spare the reader a review of the Standard Model (SM)
and the reasons to believe that there is some sort of hidden structure beyond
the Standard Model. I'll also spare the reader a survey of the some of the
dominant ideas that high-energy physicists have developed over the past
30 years for what might lie beyond the Standard Model at higher energies,
such as supersymmetry (SUSY) and grand-unified theories (GUT's).

It is important to stress, however, that neither the SUSY idea nor the
GUT idea present us with a single model with sharp predictions. Rather,
they represent certain types of quantum field theories within which the act
of model-building takes place, and therefore lots of distinct phenomenologies are possible within these frameworks. For example, on the SUSY side,
numerous questions remain unanswered, chief among them the question of
how SUSY is broken. The typical approach is to introduce SUSY-breaking
by hand, which in turn requires the introduction of many additional phenomenological parameters (masses and mixing angles) even beyond those
of the Standard Model. Similarly, even within the GUT framework, there
are many unanswered questions. For example, what is the larger symmetry group underlying the GUT force? $SU(5)$? $SO(10)$? E_6? Other
groups? How do the different particles join together under these groups?
What kinds of interactions are allowed while respecting these enlarged symmetries? Might there exist sequences of successive GUT embeddings at
higher and higher energy scales? Clearly, many questions still remain!

Both the SUSY idea and the GUT idea are very compelling. They and
their low-energy effects will be the focus of experimental high-energy physics
over the next 20 years. But high-energy theorists have plenty of work to do:
We must be able to build theories in order to be able to interpret data! But
the above questions remain. How do we build realistic SUSY theories? How
do we build realistic GUT theories? How can we make sense of alternate
proposals for physics beyond the SM, such as alternative/extended Higgs
structures, large extra spacetime dimensions, strongly coupled (Randall-
Sundrum-type) scenarios, and so forth? How do we incorporate gravity?

The possibilities clearly seem endless. It is therefore natural to seek guidance from some deeper framework. And this is where string theory comes in.

I will not take the time here to introduce string theory. However, in a nutshell, the main characteristic of string theory is that at the natural energy scale for string theory (which is parametrically identified with the Planck scale $\sim 10^{19}$ GeV), each elementary particle is viewed as a specific vibrational or rotational mode of a one-dimensional closed loop of energy called a string. Thus, in string theory, we simply determine the spectrum of allowed vibrational or rotational excitations for our fundamental strings (subject to certain geometric constraints), and then interpret these allowed excitations at lower energies as different particles: electron, neutrino, quark, photon, W, Z, gluon, Higgs, and even graviton. In this way, string theory makes predictions about the nature of "low-energy" world, and moreover "unifies" all of this information as coming or emerging from a single underlying entity, namely the string itself. Both the spectrum and allowed interactions are specified. Furthermore, because string theories naturally exist in ten dimensions, this rich geometric structure must be somehow compactified to four dimensions. Depending on how this compactification is performed, closed strings can then wrap or "wind" around these compactified dimensions.

The story is even richer if we consider open strings. Open strings can be useful tools for understanding the non-perturbative physics associated with closed strings, but can also serve as candidate fundamental strings themselves. Unlike closed strings, open strings have endpoints, and these can end on membrane-like surfaces of various dimensionalities called D-branes. Moreover, these D-branes can intersect each other and have strings stretching between them. Likewise, both the strings and the D-branes can wrap around compactified spacetime dimensions in many different ways, resulting in huge numbers of highly non-trivial geometric configurations.

This geometric richness is tremendously important because all of these features have profound implications for the allowed excitations of the fundamental strings and branes in the theory. Thus, the choices inherent in this geometric richness have tremendous impact on on the predicted "low-energy" spectrum of particles and forces to which these excitations correspond. Consequently, through studies of these possible compactifications, it has been the hope and expectation that string theory would make specific, detailed predictions about the world of the Standard Model and beyond. Indeed, extracting this information from string theory has been the primary goal of *string phenomenology* ever since the earliest days of string theory.

It of course goes without saying that over the past 30 years, string theory has come to occupy a central place in high-energy physics. It has had a profound impact in many branches of theoretical physics and mathematics, and has led to many new ideas and insights concerning the structure of field theory, gauge theory, supersymmetry, and their relations to gravity. Indeed, as early as the 1980's, it has even been called (reputedly by Edward Witten, a professor at the Institute for Advanced Study in Princeton and 1990 Fields Medalist) "a piece of 21st-century physics that fell by chance into the 20th century".

But in order for string theory to actually fulfill its phenomenological promise as a guide to physics beyond the Standard Model, it must actually make unique statements about "low-energy" physics. This uniqueness is critical.

Now, string theory *does* make detailed, specific statements about the low-energy world. However, these statements do not (yet?) rise to the level of unique predictions.

Ultimately, the reason is that string theory gives rise to a multitude of self-consistent vacua. Each one is called a different "string vacuum" or a different "string model". It is like having a big master equation with many possible solutions, each with different properties. Roughly speaking, each of these different string vacua corresponds to a different way of compactifying the theory from ten dimensions down to four dimensions. The different vacua correspond to different choices of compactification manifolds and D-brane wrappings, different Wilson lines, different vacuum expectation values for unfixed moduli fields, different choices of fluxes, and so forth. These features are all part of the geometric richness we discussed above, but string theory apparently provides no theoretical basis on which to declare that one set of choices — *i.e.*, one string model or string vacuum — is favored over another.

Unfortunately, this richness in the numbers of string models produces a corresponding richness in the different possible low-energy phenomenologies that are "predicted" by the string. For example, one set of choices might lead to a low-energy world exhibiting the Standard-Model gauge group $SU(3) \times SU(2) \times U(1)$, three chiral generations of fermions with certain charges and transformation properties under this gauge group, and $\mathcal{N} = 1$ supersymmetry. However, another might lead to an $SO(10)$ gauge group with five generations and two anti-generations of **16** representations, but without supersymmetry. Still another might lead to a low-energy theory whose properties have already been ruled out by experiment. Indeed, the possibilities for the resulting spacetime physics are quite rich and varied, and string theory does not appear to give any mechanism for selecting between them.

Of course, we do know certain things. We know, for example, that these low-energy predictions must be consistent with the Standard Model. (Or at least, we must *demand* that this be true.) Imposing this phenomenological requirement already truncates the set of self-consistent string compactifications quite significantly, and one might then hope that by studying only the surviving compactifications, string theory might still manage to provide theoretical guidance concerning the many possible *extensions* to the Standard Model — guidance that might fill in the gaps in our knowledge as we move upwards in energy scale. However, it turns out that even this truncation does not restrict the space of remaining string models enough that one might be able to make predictions for physics beyond the Standard Model on the basis of the string models that are left. There are simply too many remaining possibilities.

Thus, we see that although a given string model may be completely predictive, string theory as a general framework is not.

3.2. *Huge numbers of string models: A bit of history*

That there are so many self-consistent ways of compactifying the theory has been known since the mid-1980's. Indeed, ever since these earliest days of modern string theory, it was understood by workers in the field that the space of possible string theories is quite large and perhaps even infinite, with each string theory corresponding to a potentially distinct low-energy phenomenology. Different formalisms had been developed for constructing (often overlapping) classes of strings — orbifold constructions, bosonic constructions, fermionic constructions, constructions based on other worldsheet conformal field theories (CFT)'s, and so forth. Some of these constructions even went beyond the possibility of being interpreted as purely geometric compactifications. Each construction technique gave rise to its own "moduli space" of possible string vacua, and the number of different construction techniques seemed bounded only by the limitations of imagination or cleverness. Thus, throughout the late 1980's and 1990's, workers in the field already understood that there was a large space of possible string theories and were already grappling with the unpleasant consequences of this fact insofar as the ultimate low-energy predictivity of string theory was concerned.

In June 1998, I was invited to give a set of lectures on string phenomenology and string model-building at the TASI Summer School. After two weeks of lectures focusing on different construction techniques concerning one particular class of string theories (those which were perturbative and heterotic), I then devoted a major part of the final lecture to providing an "assessment" of the position in which we now found ourselves. The following is a direct quote from the Proceedings of my 1998 TASI lectures, as written at that time:

> At this point, it is perhaps useful to assess the position in which we now find ourselves. Clearly, through these constructions, we are able to produce *many* string models. In fact... the number of self-consistent string models in $D < 10$ is virtually infinite, and there exists a whole space of such models... Moreover, each of these models has a completely different spacetime phenomenology. What, then, is the use of string theory as an "ultimate" theory, if it does not lead to a single, unique model with a unique low-energy phenomenology?
>
> To answer this question, we should recall our discussion at the beginning of these lectures. Just as field theory is a language for building certain models (one of which, say, is the Standard Model), string theory is a new and deeper language by which we might also build models. The advantages of using this new language, as [we] discussed, include the fact that our resulting models incorporate quantum gravity and Planck-scale physics. Of course, in field theory, many parameters enter into the choice of model-building. These parameters include the choice of fields (for example, the choice of the gauge group, and whether or not to have spacetime supersymmetry), the number of fields (for example, the number of generations), the masses of particles, their mixing angles, and so forth. These are all *spacetime* parameters. In string theory, by contrast, we do not choose these spacetime parameters; we instead choose a set of *worldsheet* parameters [*e.g.*, parameters corresponding to the specific compactification of the worldsheet of the string, or to a particular internal string-construction technique]... All of the phenomenological properties in space-

time are then derived as consequences of these more fundamental choices. But still, just as in field theory, we are faced with the difficult task of model-building.

Is this progress, then? While opinions on this question may differ, one can argue that the answer is still definitely "yes". Recall that quantum gravity is automatically included in these string models. This is one of the benefits of model-building on the worldsheet rather than in spacetime. Also recall that string theory is a finite theory, and does not contain the sorts of ultraviolet divergences that plague us in field theory. This is another benefit of worldsheet, rather than spacetime, model-building. Moreover, worldsheet model-building ultimately involves choosing *fewer* parameters than we would have to choose in field theory — for example, we have seen that an entire infinite tower of string states, their gauge groups and charges and spins, are all ultimately encoded in a few underlying worldsheet parameters... Furthermore, because of this drastic reduction in the number of free parameters, string phenomenology is in many ways more tightly constrained than ordinary field-theoretic phenomenology. Thus, it is in this way that string theory can guide our choices and expectations for physics beyond the Standard Model. Indeed, from a string perspective, we see that we should favor only those patterns of spacetime physics that can ultimately be derived from an underlying set of worldsheet parameters... These would then serve as a "minimal set" of parameters which would govern all of spacetime physics!

Of course, at a theoretical or philosophical level, this state of affairs is still somewhat unsatisfactory. After all, we still do not know *which* self-consistent choice of string parameters ultimately corresponds to reality... Nonperturbative insights have thus far changed our understanding of the size and shape of this moduli space, they have not yet succeeded in leading us to an explanation of which points in this moduli space are dynamically selected.

So where do we stand? As string phenomenologists, we can do two things. First, we can pursue *model-building*: we can search through the moduli space of self-consistent string models in order to determine how close to realistic spacetime physics we can come. This is, in some sense, a direct test of string theory as a phenomenological theory of physics. Of course, this approach to string phenomenology is ultimately limited by many factors: we have no assurance that our model-construction techniques are sufficiently powerful or general to include the "correct" string model (assuming that one exists); we have no assurance that our model-construction techniques will not lead to physically distinct models which nevertheless "agree" as far as their testable low-energy predictions are concerned; and we have no assurance that the most important phenomenological features that describe our low-energy world (such as the pattern of supersymmetry-breaking) are to be found in perturbative string theory rather than in nonperturbative string theory. For example, it may well be (and it has indeed been argued) that the true underlying string theory that describes nature is one which is intrinsically non-perturbative, and which would therefore be beyond the reach of the sorts of approaches typically followed in studies of string phenomenology.

Another option, then, is to temporarily abandon string model-building somewhat, and to seek to extract general phenomenological theorems or correlations about spacetime physics that follow directly from the general structure of string theory itself. Clearly, we would wish such information to be *model-independent*, *i.e.*, independent of our particular location in moduli space... For example, if some particular configuration of spacetime physics (some pattern of low-energy phenomenology) can be shown to be inconsistent with being realized from an underlying set of [worldsheet] parameters, and if such a demonstration can be made to transcend the particular construction in use so that it relies on only the primordial string symmetries themselves, then such patterns of phenomenology can be ruled out. In this way, one can still use string theory in order to narrow the list of possibilities for physics at higher energies, and to correlate various seemingly disconnected phenomenological features with each other. Such correlations would then be viewed as "predictions" from string theory...

Despite recent advances in understanding various non-perturbative aspects of string theory, our inability to answer the fundamental question of vacuum selection persists. Until this challenge is overcome, string phenomenology therefore must content itself with answering questions of a *relative* nature (such as questions concerning relative *patterns* of phenomenology) rather than the sorts of absolute questions (such as calculating the mass of the electron) that one would also ideally like to ask. Nevertheless, as we shall see, string theory can still provide us with considerable guidance for physics beyond the Standard Model.

My purpose in providing this lengthy quote from 1998 is not to claim credit for any sort of unique idea here. As indicated above, ideas like this were already widely acknowledged and discussed by workers in this field long before this point, and I was hardly unique or especially clairvoyant. Rather, my purpose is merely to provide illustration of the kinds of thinking that were prevalent in the late 1990's, and the approaches that people were then pursuing. Indeed, as indicated above, many string phenomenologists at that time were pursuing traditional model-building, imagining that at least they were learning about certain classes of promising string theories and might ultimately come across (by dint of hard work and/or clever construction methodology) the "correct" string model. This activity continues today. Others were instead pursuing global theorems — statements (usually taking the form of correlations between different low-energy observables) that rested on fundamental string symmetries such as conformal or modular invariance and which would therefore hold in a general sense *across* the space of all models. However, the important point is that even during those "early" years of the late 1980's and 1990's, it was well understood that there existed huge numbers of apparently self-consistent string models, and that this was ultimately critical to the future of string phenomenology and the predictivity of string theory.

3.3. *The emergence of the modern string landscape*

Although many string phenomenologists were already grappling with the large space of apparently self-consistent string models, I think it is fair to say that during the late 1980's and 1990's the bulk of the string community was not directly focused on this issue. Although it is always difficult (and indeed somewhat dangerous) to speculate on the sociological reasons why a particular scientific community might focus on a particular set of problems at one time but not another, it is my sense that there were three fundamental reasons why the larger string community was not worrying about the large plethora of existing string models:

- These string models usually had flat directions — *i.e.*, there were no dynamical ways of fixing many of the continuous parameters that could be freely adjusted. Flat directions are in direct conflict with experiment, since they correspond to extra (unobserved) massless particles and unseen forces.
- These models were usually supersymmetric, yet the real world is non-supersymmetric.
- These models were usually formulated in flat space or anti-de-Sitter space (negative cosmological constant). This was also not realistic.

It was therefore tacitly assumed that some sort of vacuum-selection mechanism would eventually be found (presumably relying on the mysterious non-perturbative aspects of string theory), and that this stabilization mechanism would lead to a unique vacuum that would solve the other problems (namely, break SUSY and introduce de Sitter space). In other words, even though no vacuum-selection principle had yet been found, there was a very strong faith within the formal string community that such a principle *would eventually* be found, and that the whole problem of having to deal with large spaces of apparently self-consistent string models would disappear on its own.

So what changed? Starting in the mid-1990's, and more urgently since 2003, there has been an increasing realization that this is *not* what is going to occur. First, in the mid-1990's, formal string theorists gained considerable insight into the non-perturbative behavior of these theories, and discovered that these theories continue to be self-consistent even when these parameters are left unfixed and even at strong coupling. Second, starting around 2003, various proposals suggested the existence of controlled methods of stabilizing vacua, breaking SUSY, and realizing de Sitter space in string theory. Unfortunately, none of these ideas led to a vacuum-selection principle. Instead, they showed that a plethora of self-consistent string compactifications is likely to continue to exist, even after vacuum stabilization and other problems are solved.

What emerged, then, was the dawning realization that string theory really might contain an entire multitude of solutions — *i.e.*, a multitude of stable ground states — without a dynamical or symmetry argument to select amongst them. Such ground states can be viewed as local minima in a complex theoretical terrain of hills and valleys. It is this terrain which was then somewhat whimsically dubbed the string-theory "landscape".

A natural question that arises, then, is the total number of these solutions across the landscape. How many are there? The number most often bandied around comes as the result of a heroic analysis by Michael Douglas (`hep-th/0303194`), who showed that this number could be as large as 10^{500}. I often like to think that even writing this number in this form is an abuse of scientific notation, since it masks how truly large this number is. The digit '1' followed by five hundred zeros will easily fill a page. Or, to make this even more dramatic, we could write the digit with as many as 491 zeros after it, and we would still have only described only *one billionth* of the total number of solutions in this landscape. This is a large number indeed.

Unfortunately, the choice of which vacuum we live in *does matter*. As we have discussed above, the low-energy phenomenology that emerges from the string depends critically on the particular choice of vacuum state. Thus, detailed quantities such as the choice of gauge group at unification, the number of chiral generations, the method of and energy scale associated with SUSY-breaking, even the value of the cosmological constant, all depend on the particular vacuum state selected!

Thus, starting around 2003, it became accepted by a large fraction of the string community that the landscape might be real after all, and not merely reflect our ignorance concerning a possible as-yet-missing piece of string theory. And if so, string theory and string phenomenology will really need to find a way to deal with it.

3.4. *Living without a vacuum-selection principle: Statistics!*

How then can we make progress in the absence of a vacuum-selection principle? One natural idea (proposed by Douglas in `hep-th/0303194` and perhaps also by others) is to examine the landscape *statistically*, and look for correlations between low-energy phenomenological properties that would otherwise be unrelated in field theory. This would then provide a new method for extracting phenomenological predictions from string theory.

As I indicated above, there had been many previous proposals by string phenomenologists to seek correlations across the landscape of possible string vacua as a way of extracting predictions from string theory in the absence of a vacuum-selection principle. However, what was usually envisioned was the extraction of *theorems* — mathematical relations which followed directly from underlying string symmetries. What was now being proposed, by contrast, was to find *statistical* correlations — correlations which might hold *frequently* although not always. Such correlations might therefore emerge out of the hidden complexity of string theory without being a direct mathematical consequence of some underlying string symmetry.

In some sense, this proposal represents the most direct possible frontal assault on the string landscape. One takes its existence at face value, and probes it almost as an experimentalist would, gathering statistical data and looking for noteworthy features.

This idea has triggered a surge of activity examining the statistical properties of the landscape. Indeed, string theorists have undertaken detailed statistical studies of numerous phenomenological features on the landscape, focusing on issues such as the SUSY-breaking scale, the cosmological constant, the ranks of predicted gauge groups, the prevalence of actual SM gauge group $SU(3) \times SU(2) \times U(1)$, the numbers of predicted chiral generations, and so forth. Moreover, this statistical line of attack has also led to various paradigm shifts. For example, based on statistical studies of the landscape, string theorists have advanced alternative notions of naturalness, new cosmological/inflationary scenarios, new kinds of anthropic arguments, and even field-theory analogues of the landscape. There has even been work attempting to understand the distinction between the landscape and the so-called "swampland" (*i.e.*, the space of theoretical phenomenologies allowed by field theory but not by string theory), as well as outright negative assertions (often called "landskepticism") concerning the existence of the landscape itself and the methods by which it can be explored.

Indeed, such work on the landscape has even led in some quarters to a formal organized attempt (the so-called "String Vacuum Project", or SVP) to bring systematic methods to bear on the constructions and analysis of compactifications of string theory. The SVP is a large, multi-year, multi-institution, inter-disciplinary collaboration to explore the space of string vacua, compactifications, and their low-energy implications through the enumeration and classification of string vacua, through the detailed analysis of those vacua with realistic low-energy phenomenologies, and through statistical studies across the landscape as a whole. As such, the SVP involves intensive research at the intersection of particle physics (string theory and string phenomenology), mathematics (algebraic geometry, classification theory), and computer science (algorithmic studies, parallel computations, database management).

At first glance, this might seem like massive over-reach, especially given that the landscape is so huge, consisting of $\sim 10^{500}$ vacua. However, if the landscape exists, then explore it we must. Lots of branches of science, from astrophysics and botany all the way to zoology, begin with large data sets. For each, science then proceeds through well-defined stages: enumeration, classification, and pattern-hunting. Why should the string landscape be any different?

4. Examples of landscape statistical analyses

To date, there has been considerable work in studying the properties of the string landscape. Collectively, this work has focused on different classes of string models, both closed and open, employing a number of different underlying string constructions and formulations. However, regardless of the particular string model or construction procedure utilized, any such statistical analysis can be characterized as belonging to one of four different classes:

- **Abstract studies:** First, there are abstract mathematical studies that proceed directly from the construction formalisms (*e.g.*, considerations of flux combinations). Although large sets of specific string models are not enumerated or analyzed, general expectations and trends are deduced based on the statistical properties of the parameters that are relevant in these constructions.
- **Direct-enumeration studies:** Second, there are statistical studies based on direct enumeration of finite subclasses of string models. Within these well-defined subclasses, one enumerates literally all possible solutions and thereby collects statistics across a large but finite tractable data set.
- **Random-search studies:** Third, there are statistical studies that aim to explore a data set which is (either effectively or literally) infinite in size. Such studies involve randomly generating a large but finite sample of actual string models and then analyzing the statistical properties of the sample, assuming the sample to be representative of the class of models under examination as a whole.

- **"Fertile-patch" studies:** Finally, there have been studies which concentrate on some particular region of the landscape which one has reason to believe is "fertile" — *i.e.*, likely to give rise to a particularly attractive low-energy phenomenology.

Indeed, all four types of studies have been undertaken in the literature.

Certain difficulties are inherent to all of these approaches. For example, in each case there is the over-arching problem of defining a measure in the space of string solutions. We shall discuss this problem further below. However, for simplicity, most researchers simply assume that each physically distinct string model is to be weighted equally in any counting or averaging process.

By contrast, other difficulties are specifically tied to individual approaches. For example, the first approach has great mathematical generality but often lacks the precision and power that can come from direct enumerations of actual string models. Likewise, the second approach is fundamentally limited to classes of string models for which a full enumeration is possible — *i.e.*, string constructions which admit a number of solutions which is both finite and accessible with current computational power. Finally, the fourth approach is perhaps the most efficient if one's goal is the traditional goal of string phenomenology — namely to construct realistic string models with the hopes of finding "the" correct string model. However, this approach can never teach us about the properties of the landscape as a whole, such as its structure and overall global properties.

For this reason, a number of researchers have pursued the third direction — that of undertaking random search studies. However, even within this approach, one must choose a particular class of string models to study as well as a particular formalism within which to generate models in this class. Amongst such classes of models might be, for example, Type I models or heterotic models. Depending on the choices of class, different construction techniques are also available. For example, heterotic string models might be constructed through so-called orbifold-based techniques, or through bosonic or fermionic formulations. Likewise, Type I models are even richer — they can involve different sorts of flux vacua, intersecting D-branes, *etc.* Moreover, the models that one chooses for examination might be further narrowed in other *phenomenological* ways: they might all have $\mathcal{N} = 1$ SUSY, for example, or be non-supersymmetric but tachyon-free. They might all have the Standard-Model gauge group as a precondition (a so-called "prior"), or satisfy some other phenomenological requirements.

Finally, such models can vary in their levels of sophistication. For example, may exhibit various degrees of moduli stabilization — *i.e.*, they may truly represent local "minima" within an actual dynamical landscape, or they might continue to have unfixed moduli. Indeed, the vast majority of random searches involve models with unfixed moduli — this is often the state of the art for such searches. Thus, it is clear that many different types of random landscape searches can be contemplated and executed.

In the rest of this section, I will provide a few illustrative results from my own research which has focused on that portion of the landscape corresponding to perturbative heterotic strings. These models have all been constructed through the so-called "free-fermionic" construction, and as such, they contain unfixed moduli. However, experience has shown that this is a class of models which are particularly amenable to rapid computer generation and analysis, but for which a great deal of complexity is nevertheless possible.

One kind of investigation one can perform concerns the gauge groups associated with these string models. For example, how likely are different gauge groups to appear? In `hep-th/0602286`, we showed that across all string models in our sample, 10.65% contain $SU(3)$ factors, while 95.06% contain $SU(2)$ factors and 90.80% contain $U(1)$. Approximately only 10% contain all three factors simultaneously, thereby comprising the Standard-Model gauge group. In fact, we have found that 99.81% of all heterotic string models in our sample which contain one or more $SU(n)$ factors also exhibit an equal or greater number of $U(1)$ factors. This is an example of an extremely strong correlation between two spacetime quantities that would otherwise be completely unrelated in field theory.

Another useful piece of information concerns *cross-correlations* between all possible gauge groups of interest? What are the joint probabilities that two different gauge-group factors will appear within the same string model simultaneously? This is especially useful to know if one factor is observable, the other hidden. Such information can be found in `hep-th/0602286`.

Another important quantity which string theory is in a unique position to predict/evaluate is the one-loop vacuum energy (cosmological constant) Λ. A histogram of results found across our sample set is shown in Fig. 1(a). Note that our conventions are such that $\Lambda > 0$ corresponds to *anti* de Sitter space. By contrast, in Fig. 1(b), we see a striking correlation between *cosmological constants* and *gauge groups* — models that have greater numbers of twists, thereby breaking their gauge groups into smaller and smaller pieces with smaller and smaller ranks have smaller average cosmological

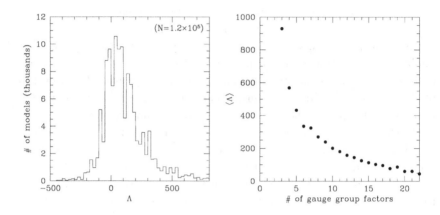

Fig. 1. *Left panel:* Histogram showing the one-loop vacuum amplitude Λ across our sample of $N \gtrsim 10^5$ tachyon-free perturbative heterotic string vacua with string-scale supersymmetry breaking. Both positive and negative values of Λ are obtained, with over 73% of models having positive values (corresponding to anti de-Sitter spacetimes). *Right panel:* A striking correlation between cosmological constants and gauge groups: The average value of the cosmological constants obtained from models with a fixed number of gauge-group factors is plotted as a function of this number.

constants. Again, these are features which would be completely unrelated in quantum field theory. Both are discussed further in in hep-th/0602286.

These examples are drawn from my own work, but I do not mean to suggest that my work is unique in any way. By now there is a relatively large literature providing statistical analyses of all sorts of phenomenological quantities including chirality, numbers of fermion generations, hypercharge normalizations, gauge-coupling unification, Yukawa couplings, string threshold corrections, intermediate-scale physics (SUSY-breaking, new gauge structures, *etc.*). Moreover, such analyses have been performed across a wide variety of string models (Type I, heterotic, perturbative versus non-perturbative, supersymmetric versus non-supersymmetric, and so forth).

5. Two cautionary tales

On the one hand, it is incredible that string theory enables statistical calculations of the sort discussed above. After all, these are literally statistical calculations regarding probabilities that one set of laws and fundamental constants for the universe are favored over another!

On the other hand, there are numerous subtleties that emerge when trying to perform analyses of this type, and new methods need to be developed in order to extract phenomenological predictions in a meaningful way. In order to illustrate this point, I'll now tell two cautionary tales. Both must be borne in mind when attempting to extract statistical information from studies of the string landscape.

5.1. *Cautionary tale #1: Counting is hard*

Our first complication has to do with the problem of *floating correlations*. This was first discussed in `hep-th/0610319` (with my collaborator Michael Lennek), and is a generic problem which affects any random search through the landscape. In particular, this problem turns out to play a huge role in obtaining meaningful statistical results from any data set to which one has only limited computational access.

The problem of floating correlations is the observation that some statistical correlations are *unstable* — they "float" (or evolve) as the sample size increases. Why does this happen? Essentially, as we continue to randomly generate models, it gets harder and harder to find new (*i.e.*, distinct) models. Thus, physical characteristics which were originally "rare" are often forced to become less "rare" as the sample size increases and as we probe more deeply into the space of models.

To see this more explicitly, let us consider the process of randomly generating string models. In general, one must generically employ a model-construction technique which specifies models according to some set of internal parameters (*e.g.*, fluxes, orbifold twists, boundary conditions or phases, Wilson lines, *etc.*) Each set of parameters maps to a single model, but the mapping is rarely unique! Thus, as illustrated in Fig. 2, some models are much more likely to be generated than others! (An example of this would be Model A in Fig. 2.) This feature is essentially unavoidable.

As a result, any random search of the parameter space is not a random search in the space of corresponding models. The implication of this, as illustrated in Fig. 3, is that we are not actually probing the space of models directly; we are instead actually probing a *deformed* version of this space, the so-called *probability* space in which different models occupy volumes proportional to their odds of being realized through the chosen construction technique.

Unfortunately, this difference is critical when we are trying to extract statistical correlations between physical observables. In general, the phys-

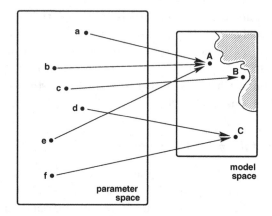

Fig. 2.   Random searches of the landscape typically explore the space of parameters which define string models. However, this mapping is rarely one-to-one. Thus a random search of the parameter space is not a random search in the space of corresponding models.

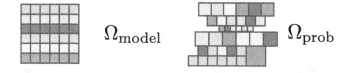

Fig. 3.   The consequence of the situation sketched in Fig. 2 is that we don't have access to the space of models directly; instead we are actually probing a deformed version of this space, the so-called *probability* space, in which different models occupy volumes proportional to their odds of being realized through the chosen construction technique.

ical properties of our models will be correlated with these probability deformations. Thus, even though our goal is to extracting statistical correlations within the model space (where all models are weighted equally), all we really have access to is the probability space, with no knowledge of the deformations that have transformed the former to the latter.

In order to proceed, we can begin by revisiting our model-generation methodology. A partial solution is to avoid counting a newly-generated model if it has already been generated once before and is therefore already in our data set. Rather, we can consider it a "failed attempt", disregard this model, and try again. However, we are still not finding the very "rare" models (such as Model B in Fig. 2) which are not often generated. It will take a considerably larger data set before will stumble across such rare models, and in principle we have no information about where they are, how common they are, or whether they even exist!

The solution is to restrict our attention to the relative *ratios* of probabilities of models with different characteristics — but equally importantly, to only calculate those ratios when the spaces of models with these characteristics are *equally explored*.

Of course, we need a measure for the notion of "equally explored". How can we judge how deeply we have penetrated into a particular model space? The solution is to look at the number of attempts it takes to randomly generate a new model with a specified characteristic. If it is easy to generate new models of a given type, then the corresponding space of models of that type is relatively unexplored. As we progress, however, it gets much harder to find new models of that type and the number of failed attempts per new model increases. Thus, by measuring numbers of models found against numbers of attempts to generate new models, and by comparing this ratio for two different groups of models, we can extract information about the relative volumes of their corresponding model spaces and thereby deduce their true relative probabilities.

To see how this works in practice, let us consider a simple example: An urn contains 300,000 balls of different colors. One third of the balls are red. We seek to know what fraction of balls in the urn are red, and we try to determine this by choosing a ball randomly from the urn, noting its color, marking it for future identification, replacing the ball in the urn, mixing, and then repeating over and over. Clearly, if all balls are treated equally (no bias), approximately one third of all balls selected will be red. This will not vary significantly with sample size. However, let us now suppose the red balls have a different size than the others, so that the probability of picking a red ball from the urn on a given try is γ times that of picking a ball of any other color. What fraction of selected balls will be red? Clearly at the beginning of the process, we will find a probability $\gamma/3$ of finding red balls. However, as we continue, this fraction will "float" with the sample size, only reaching the true value $1/3$ once we have fully explored the urn. This is the problem of floating correlations. But suppose we don't have enough time/ability to wait that long, and we don't know γ. What can we do?

Our solution, as outlined above, is the following. As we select balls from the urn, one after another, we simply keep a tally of two numbers: X_{red} (the number of failed "red" attempts to find the last new red ball — *i.e.*, the number of red balls that must be drawn before finding a new red ball) and X_{other}. At any moment in the search, these X-variables measure how deeply into the corresponding spaces of red (and "other") balls we have already probed. We also record $N_{\text{red}}(X_{\text{red}})$, the total number of new red balls which have been found by the time our search has reached a penetration level X_{red}. We also record the analogous quantity $N_{\text{other}}(X_{\text{other}})$. Our solution is then to identify the desired quantity (in this case the number of red balls in the urn divided by the number of "other" balls in the urn) as $N_{\text{red}}(X_{\text{red}})/N_{\text{other}}(X_{\text{other}})$ for any sufficiently large but *equal* values of X_{red} and X_{other}. Here the condition "sufficiently large" can be considered to be satisfied when the corresponding ratio ceases to float. This happens relatively quickly under this prescription. Further details and examples of this prescription in action are given in `hep-th/0610319`.

In fact, the true computational situation we face for the landscape is even more complicated than in this simple example. There can be a whole spectrum of different sizes (intrinsic probabilities) for the different balls (string models). Likewise, there is no guarantee that the sizes (intrinsic probabilities) of the balls (models) are in any way correlated with their colors (physical characteristics). In general, there can be a huge "CKM matrix" between colors and sizes, all of whose entries are essentially unknown! Thus, one requires methods of extracting meaningful statistical information even for such general situations. These issues are discussed further in `hep-th/0610319`.

5.2. *Cautionary tale #2: Not all vacua are vacua*

Another possible complication when calculating landscape statistics is the following. All of our previous discussions assume that the low-energy limit of a given string model has a relatively simple field-theoretic structure: a single vacuum (the ground state), and a tower of excited states built on that vacuum. In such cases, the resulting phenomenology associated with each string model is uniquely determined, and each string model corresponds to a unique possible ground state for the universe. In other words, one string model corresponds to one vacuum, and thus counting models corresponds to counting vacua.

In recent years, however, there has been increasing recognition that many models also contain additional *metastable* vacua whose lifetimes can easily exceed cosmological timescales. Moreover, the phenomenological properties of the metastable vacuum can be completely different than those of the true ground state! For example, supersymmetry and R-symmetries may be preserved in the true vacuum but broken in the metastable vacuum;

these different vacua may have different gauge groups and particle contents; and so forth. As a result, the one-to-one connection between models and vacua need not apply! As a result, the full landscape of string theory can be even richer than previously imagined, since all long-lived metastable vacua must also be included in the analysis.

This effect can be extremely dramatic and can completely alter our perspective on the sorts of physics which might dominate the landscape. This is because many string vacua take the form of so-called "flux compactifications", and these theories have "deconstructed" low-energy versions which correspond to supersymmetric abelian gauge theories with very specific particle contents. In the presence of kinetic mixing, however, it has recently been shown (see arXiv:0811.3335, in collaboration with B. Thomas) that these theories give rise to infinite towers of metastable vacua with higher and higher energies! Indeed, as the number of vacua grows towards infinity in these models, the energy of the highest vacuum remains fixed while the energy of the true ground state tends towards zero. Thus, even if such models are relatively rare across the landscape, the fact that they give rise to infinitely many vacua means that they could completely dominate the statistical properties of the landscape as a whole! Clearly, such effects must also be taken into account in any exploration of the string landscape.

6. Naturalness and SUSY: A case study

It turns out that the existence of the landscape allows us to reformulate many of our usual theoretical notions in hitherto-unimaginable ways. For example, let us ask a simple question: *Is SUSY natural?*

This is an important question. For very compelling phenomenological and theoretical reasons, most theoretical frameworks for physics beyond the SM involve the introduction of SUSY. As a result, SUSY is truly ubiquitous across the landscape of theoretical particle physics. However, in the past 15 years, many competing theories have emerged — theories with large extra dimensions, small extra dimensions, strong dynamics, and so forth. Some of these theories are quite arcane, and require that we are made of open strings, that we live on a brane, that the brane lives in extra dimensions, that the brane is wrapped and intersects other branes, that the extra dimensions are warped, that the warping is severe and forms a throat, that the brane is falling into the throat, and so forth. All of this may sound highly unnatural. But is SUSY itself truly natural? What does it mean to be "natural", anyway?

There are many different notions of "naturalness" which are often bandied about. For example, EFT (Dirac) naturalness posits that an EFT (an Effective Field Theory) is "natural" if the dimensionless coefficients of all operators are $\sim \mathcal{O}(1)$ — no unnaturally small numbers are allowed. Under this criterion, the gauge hierarchy is unnatural (and is thus the biggest motivation for SUSY). Another example is 't Hooft naturalness, which posits that even if a number is small, it can be "natural" if protected by a symmetry. But neither of these addresses the question as to whether a theory, even if "natural" in the above sense, is likely to be right. How *likely* is SUSY to be the correct theory?

"Likely"? Even though we constantly judge theories in this way, we don't say it aloud because the question seems more philosophical than scientific. How likely relative to what? All other theories that one can imagine? Who is doing the imagining? How can one compare the likelihood of one theory against another?

As we have seen, string theory provides a framework in which this question can be addressed in a meaningful way. Thanks to the landscape, we can reformulate this question as follows: In the landscape of possible string solutions, how many of these solutions are supersymmetric? Is SUSY "natural" on this landscape, or relatively rare?

Using the statistical techniques we have developed (as discussed, above), we have investigated this question within the space of perturbative four-dimensional tree-level free-fermionic heterotic string models. Our results (derived in collaboration with M. Lennek, D. Sénéchal, and V. Wasnik in arXiv:0704.1320) are shown in Table 1.

Table 1. Classification of the four-dimensional tree-level heterotic landscape as a function of the number of unbroken spacetime supersymmetries and the presence/absence of tachyons at tree level.

SUSY class	% of heterotic landscape
$\mathcal{N}=0$ (tachyonic)	32.1
$\mathcal{N}=0$ (tachyon-free)	46.5
$\mathcal{N}=1$	20.9
$\mathcal{N}=2$	0.5
$\mathcal{N}=4$	0.003

We thus see that nearly half of the heterotic landscape is non-SUSY but tachyon-free! Indeed, the SUSY portion of the heterotic landscape represents less than 1/4 of the full landscape, even at the string scale! Moreover, we see that models exhibiting extended ($\mathcal{N} > 1$) SUSY are exceedingly rare, representing less than 1% of the full landscape. In fact, the SUSY fraction of full landscape may be even smaller than this. Free-field constructions probably tend to favor models with unbroken SUSY and large gauge groups. Moreover, even when stabilized models exhibit SUSY at the string scale, it's statistically unlikely that SUSY will survive down to the weak scale.

Thus, we conclude that weak-scale SUSY is rather unnatural from a string landscape perspective. Is this a problem? Not at all — it could even be considered good news — for it implies that we will actually learn something about string theory and its preferred compactifications if/when weak-scale SUSY is discovered in upcoming collider experiments!

One can extend such an analysis even further. For example, one can ask how the degree of supersymmetry of a given string model is correlated with its gauge group. Within the same class of perturbative heterotic strings discussed above, this question was analyzed in `arXiv:0804.4718`, with results shown in Table 2. From this table we see that the Standard Model prefers to remain non-supersymmetric, while GUT's apparently have greater preference for SUSY than does the SM alone. Indeed, exceptional groups such as E_6, E_7, or E_8 almost *require* SUSY! Thus, we see that such strings favor either the non-SUSY Standard Model or SUSY GUTs, but not the MSSM!

Table 2. The likelihood that a given perturbative heterotic string model with gauge group G will exhibit various levels of unbroken SUSY, within the sample set studied. Note that smaller (larger) gauge groups are strongly correlated smaller (larger) amounts of SUSY.

SUSY	U_1	SU_2	SU_3	SU_5	$SU_{>5}$	SO_8	SO_{10}	$SO_{>10}$	$E_{6,7,8}$
$\mathcal{N} = 0$	69.80	58.41	68.79	45.29	17.33	37.98	43.68	16.21	1.85
$\mathcal{N} = 1$	29.68	40.94	30.51	52.78	71.56	56.66	46.75	55.38	83.00
$\mathcal{N} = 2$	0.51	0.65	0.69	1.92	10.65	5.25	8.95	26.84	10.59
$\mathcal{N} = 4$	0.004	0.002	0.002	0.006	0.44	0.11	0.63	1.57	4.57

7. Is string theory predictive?

Needless to say, the existence of the landscape also prompts a number of questions of a more philosophical nature. For example, given the existence of the landscape, one natural question concerns the extent to which string theory is predictive. In some sense, this goes to the heart of what it means to be doing science. As such, there can be no more critical question for string theory than this!

It is easy to imagine how a debate on this topic might proceed. P will

start by remarking that predictivity is not an absolute necessity for all
aspects of science — indeed, good science often begins with observation
and classification. Q will retort that this is true, but while observers and
experimentalists need not be primarily concerned with making predictions,
theorists must be. Theories of science must incorporate the ability not only
to explain, but also to predict. P will then protest that the most direct
experimental consequences of string theory lie at inaccessible energy scales!
Is it fair, then, to hold string theory to normal standards of predictivity? Q
will then state that even though many of the direct consequences of string
theory lie at presently inaccessible energy scales, not all will be. And even
if all of the firm experimental consequences of string theory were somehow
proven to lie at scales exceeding those reachable by current accelerator
technology, this would not free string theory from its obligations to make
predictions which are testable at those higher energy scales — *i.e.*, testable
in principle, if not in practice.

At this stage, P might respond that "string theory" is not a model like
the Standard Model — it's a *language* (like QFT) within which the subsequent act of model-building takes place! QFT does not make predictions
on its own — why hold string theory to such a standard? But of course Q
could claim that this misses a critical point. While quantum field theory
tolerates many free parameters, string theory does not: generally all free
parameters in string theory (such as gauge couplings, Yukawa couplings,
etc.) are determined by the vacuum expectation values of scalar fields and
thus are expected to have dynamical origins within the theory itself. String
theory should determine its own parameters!

Given the existence of the landscape, it is certainly too much to demand
that string theory give rise to predictions for such individual quantities as
the number of particle generations. However, as we've seen, it is perhaps
not too much to ask that string theory manifest its predictive power through
the existence of *correlations* between physical observables that would otherwise be uncorrelated in quantum field theory. Such correlations would
be the manifestations of the deeper underlying geometric structure that
ultimately defines string theory and distinguishes it from a theory whose
fundamental degrees of freedom are based on point particles.

Thus, our question concerning the predictivity of string theory boils
down to a single critical question: *To what extent are there correlations
between different physical observables across the string-theory landscape as a
whole?* The existence of such correlations would imply that string theory is
predictive, while the absence of such correlations would imply the opposite.

Unfortunately, the true picture is likely to be much more complicated, lying somewhere between these two extremes, with different regions of the landscape exhibiting different correlations. Such regions may have different sizes, and moreover are likely to exhibit non-trivial overlaps.

This then leads to a highly non-trivial pattern of correlations. To see why, let us suppose that our landscape consists of only three regions, as illustrated in Fig. 4, and let us further suppose that each of these regions exhibits a correlation between only two physical observables. In particular, let us imagine that Region I exhibits a correlation between two observables X and Y, while Region II exhibits a correlation between Y and Z and Region III exhibits a correlation between W and Z.

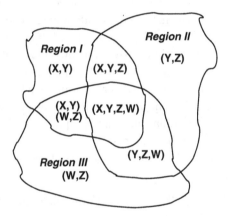

Fig. 4. Sketch of a landscape in which different regions exhibit different correlations between phenomenological observables X, Y, Z, and W. As discussed in the text, the overlaps between these regions can then exhibit correlations amongst larger subsets of observables or multiple independent correlations involving smaller subsets of observables.

This then leads to a highly non-trivial pattern of correlations in the different overlap regions! Indeed, the overlap region between Regions I and II exhibits a single *three*-quantity correlation among (X, Y, Z), while that between Regions II and III exhibits a single *three*-quantity correlation among (Y, Z, W) and that between Regions I and III exhibits two *two*-quantity correlations (X, Y) and (W, Z). Moreover, the overlap region between all three regions exhibits a single *four*-quantity correlation amongst (X, Y, Z, W). This is clearly a very complex structure!

How then might we proceed? One evidently requires practical statistical methods of probing such a non-trivial correlation structure "experimentally" through the random generation and analysis of string models drawn across the landscape as a whole! In this way, one might hope to develop and quantify a practical notion of "predictivity" for such a system. Initial steps in this direction are discussed in `arXiv:0809.0036`.

8. The Multiverse, the A-word, and Captain Kirk

Thus far, we have treated the landscape in a rather simplistic manner: There are many possible states, and the universe chooses one. But of course, from a quantum cosmological standpoint, it is more likely that all possibilities are realized, and that our universe is only one "bubble" in a such a larger multiverse (or megaverse). In accordance with the string landscape, each universe in the multiverse would have its own physical laws and its own constants of nature. Welcome to the Multiverse!

If this is the true nature of things, then our own universe is not special at all, and there would be many other parallel universes whose properties need not resemble those of our own universe in any way! Indeed, one can further imagine that these different universes are continually being spawned in a process dubbed "eternal inflation", first proposed by Andrei Linde in a more general context more than 25 years ago.

If these ideas are correct, then entirely new sets of questions are spawned as well. For example,

- Is the number of possible universes finite or infinite? Is this even knowable? Does it matter?
- How are these universes generated? Through vacuum decay/tunneling? In the interiors of black holes? Do they decay?
- Are all possible universes created equal? Or are some favored in a Darwinian sense, having characteristics that will cause them to inflate more rapidly than others and thereby come to dominate the landscape of universes?
- How did we land in this universe? Are the fundamental laws of physics (as we know them) destined to become nothing more than environmental accidents of initial conditions?
- Are quantities such as the electron/proton mass ratio now going to be viewed as having no deep theoretical underpinnings, but instead like the planetary orbital radii, valid just "by accident" in this universe?

Or, to summarize all of these questions in the most dramatic of ways:

- *Is theoretical particle physics destined to become a branch of cosmology, a study of randomly chosen cosmological initial conditions?*

Along with this comes perhaps the biggest question of all:

- Why are we *here*? Is there anything special about our own universe whatsoever, any tool that remains by which we can hope to develop insight into our universe and make predictions?

[Obviously, the astute reader recognizes that we are heading towards very dangerous waters. We now take the plunge...]

There is, of course, one way of answering all of these questions: the Anthropic Principle. There are many ways of expressing this principle, some of which sound silly and others of which may be profound. Here's one way: *The universe takes the form that it does so as to allow observers to observe it.*

My personal verdict on this formulation of the anthropic principle is that it is silly. In my opinion, the universe doesn't care about me or you, and it doesn't exhibit narcissistic or exhibitionist tendencies that make it want to be observed. (This would not be anthropic, but *anthropomorphic*!) This form of the anthropic principle essentially asserts that the universe may seem random and spontaneous, but that there's really a hidden "script" behind the scenes which is fine-tuned towards the single goal of being witnessed by an "audience". Now, some universes are indeed known to operate this way (I'm here thinking of the "WWE Universe") — but some have argued that the anthropic principle is really about *intelligent* life...

But there are other versions of the anthropic principle which are perhaps *not* so silly. Here's a pop quiz: *The Federation Star Ship "Enterprise" enters an uncharted solar system with 10^{500} planets. In order to survey the planets quickly, Captain Kirk sends a landing party down to each planet simultaneously. After an hour, he puts out a general call for survey reports to be sent back to the ship. Question: What percentage of surveys will report an M-type planet (i.e., capable of supporting life)?*

A moment's thought will convince you that there is only one answer: 100%! All other teams will be dead, and won't be able to file any reports. Indeed, there is a profound lesson here: Certain outcomes about the universe are guaranteed, because otherwise we couldn't have even asked the question.

This form of the anthropic principle has had some successes, most notably the prediction of a non-zero cosmological constant Λ of approximately the right size, a full decade before it was observed! Indeed, in 1987, Weinberg observed that Λ cannot be too big, or else the universe would have expanded too rapidly to allow the formation of structure (galaxies, stars, ...) as needed to generate life. This gives an upper value for Λ. Now, this alone is not the anthropic principle. This is just an upper bound on Λ. In particular, $\Lambda = 0$ would still be allowed. The anthropic principle which Weinberg then used is to say that since there is no *other* argument concerning the size of Λ, there is nothing else to suppress Λ further. Consequently the value of Λ should be at or near this critical value (and hence not zero). And indeed, experimental observations ten years later have borne this out!

If the multiverse is real, the only hope we may have for understanding the properties of our own universe is through these sorts of applications of the anthropic principle. But it should be noted that there is also fierce opposition to this idea. Some physicists assert that this is not the way science should be done, that the anthropic principle represents a surrendering of the idea that the fundamental laws of physics are unique and not tuned for particular outcomes — especially not an outcome such as life. Indeed, as Burt Richter, the former director of SLAC, commented during a public debate during the SUSY 2006 Conference at UC Irvine, "The anthropic principle is not an *explanation*; it's an *observation*."

In science, we normally accept various "priors" (inputs, assumptions, axioms), and seek to use those inputs in order to derive new results. The question that emerges, then, is whether it is fair to use our own existence as such an input. How we decide which input "data" are acceptable, and which "data" aren't? And what is the data: intelligent civilizations? Life in general? What kinds of life can we imagine? Indeed, are universes which are capable of supporting life somehow preferred (in a Darwinian sense) over those that don't? Obviously, there are no easy answers to these questions.... just a raging debate.

Closely related to these questions is another: How can we *test* the Multiverse idea? Is it even *falsifiable*?

I have found no better discussion of this issue than that from Lenny Susskind. His words are eloquent, so I will simply quote them here:

> Throughout my long experience as a scientist, I have heard unfalsifiability hurled at so many important ideas that I am inclined to think that no idea can have great merit unless it has drawn this criticism. I'll give some examples..

In the early days of the quark theory, its many opponents dismissed it as unfalsifiable. Quarks are permanently bound together into protons, neutrons and mesons. They can never be separated and examined individually. They are, so to speak, hidden behind a veil... But by now, although no single quark has ever been seen in isolation, there is no one who seriously questions the correctness of the quark theory. It is part of the bedrock foundation of modern physics.

Another example is Alan Guth's inflationary theory. In 1980 it seemed impossible to look back to the inflationary era and see direct evidence for the phenomenon. Another impenetrable veil called the "surface of last scattering" prevented any observation of the inflationary process...

Good scientific methodology is not an abstract set of rules dictated by philosophers. It is conditioned by, and determined by, the science itself and the scientists who create the science. What may have constituted scientific proof for a particle physicist of the 1960's — namely the detection of an isolated particle — is inappropriate for a modern quark physicist who can never hope to remove and isolate a quark. Let's not put the cart before the horse. Science is the horse that pulls the cart of philosophy.

In each case that I described — quarks, inflation, ... — the accusers were making the mistake of underestimating human ingenuity. It only took a few years to indirectly test the quark theory with great precision. It took 20 years to do the experiments that confirmed inflation... What people usually mean when they make the accusation of unfalsifiability is that they, themselves, don't have the imagination to figure out how to test the idea. Will it be possible to test eternal inflation and the Landscape? I certainly think so, although it may be, as in the case of quarks, that the tests will be less direct, and involve more theory than some would like. — *Lenny Susskind*

Indeed, several ideas along these lines have already been proposed. Some tests are possible only in principle...

- in the long-distance future, if our horizon expands sufficiently,

- if/when our universe tunnels into another vacuum state,
- through signatures of physics at or near a domain wall,

while others are potentially more realistic, relying on

- traces of stringy physics and/or inflationary history imprinted on the Cosmic Microwave Background (CMB),
- evidence for strings through deviations from general relativity,
- direct observation of string theory at the LHC (which is possible if the string scale M_{string} is in the TeV-range), and/or
- observation of spatial variation of the fundamental constants.

But these issues are far from settled:

> I am very glad that Susskind has been able to give these issues much more visibility. But it would be very unfortunate if string theorists finally accept there is an issue with predictability, only to fall for the easy temptation of adopting a strategy towards it that cannot yield falsifiable theories. The problem with non-falsifiable theories is nothing other than that they cannot be proven wrong. If a large body of our colleagues feels comfortable believing a theory that cannot be proved wrong, then the progress of science could get stuck, leading to a situation in which false but unfalsifiable theories dominate the attention of our field.
> — *Lee Smolin, Perimeter Institute, Waterloo, Canada*

Finally, one can ask why we should go through all of this worry. Even if the multiverse exists, why not just focus on our own universe?

In my opinion, there are three fundamental reasons why we should care. First, it is part of the compelling nature of scientific challenge. As the old joke says, "Why climb Mount Everest? Because it is there." Second, if there are really 10^{500} vacua, it is very unlikely that we will determine which one describes our universe, exactly. Many will satisfy current experimental constraints. So our need to make predictions still requires that we understand something of the more global structure. But finally, I believe that such an understanding provides the only way of answering the "why" questions associated with the Standard Model. *Why* are there three generations? *Why* are there three types of non-gravitational forces? As A.N. Schellekens of NIKHEF in the Netherlands has written in 0807.3249, "If the Standard Model is part of a huge ensemble, then the only way to answer such questions is to understand the distribution of that ensemble. We simply *have* to care about more than just our own universe, and... it is inevitable that anthropic arguments will play a role in addressing such questions..."

9. Conclusions and perspectives

Clearly, a statistical analysis of the string landscape has the potential to address questions of relevance to string phenomenology — *even without a vacuum-selection principle*. However, more work along these lines remains to be done. Perhaps most importantly, one needs to develop methods of generating and analyzing large classes of *stable* string models without un-fixed moduli. Concurrently, one also needs to develop new algorithmic and statistical tools with which to conduct analyses of the types that will be required. The landscape is a huge place, and it will not be possible to visit all corners and perform the types of complete studies that one would other-wise like to perform. However, this does not mean that certain sections of the landscape cannot be examined, with meaningful information extracted.

That said, one must be aware of certain dangers inherent in conducting such studies. There are three, in particular, that I consider paramount. First is what may be called the "lamppost" effect — the danger of re-stricting one's attention to those portions of the landscape where one has control over calculational techniques. This does not guarantee that one is looking at the most interesting regions of the landscape at all. Second is what may be called the "Gödel" effect — the landscape is so large that it is possible that no matter how many input "priors" one demands, there will always be another observable which cannot be uniquely predicted. Fi-nally, there is also the "bull's-eye" effect — we don't always know what our "target" is (*i.e.*, how to define success), since we are not certain how our low-energy world ultimately embeds into the fundamental theory (SUSY? GUTs? technicolor? something else?).

Nevertheless, despite these dangers, I believe that direct examination of actual string models can uncover features and behaviors that might not otherwise be expected. Moreover, through direct enumeration, we gain valuable experience in the construction and analysis of phenomenologically viable string vacua. Finally, as string theorists, I believe we must ultimately come to terms with the landscape. Just as in astrophysics, botany, and zoology, the first step in the analysis of a large data set is enumeration and classification. Thus, properly interpreted, I believe that statistical landscape studies can be useful and relevant in this overall endeavor.

So where does all of this lead us? Unfortunately, the answer is not clear. Perhaps more than any other, the landscape idea has spawned a raging debate within the string community. Let us recall the quote from Lenny Susskind with which we began this Chapter. However, not all physicists share in his excitement:

> I love Lenny, but I hate this recent landscape idea and I am hopeful it will go away. — *Paul Steinhardt, Albert Einstein Professor of Science, Princeton University*

Likewise,

> When I hear Lenny say that "this theory is going to win, and physicists who are trying to deny what is going on are going to lose", then to my opinion he is going too far... This is not the way physics has worked for us in the past, and it is not too late to hope that we will be able to find better arguments in the future. — *Gerardus 't Hooft, University of Utrecht, the Netherlands, Nobel Prize in Physics, 1999*

Lenny responds,

> That's hard to argue with. I consider myself to be a cautious, rather conservative physicist. I really don't like new ideas. But I also find wisdom in a quote from Sherlock Holmes: "When you have eliminated all that is impossible, whatever remains must be the truth, no matter how improbable it is."

Many others have joined this debate:

> I feel the views of some, that such a picture is unscientific, or a cop-out, are extreme. In particular, understanding the laws that give rise to the megaverse is a very scientific question, and one that I think is well worth studying further. — *Steve Giddings, University of California, Santa Barbara*

Indeed, many statements are quite provocative:

> Finally, after 15 years, the debate has started that should have started around the mid-80's, but was stifled

by irrational opposition against the notion that our observation of the Standard Model could be biased by our own existence... To me, at least one thing seems absolutely obvious: the idea that the Standard Model is (even approximately) unique will eventually find its place in history next to Kepler's attempt to compute the orbits in the solar system: understandable at its time, but terribly anthropocentric. — *A.N. Schellekens, NIKHEF, Amsterdam, the Netherlands, in* 0807.3249

However, I'd like to close with a final quote which I think sums up the debate in the most succinct way possible:

We now believe we live on an ordinary planet, one of many, circling an ordinary star, one of many, in an ordinary galaxy, one of many. Perhaps we need to take the next step, admittedly a revolutionary one, of saying we live in an ordinary universe, a very small part of an enormous megaverse. — *Gino Segre, University of Pennsylvania*

Indeed, perhaps this debate about the meaning of the string landscape can be viewed as nothing more than a 21st-century continuation of the conflict between the Copernican and Ptolemaic world-views. If so, then the current debate about the string landscape is not new at all — it is actually an ancient one, taken to what might be considered its logical and most complete conclusion. This is not entirely unexpected. A generation goes and a generation comes, but the debate concerning the universe and our true place within it remains forever.

Acknowledgments

I wish to thank my collaborators E. Dudas, T. Gherghetta, M. Lennek, D. Sénéchal, B. Thomas, and V. Wasnik on much of the work described here. I also wish to thank G. Kane for the invitation to transcribe what was originally a Colloquium and turn it into a Chapter for this "Perspectives" volume. My work on this subject through the years has been funded by the National Science Foundation and/or the Department of Energy, although my preparation of this Chapter took place independently of these agencies and no conclusions or opinions expressed herein should be associated with any funding agency whatsoever.

6

Mathematics for String Phenomenology

Michael R. Douglas

Simons Center for Geometry and Physics
Stony Brook University
Stony Brook, NY 11794 USA
mdouglas@scgp.stonybrook.edu

We survey some of the basic mathematical ideas and techniques which are used in string phenomenology, such as constructions of Calabi-Yau manifolds, singularities and orbifolds, toric geometry, variation of complex structure, and mirror symmetry.

1. Introduction and overview

String theory and M theory predict that our universe has six or seven extra dimensions, which form a small compact space. In principle, all of the fundamental laws we observe – the Standard Model, and whatever extensions of it we will someday discover – can be derived from the properties of this space and the fields and objects contained within it.

If we believe this, then the problems of classifying the relevant six and seven dimensional manifolds, and making the relevant computations, become central problems of fundamental physics. While these problems are much discussed elsewhere in the volume, let us begin with a brief summary of their mathematical content. To a first approximation, one can think of string and M theory as leading to supergravity in ten or eleven dimensions. This theory contains general relativity and the simplest class of quasi-realistic solutions are a product of Minkowski space-time and a Ricci-flat internal manifold. Thus to get started, we would like to classify the six and seven-dimensional Ricci flat manifolds. Some of the string theories also contain ten-dimensional Yang-Mills theory, so we would also like to classify solutions of the Yang-Mills equations of motion on these manifolds.

These are already hard problems, but here they are only a starting point. Once we have a solution, we would like to understand the small fluctuations around the solution, particularly those which lead to massless fields in four dimensions. We would then like to compute overlaps between these modes, meaning the integral over the extra dimensions of products of three or more eigenfunctions, to get low energy couplings. Once we know how to do this, we are ready to face up to the problem that supergravity was only a first approximation, and the real string and M theories involve many corrections, controlled by the length scales of the extra dimensions in units of the string or Planck scale, and by Planck's constant. Furthermore, string and M theory allow singularities which would not have made sense in general relativity, such as metric degenerations, and branes which carry gauge fields and other matter. There are even further generalizations of the problem we will not discuss, such as nongeometric backgrounds. Understanding all of this is a tall order and it is rather amazing that string theorists have gone as far as we have in this direction.

The initial problems on our list, of classifying manifolds and solutions on them of nonlinear PDE's such as the Einstein and Yang-Mills equations, are particular cases of more general problems which have been actively studied by mathematicians for about fifty years. The classification of spaces (or manifolds) is part of topology and geometry. Spectral theory on Riemannian manifolds was originally studied as a generalization of the theory of vibrating media. The combination of these theories leads to index theory, which relates the number of zero modes of linear operators to the topology of the manifolds and bundles on which they are defined. Index theory is one of the most powerful and general theories of modern mathematics. Since in most compactifications the particles we can produce and study are far lighter than the scale set by the size of the extra dimensions, they correspond to approximate zero modes, and thus index theory is the usual starting point for studying compactification. An introduction from this point of view, still well worth readng, can be found in the second volume of Green, Schwarz and Witten's famous textbook (GSW [1]). Still, index theory is not able to answer important questions such as the stability of a compactification, or whether particular wavefunction overlaps are zero or nonzero. Of course, eventually we need quantitative results to do string phenomenology.

At present, the most powerful and successful mathematical framework for getting a quantiitative handle on string compactification is algebraic geometry. Originally, algebraic geometry was the study of spaces defined by simple equations, such as spheres, Riemann surfaces, and the like. Typically, one starts from a very simple higher dimensional space, say real or complex Euclidean space, and describes the space of interest as the set of solutions to one or more equations formulated in the higher dimensional space. For the simplest case of the unit N-sphere, rather than define it in terms of coordinate patches or symmetry properties, the algebraic geometric definition is the subset of real $N+1$-dimensional space consisting of solutions of the equation

$$1 = \sum_{i=1}^{N+1} X_i^2. \tag{1}$$

The idea of algebraic geometry is then to relate properties of the defining equations to properties of the space. For example, the degree 2 of this equation, can be related to the fact that the N-sphere is simply connected.

As physicists, this might already give us pause – while such relations are certainly interesting, it would be rather odd for the laws of physics to know or care about whether physical space and time could be embedded in another, still higher-dimensional space, by equations which take a simple form. And indeed, algebraic geometry was not much used in fundamental physics before the first superstring revolution of the 80's. One of the main goals of this article will be to explain how it became not just relevant to fundamental physics but one of the most basic parts of the discussion. In particular, one can show in some generality that the ability to describe the extra dimensions as solutions to polynomial equations in a higher dimensional space is not an additional assumption. Rather, it follows from the other physical and mathematical constraints on the problem, in an indirect way which will take us some time to explain.

One of these physical constraints, justified elsewhere in this volume, is to restrict attention to solutions with low energy supersymmetry.[a] Doing this has a huge impact on the mathematical nature of the problem, focusing our attention on complex geometry and manifolds of special holonomy. Of course, the general theory of four-dimensional supersymmetry requires the scalars to group into complex fields, and from this point of view it seems natural that the extra dimensions should be a complex manifold. Actually this is oversimplified as there are other possibilities for the extra dimensions, such as G_2 manifolds, which are not complex. We will outline the broader picture, before focusing on the case of complex extra dimensions.

---

[a]Note that "low" here means relative to the string, Planck and compactification scales, so this assumption is not based solely on bottom-up considerations. Some top-down arguments for supersymmetry can be found in [4–6].

The starting point for supersymmetric compactification, as discussed in GSW and many textbooks, is to find a covariantly constant spinor on the extra dimensional space, as this implies an unbroken supersymmetry in four dimensions. This leads us to the study of holonomy and generalized holonomy. We review this topic and some of the mathematics it leads to in §2. There is a classification of holonomy groups, and an emerging classification of generalized holonomy,[b] which we expect to provide a fairly short list of around ten or twenty local geometries which are compatible with supersymmetry.

By far the best-studied local geometry is $SU(3)$ holonomy, which leads to Calabi-Yau compactification. We discuss a variety of constructions of these manifolds in §3, such as hypersurfaces in projective space and in toric varieties, and the theory of orbifolds. We give only the most basic examples, instead focusing on the concepts in algebraic geometry which are used to study them, such as line bundles, characteristic classes, and resolution of singularity. Along the way, we will explain why six-dimensional Calabi-Yau manifolds, which *a priori* need have no simple definition as the solution of equations in a higher dimensional space, nevertheless always do. This justifies the central role of algebraic geometry in the discussion.

In §4 we discuss how the definition of complex moduli space, which the reader should be acquainted with from its use in string world-sheet perturbation theory, is made for Calabi-Yau threefolds. A number of explicit one and two-parameter examples have been worked out in detail, originally as tests of mirror symmetry and string duality. These results are much used in quantitative work on string phenomenology, such as moduli stabilization and the computation of coupling constants.

In §5, we survey some more advanced topics, including construction of bundles and mirror symmetry. We close with some comments on longer range questions.

2. Holonomy and complex geometry

An unbroken supersymmetry in a compactification is a spinor ϵ for which the supersymmetry variation of the gravitino and other fermionic fields vanishes, schematically

$$0 = \delta\psi_I = D_I\epsilon + \sum G^{(p)}_{IJ...K}\Gamma^{J...K}\epsilon \tag{2}$$

---

[b]At the time this was written, this theory was under active development, but the general classification had not been finished.

where D_I is a covariant derivative and the $G^{(p)}_{IJ...K}$ are p-form field strengths, often called "flux."

Let us begin with the case $G = 0$, so we are looking for a covariantly constant spinor ϵ. To understand the nature of this problem, consider the analogous problem of finding a covariantly constant vector field, one preserved by parallel transport along any path. This is a very strong condition which forces the Riemann curvature tensor to be zero in every component along the vector field, effectively restricting the curvature to one lower dimension. A more geometric way to say this is to define the holonomy group of the manifold as the group of all rotations of an orthonormal frame which can be achieved by parallel transport (to be precise, we choose a basepoint and consider parallel transport around a loop starting and ending at this basepoint). The condition for a covariantly constant vector field is that the holonomy group, which for a general D-dimensional manifold would be $SO(D)$, must be $SO(D-1)$ or a subgroup thereof. Equivalently, if we restrict the fundamental representation of $SO(D)$ to the holonomy group, it must contain an invariant subspace, or singlet representation.

Now, the parallel transport of a spinor can be derived from that of a vector, by taking the associated rotation acting on an orthonormal frame, considered as an element of $SO(D)$ acting in the fundamental representation, and re-expressing it in the spinor representation. Thus, the condition that there exists a spinor which is preserved by parallel transport, is that the spinor representation of $SO(D)$, when restricted to the holonomy group, must contain a singlet. If we do not want extended supersymmetry, the singlet must be unique.

What is the simplest way this can happen? Suppose our manifold admits complex coordinates, in other words we can group the coordinates in pairs $z^n = x^{2n} + ix^{2n+1}$ such that the transition functions between coordinate patches are holomorphic. Suppose further that the metric is hermitian, so that it takes the form $g_{m\bar{n}}dz^m d\bar{z}^n$ in each patch. In this case, the subgroup of frame rotations which do not mix up the z's and $\bar{z}$'s will preserve the metric – so it is the unitary group $U(N)$ with $N = D/2$. The spinor representations of $SO(2N)$ then decompose into a sum over the n-index antisymmetric representations of $U(N)$, with n even or odd depending on their chirality. Since the case $n = 0$ is a singlet of $SU(N)$, if we can find a metric with $U(N)$ holonomy group, this gets us most of the way.

It turns out that $U(N)$ holonomy is the condition that the metric be Kähler. As will be familiar from $N = 1$ supersymmetry, whereas a general

metric tensor has $D(D + 1)/2$ components, a Kähler metric is determined
(locally) by a single function. This is an example of how the constraints of
supersymmetry will dramatically simplify our problems.

The subrepresentation with $n = 0$ will still be charged under $U(1) \subset U(N)$ and thus to have a covariantly constant spinor, the holonomy group
must in fact be $SU(N)$. One consequence of this is that there must exist
a covariantly constant N-index antisymmetric tensor, because this representation is also a singlet of $SU(N)$.[c] It is not hard to show that a metric
with $SU(N)$ holonomy is Ricci-flat, [1] and thus we can get $N = 1$ supersymmetry in $d = 4$ Minkowski space-time by taking the internal space to
be a six-dimensional complex manifold with a metric of $SU(3)$ holonomy.

Showing that such a metric exists still sounds hard. Now there is a
necessary condition on the $U(1)$ part of the holomony, which is easy to
analyze. As we will discuss in the next section, the integrals of the $U(1)$ part
of the curvature over two-cycles give a topological invariant of the manifold,
the first Chern class. If this is nonzero, any metric on the manifold must
have nontrivial $U(1)$ holonomy, and therefore a manifold of $SU(3)$ holonomy
must have zero first Chern class. Famously, it was conjectured by Calabi
and proven by Yau that this necessary condition is also sufficient. Although
we have no closed form expression for the resulting Ricci flat metrics (and
there is a strong belief that none exists), they have been proven to exist,
and one can say a fair amount about them if needed.

It turns out that the Ricci-flatness condition can be written as a single
PDE for a single unknown function, and thus one might expect its solution
to be unique. This is true with the caveat that reducing it to a single
equation requires specifying additional parameters, called moduli, and thus
the Ricci flat metrics come in a family parametrized by a moduli space.
One way to see that additional parameters must come in is that, since the
Ricci flatness condition is scale-invariant, the overall scale of the metric
will always be a modulus. We will discuss the moduli further below and
see that they come in two types – the Kähler moduli (as we explain in
§3) and the complex structure moduli (§4). The moduli correspond to
fields in the $D = 4$ effective field theory, and computing those parts of the
effective potential which stabilize them is one of the key problems which a
quantitative treatment of compactification must solve.

One can look at group theory tables to find other subgroups of $SO(D)$
for which the spinor contains a singlet. However, it turns out that not all

---

[c]For N odd, it corresponds to the same covariantly constant spinor.

subgroups of $SO(D)$ can appear as holonomy groups.[d] The possibilities were classified by the mathematician Berger in the 50's, and are called the special holonomy groups. The possibilities which admit covariantly constant spinors are $SU(D/2)$, $Sp(D/4)$ (the hyperkähler metrics), G_2 in seven dimensions, and $Spin(7) \subset SO(8)$ in eight dimensions. Of these, the G_2 case leads to four-dimensional compactifications of M theory with $N = 1$ supersymmetry, a number of phenomenological implications of which have been studied in [2]. While G_2 metrics have been proven to exist, starting with Joyce [3], this theory is still in its infancy.

Let us turn now to the case with flux, $G \neq 0$. The particular list of field strengths G depends on the supergravity theory under discussion. The first example to be discussed was the $D = 10$, $N = 1$ theory which is the low energy limit of the heterotic string. It has a Yang-Mills field strength F_{IJ}^a and a three-form NS field strength usually denoted H_{IJK}. In a precise discussion, we would also have to take nonconstant dilaton and warp factor, leading to additional functions multiplying the terms in Eq. (2), but this will not be needed for our qualitative discussion.

To see how these field strengths fit into the holonomy analysis, we need to decompose the field strengths into representations of the holonomy group. In the case at hand of $SU(3)$, this amounts to the reduction $SU(3) \subset SO(6)$, under which a fundamental $6 = 3 + \bar{3}$. Introducing complex coordinates, the fundamental and antifundamental of $SU(3)$ have local bases dZ^I and $d\bar{Z}^{\bar{I}}$ respectively, and are usually called holomorphic and antiholomorphic indices. The space of n-forms can then be decomposed into (p, q)-forms with p antisymmetrized holomorphic indices, q antisymmetrized antiholomorphic indices, and $p + q = n$. Normally a background flux will break supersymmetry unless its nonzero terms are singlets under this decomposition.

Let us begin with the three-form NS field strength. A real-valued triple antisymmetric representation of $SO(6)$ contains a unique singlet of $SU(3)$, a combination of $(3, 0)$ and $(0, 3)$ forms which are complex conjugates of each other. Thus, it is possible to have a non-zero NS field, but for a given Calabi-Yau metric it will be unique up to overall scale. Since these field strengths are quantized, this can be used to stabilize some of the moduli, as we will discuss in §4.

A similar story can be told about M theory compactification on G_2

---

[d] They can all appear as holonomy groups of homogeneous spaces, but these spaces have too much symmetry to lead to realistic compactifications.

manifolds. Since this geometry has an invariant four-form, the four-form field strength $G^{(4)}$ can be nonzero, and will stabilize metric moduli.

As for the Yang-Mills field, to be consistent with supersymmetry it must satisfy

$$F_{I\bar{J}}^a = \sum e^a g_{I\bar{J}}, \tag{3}$$

where the e^a are constants taking values in $U(1)$ factors, and $g_{I\bar{J}}$ is the Kähler metric. These are known as the Donaldson-Uhlenbeck-Yau equations. It is not hard to show that they imply the Yang-Mills equations, and thus supersymmetry again implies the equations of motion. One can also show that, like the $SU(3)$ holonomy condition, these are as many equations as unknowns (the local structure of the connection is determined by a single function of matrices) and thus the solutions will come in finite-dimensional moduli spaces. These moduli correspond to $D = 4$ matter multiplets.

The most important attribute of a background Yang-Mills connection is its holonomy group, again referring to the group of gauge rotations which can be induced by parallel transport around an arbitrary loop, but now acting on the gauge charges. Physically, a compactified theory with gauge group G and a background with holonomy H, will have unbroken lower-dimensional gauge symmetry H', the commutant of H in G. Working these out for E_8 leads to the well known result that heterotic compactification will lead to the GUT groups E_6, $SO(10)$ and $SU(5)$, if H is one of $SU(3)$, $SU(4)$ or $SU(5)$ respectively. In fact, the generic solution of Eq. (3) will have $H \cong G$ and break all gauge symmetry.

From the point of view of a solution which preserves some gauge symmetry, many of the moduli will enlarge H and reduce H'. Physically, they are charged. The discussion of charged moduli is intricate as their $N = 1$ effective field theory is relatively unconstrained, leading to complicated superpotentials and symmetry breaking patterns. In practice, this discussion is often done in the higher dimensional language, considering families of bundle and brane configurations and reformulating the equations of motion as conditions in algebraic geometry, as we explain in §5.1. This has the advantage that (at least some) quantum corrections can be obtained from the higher dimensional picture.

The type II supergravities have several field strengths and thus the discussion becomes more involved. Now it is possible for the covariant derivative and the field strength terms in Eq. (2) to cancel in a nontrivial way. To treat all the terms on a more equal footing, one can regard the field strength terms as additional components to add to the metric connection, to obtain a new connection on the spinor bundle. This new connection will have a holonomy, which is referred to as generalized holonomy. Unbroken supersymmetries will again correspond to a reduction of the generalized holonomy group which preserves one or more singlets, and in principle the discussion could be made in a parallel way. In practice, the analysis is usually done by introducing "torsion classes." These are defined by considering the field strengths and their first derivatives, and decomposing these under one of the standard holonomy groups. The equations of motion then lead to constraints on the nonzero torsion components, which can be worked out to find the possible local geometries with flux.

3. Constructing complex manifolds

We will go back and forth between the many techniques for constructing and analyzing the spaces used in string phenomenology, and the concepts which are being used and which tell us to what extent these techniques actually do cover all the possibilities. Of course, we will barely be able to scratch the surface; a number of introductory articles as well as mathematical articles exist on the subject.

As we reviewed, the starting point for Calabi-Yau compactification is to construct a Kähler manifold with zero first Chern class. There are various recipes in the literature for this. The simplest is to use the fact that a submanifold of a Kähler manifold, defined as a set of solutions to a set of complex equations which is nonsingular at every point, is itself a Kähler manifold. We want a three complex dimensional manifold, so the simplest possibility is to take a simple four-dimensional Kähler manifold and impose a single complex equation, to get a "hypersurface."

Now the simplest Kähler manifolds are the complex projective spaces $\mathbb{CP}^N$, defined by taking $N + 1$ complex coordinates Z_i, and identifying two points $\vec{Z} \sim \vec{Z}'$ if they are related by an overall rescaling $Z_i = \lambda Z_i'$ with λ a complex number. Excluding the case of all $Z_i = 0$, the resulting set is a compact complex manifold with $b^2 = 1$. To get an explicit description in terms of coordinate patches, we define the j'th patch by setting $Z_j = 1$ and using the Z_i for $i \neq j$ as its coordinates.

Next, we need to pick an equation $f = 0$ on $\mathbb{CP}^4$. At this point a reader truly new to the subject might ask how we could do this, as the only holomorphic functions on a compact manifold are the constant functions. The answer is to work with local coordinate patches and postulate a different defining function, call it f_j, in each patch. All we need is for pairs of equations $f_j = 0$ and $f_k = 0$ to have the same solution set wherever they overlap. This will be true if they are related by a nonvanishing holomorphic transition function, $f_j = \rho_{jk} f_k$.

An explicit discussion in terms of patches can be found in textbooks; let us simply note that the appearance of transition functions suggests that a better way to think about f is that it is not a function but rather a section of a vector bundle, related between patches by nontrivial gauge transformations. The special case in which the section is locally a single complex function, related by a single holomorphic transition function, is called a "line bundle" as in algebraic geometry the space parameterized by a single complex number is always called a "complex line."

Taking f to be a holomorphic section of a line bundle works and can be used on arbitrary manifolds, but for the special case of $\mathbb{CP}^N$ we can do something simpler. We take f to be a homogeneous polynomial (one of fixed degree) in the Z_i, such as the "Fermat polynomial"

$$0 = f_{N,d}(Z) \equiv \sum_{i=1}^{N+1} (Z_i)^d. \qquad (4)$$

One way to see that this makes sense is to check that its solutions agree in each of the coordinate patches $Z_j = 1$. A better way is to note that under the relation $\vec{Z} \to \lambda \vec{Z}$, we have $f_{N,d} \to \lambda^d f_{N,d}$, so the condition $f_{N,d} = 0$ is independent of λ.

We should check that our equation $f = 0$ defines a manifold and not some more singular space. Just as in real geometry, the condition for this is that the gradient $\nabla f \neq 0$ at every point of the manifold (in other words where $f = 0$). This is clearly true for the Fermat polynomial Eq. (4), and true for generic polynomials. On the other hand, by adding an additional term with a parameter, say for definiteness

$$0 = Z_1^5 + Z_2^5 + Z_3^5 + Z_4^5 + Z_5^5 + \lambda Z_1 Z_2 Z_3 Z_4 Z_5, \qquad (5)$$

one can see that it is also easy to violate this condition – doing this requires tuning one parameter and thus "singularities are codimension one."

Singularities are ubiquitous in algebraic geometry and thus one usually talks not about manifolds but about "varieties," which can have singularities. Analogous to the definition of a manifold which is made by patching together small regions which each look like $\mathbb{R}^N$ or $\mathbb{C}^N$, the definition of a variety is made by patching together small regions which each look like the set of solutions of some set of equations in $\mathbb{C}^N$. Another of the discoveries that makes algebraic geometry so useful for string/M theory is that many of these singular varieties still define sensible compactifications – this is usually what is meant when one says that "string/M theory resolves singularities of conventional geometry."

3.1. Characteristic classes

Although one can continue the discussion in this concrete vein, as the next step is to determine the topology of our manifold and find a case with vanishing first Chern class, we have arrived at the point where some elementary mathematical formalism will prove its worth. We leave a systematic discussion to the references, and simply introduce a few of the concepts we would need to develop, beginning with the theory of characteristic classes of vector bundles. Its basics will be familiar to many readers from its use in the theory of monopoles and in the discussion of nonperturbative effects in gauge theory. Consider the definition of a Dirac monopole of magnetic charge n. Rather than talk about a connection which is singular on a Dirac string, the modern discussion takes the vector potential to be a $U(1)$ connection on a nontrivial bundle over $\mathbb{R}^3 - \{0\}$. Such a bundle can be defined concretely in terms of coordinate patches and transition functions, as is done in textbooks. Its most important attribute, and indeed its only topological invariant, is the total magnetic field, defined as an integral over any surface surrounding the origin. This is $\int F = 2\pi n$, where n must be an integer by the Dirac quantization condition.

In mathematical terms, the electron wavefunction in a monopole background is a section of a line bundle, and n is its first Chern class. More generally, given a line bundle on any manifold, we can postulate some $U(1)$ connection, compute its curvature F, and its integrals over any two-cycle define the first Chern class of the bundle. A somewhat better rephrasing of this is that the first Chern class is the cohomology class $[F]$ in $H^2(M, \mathbb{Z})$, in other words the cohomology class of a two-form whose integral over any two-cycle is an integer. It is a topological invariant of the bundle, because under any continuous change of the connection $A \to A + \delta A$, the curvature changes by an exact form, $F \to F + d\delta A$.

Connections for nonabelian groups will act on vector bundles, and we also need to discuss these. One can try to extend this definition to a vector bundle by taking the cohomology class of $\text{Tr } F$, which is invariant up to an exact form even in the nonabelian case. This will be interesting if the trace can be nonzero, which is a question of group theory; it is only true for structure groups with $U(1)$ factors, such as $U(N)$. To get characteristic classes for other groups, one must use more general invariants, such as $TrF \wedge F$, higher powers, and even traces in other representations for non-unitary groups.

For $U(N)$ structure group (so, complex vector bundles), the standard characteristic classes are the Chern classes c_n, and the Chern characters ch_n. These are defined as terms in the series expansions of

$$\det(1 + zF) = \sum_n z^n c_n = 1 + z\text{Tr } F + \frac{z^2}{2}\left(\text{Tr } F \wedge \text{Tr } F - \text{Tr } F \wedge F\right) + \ldots$$

$$\text{Tr } \exp zF = \sum_n z^n ch_n = N + z\text{Tr } F + \frac{z^2}{2}\text{Tr } F \wedge F + \ldots$$

and then taking the cohomology classes of these $2n$-forms. These are essentially two ways of packaging the same information, each more convenient for certain computations. In particular, the Chern character is additive under taking direct sums, and is usually the one which arises in physics, as in tadpole cancellation conditions and brane charges.

A good deal of the topology of a line or vector bundle is captured by these characteristic classes. Since an embedded manifold has several natural bundles associated to it – its tangent bundle, its normal bundle in the ambient space, and others – this is very useful, and often just using relations between characteristic classes one can work out all of the even cohomology of the manifold. The odd cohomology is rather different; for a Calabi-Yau threefold it is related to its complex structure as we discuss later.

To begin these arguments, let us work out the possible line bundles on $\mathbb{CP}^N$, and decide which line bundle has the defining polynomial in Eq. (4) as a section. The simplest case is the trivial bundle whose sections are functions – this is denoted $\mathcal{O}_{\mathbb{CP}^N}$ or simply $\mathcal{O}$. Of course, it has a flat connection, so $c_1(\mathcal{O}) = 0$.

The next case is $d = 1$, which includes the homogeneous coordinates Z_i. These are not coordinates on $\mathbb{CP}^N$, bur rather sections of the "hyperplane line bundle" which is denoted $\mathcal{O}(1)$ and has $c_1 = 1$.[e] Similarly, a homogeneous polynomial of degree d is a section of a line bundle denoted $\mathcal{O}(d)$, with $c_1 = d$. There is also a series of bundles whose transition functions are the inverses of those for $\mathcal{O}(d)$. These are $\mathcal{O}(-d)$, and they only have meromorphic sections (with poles). In any case, d must be an integer.

---

[e]The name comes because setting a degree one $f = 0$ defines a hyperplane in the ambient space. As for the sign of c_1, the conventions are chosen so that bundles with holomorphic sections are positive.

Then, it turns out that we have listed all topological classes of line bundle on $\mathbb{CP}^N$ – they are classified by the single integer d, the first Chern class. Furthermore, one can work out a natural (most symmetric) connection on this bundle, and check that its curvature satisfies $\int F = 2\pi d$.[f]

Given these concepts, there is a very simple computation of the first Chern class of the hypersurface M defined by Eq. (4). The point is that, since they can be expressed in terms of curvature, the Chern classes for sums, quotients and products of bundles satisfy simple additive relations. In particular, the tangent bundle for the ambient space $\mathbb{CP}^N$, is locally a direct sum of the tangent bundle of M with the normal bundle to M, and the latter must be somehow related to the bundle of which the defining polynomial is a section. Since a normal vector n to M satisfies $n\nabla f = 1$, and 1 is a function, they are in fact inverses. This leads to

$$c_1(M) + c_1(\mathcal{O}(-d)) = c_1(\mathbb{CP}^N). \tag{6}$$

Thus we need the first Chern class of the tangent bundle of $\mathbb{CP}^N$. Although we leave a convincing argument to the references, here is its essential point. The tangent bundle has as basis the vectors $\partial/\partial Z_i$, modulo the overall rescaling $\sum_i Z_i \partial/\partial Z_i$. These have degree -1 and 0 respectively, while the first Chern class is defined as a trace, so we find $c_1(\mathbb{CP}^N) = (N+1)\cdot(-1) - (0) = -(N+1)$. Combining all this, to get a Calabi-Yau threefold, we need to take $N = 4$ and $d = 5$.

One can go on to understand the even-dimensional homology of a manifold in detail using these techniques. For example, a section of the line bundle $\mathcal{O}(n)_{\mathbb{CP}^N}$ should have "n zeroes," which at least locally will define submanifolds of complex dimension $\dim M - 1$. This idea can be made precise by defining a "divisor" associated to a line bundle. Then, two such submanifolds will generically intersect in a submanifold of complex dimension $\dim M - 2$. Continuing, we can intersect $N = \dim M$ divisors, to get a collection of points. The number of these points is the intersection number I of the set of divisors, which we can write

$$I = D_1 \cap D_2 \cap \ldots \cap D_N. \tag{7}$$

Just as homology is dual to cohomology, one can compute this intersection number by taking the first Chern classes of the line bundles involved, $c_1(L_1)$, $c_1(L_2)$ and so on, and computing

$$I = \int_M c_1(L_1) \wedge c_1(L_2) \wedge \ldots \wedge c_1(L_N). \tag{8}$$

---

[f]The monopole is a special case of this, using the topological identification $S^2 \cong \mathbb{CP}^1$.

This is both a basic topological invariant of the manifold, and turns out to have direct physical applications. The original example is that, by the fact that cohomology classes such as $c_1(L)$ have harmonic representatives (which solve equations of motion), one sometimes finds that these intersection numbers directly compute wave function overlaps. In the original Candelas *et al* Calabi-Yau compactifications of the heterotic string, these numbers gave the tree level Yukawa couplings of 27 multiplets associated to elements of $b^{1,1}$.

3.2. *Calabi-Yau threefolds are algebraic*

Let us now explain how we know that, as we said many times, any Calabi-Yau threefold (of the sort which leads to $N = 1$ supersymmetry; so we leave out cases like $K3 \times T^2$), can be defined as the set of zeroes of some set of polynomials. This follows from two deep theorems. First, the Kodaira embedding theorem tells us that these manifolds are projective, meaning that they can be embedded in $\mathbb{CP}^N$ for some (possibly large) N. Second, Chow's theorem tells us that any embedding into $\mathbb{CP}^N$ can be defined by polynomial equations.[g]

The intuition behind Chow's theorem is that, once we are talking about embeddings in $\mathbb{CP}^N$, since the only holomorphic quantities at all are sections of the line bundles $\mathcal{O}(n)$, the only global equations we can impose are polynomial equations. Showing that any embedding can be defined globally by equations requires arguments from commutative algebra.

Let us say a bit more about the Kodaira embedding theorem as it can be understood to some extent using intuition from physics [7]. The idea is to embed the manifold by sections. Suppose we have some line bundle L over our manifold, which has holomorphic sections. Although a section is not a function, a ratio of sections is a function, though possibly with poles, because the action of a gauge transformation $s \to \alpha s$ will cancel out. Stated another way, a pair of sections (s_1, s_2) defines a map from the manifold into $\mathbb{CP}^1$, where the s_I are homogeneous coordinates. In the same way, a set of N sections defines a map into $\mathbb{CP}^{N-1}$. If this map is injective (never takes two distinct points of M to the same point of $\mathbb{CP}^{N-1}$), the map could be an embedding. Of course, we should check that the map is nonsingular as well. A bundle whose holomorphic sections provide an embedding is called

---

[g]The general principle behind these theorems, that complex manifolds defined "globally" and "analytically" have (under certain assumptions) a global algebraic definition, is sometimes called "GAGA" and was developed in the 1950's by Serre and Grothendieck.

"very ample."

To see that a very ample bundle exists, let us think of the sections as quantum mechanical wavefunctions for a particle moving on the manifold, in the magnetic field F defined by the connection on the line bundle. In Euclidean space, these wavefunctions would be organized into highly degenerate Landau levels, each with a basis of localized states of width $1/\sqrt{|F|}$. On a more general manifold, if the magnetic field is much larger than all of the other scales (such as the curvature), the same picture will apply.

Now, one can show that on a complex manifold, and with a magnetic field satisfying $F^{0,2} = 0$ (as is the case here), the lowest Landau level wavefunctions are related (by a complex gauge transformation) to holomorphic sections. This is simply because $F_{\bar{I},J} = [\bar{\partial}_I + A_{\bar{I}}, \bar{\partial}_J + A_{\bar{J}}] = 0$ is the integrability condition that allows us to gauge $\bar{\partial}_I + A_{\bar{I}} \to \bar{\partial}_I$. Thus, we have a picture of the holomorphic sections for a bundle with a large positive first Chern class – there are many of them, and there is a basis in which each one is localized to a small region of the manifold. One can check by explicit calculations on $\mathbb{C}^n$ that this provides an embedding locally. Furthermore, since each small region has one dominant wavefunction, no two regions can have the same values for all of the wavefunctions, so this provides a global embedding.

There is one more subtle condition which must be satisfied for this to work. It is clear from the definition of a holomorphic vector bundle that the curvature will be a $(1,1)$-form. However, the condition that the magnetic field is large everywhere might only be satisfied by a more general two-form. This is sometimes the case and that is why, for example, not all K3 surfaces are projective. However, if all two-forms are $(1,1)$-forms, or in other words $b^{2,0} = 0$, this cannot happen. This is the case for our Calabi-Yau threefolds and thus they are always projective. As we emphasized in the introduction, this is the deep reason why algebraic geometry is always applicable to Calabi-Yau compactification.

Although the possibility to embed in projective space is important for the general theory, the embeddings we get this way will often be too complicated to be useful. We would be happier if the manifold were defined by one or a few equations. More importantly, it often happens that one needs more defining equations than the codimension of the manifold, so that some of the defining equations are redundant. This makes the analysis extremely complicated.

These redundancies often arise for topological reasons. As an example, a manifold embedded into projective space will naturally have $b^2 = 1$,

since the ambient space does. While one can realize $b^2 > 1$, this is going
to involve complicated embeddings, singular embeddings, or both. Since
almost all Calabi-Yau threefolds have $b^2 > 1$, it would be better to have
ambient spaces with $b^2 > 1$ to embed in as well.

There are a variety of ways to get these. One which was much used in
the early literature was the "weighted projective space." This is defined by
weighted homogeneous coordinates satisfying an identification like

$$Z_I \cong \lambda^{n_I} Z_I' \qquad (9)$$

where the n_I are various integers. Now these spaces are singular, and the
embedded submanifold can inherit these singularities, in which case it is
not obvious *a priori* whether string theory makes sense. However, some
Calabi-Yau hypersurfaces in weighted projective space can also be defined
physically as Gepner models, so these ones do make sense. Furthermore, it
was found that, somehow, the singularities led to $b^2 > 1$.

3.3. *Orbifold singularities*

The underlying reason for this, is that singularities in complex geometry
tend to have natural smooth spaces or "resolutions" associated to them, and
string theory knows about these resolutions. The simplest and prettiest
(though by no means only) case is that of an orbifold singularity. An
orbifold, as will be familiar, is a quotient of a space by a discrete group,
call it Γ, whose action has fixed points. The example of $T^6/\mathbb{Z}_3$ is discussed
in GSW.

An isolated singularity can be understood by considering its small
neighborhood, which one can model as complex Euclidean space, say $\mathbb{C}^3$.
Let us work our way up to this. A quotient $\mathbb{C}/\mathbb{Z}_n$, say by the action
$Z \to Z \cdot \exp 2\pi i/n$, does not define a new complex space, as one could
simply make the reparameterization $W = Z^n$. However, it does define a
nontrivial singular metric – we can see this because the space is a cone with
a deficit angle, which can be measured far from the singularity. Since the
deficit angle is an integral of the curvature, which in two real dimensions
is a total derivative, we see that no resolution of this space (which keeps
the asymptotics fixed) could possibly be Ricci flat. For our purposes of
supersymmetric compactification, we can ignore it.

The same conclusion could be reached by considering the holonomy
group. If we parallel transport a tangent vector around the singularity,
even at a distance, we will rotate it by $\exp 2\pi i/n$. Thus the holonomy

group will be at least a representation of $\mathbb{Z}_n$ in $U(1)$ and, once we smooth out the singularity, perhaps larger. Again, since the holonomy is visible at infinity, it cannot be removed by resolution.

The holonomy analysis can be easily generalized to higher dimensions. Any resolution of a $\mathbb{C}^N/\Gamma$ singularity, which preserves the asymptotics, will have as holonomy group either an embedding of Γ in $U(N)$, *i.e.* an N-dimensional representation of Γ, or some larger subgroup of $U(N)$. Since for supersymmetry we know we need $SU(N)$ holonomy, we find the necessary condition that the action of Γ on a tangent vector must be a subgroup of $SU(N)$.

We are now ready for the two-complex dimensional singularities which are compatible with supersymmetry. We have seen that the group Γ must have a representation in $SU(2)$. The simplest case is the cyclic group $\mathbb{Z}_n$, which we can take to act as $(Z^1, Z^2) \to (e^{2\pi i/n}Z^1, e^{-2\pi i/n}Z^2)$. There is no other possibility among abelian groups. To get the possible nonabelian groups, since $SU(2)$ is the double cover of $SO(3)$, it is plausible that these will be the ones which have an action preserving a finite collection of points in three-dimensional space. These are the dihedral groups, obtained by adjoining the exchange of the two coordinates, and the symmetry groups of the regular polyhedra – tetrahedral, octahedral and icosahedral. This collection of cyclic, dihedral and polyhedral symmetry groups makes up the famous ADE classification of discrete subgroups of $SU(2)$.

Now, the beautiful fact, which we cannot explain in detail here, is that all of these singularities admit smooth resolutions with $SU(2)$ holonomy, which contain additional two-cycles. The two-cycles have a natural intersection form Eq. (7), which because $N = 2$ here is a number associated to a pair of cycles. This structure is encoded by the extended Dynkin diagram associated with the corresponding ADE Lie group – cycles correspond to nodes, and two cycles which intersect are connected by a link. This shows up in many ways in string duality, with the most direct being the enhanced gauge symmetry of M theory compactified on K3. A K3 manifold has a moduli space of metrics, with many singular limits. In these limits, one or more groups of two-cycles degenerates to a $\mathbb{C}^2/\Gamma$ orbifold singularity. This causes wrapped M2-branes to become massless, which provide the charged vector multiplets of ADE gauge symmetry.

The heart of the theory of these singularities, known as the McKay correspondence, is the relation between the representation theory of the group Γ, and the intersection form and Dynkin diagram. Each irreducible representation R_i of Γ corresponds to a node and thus a two-cycle. The links correspond to terms in the tensor decomposition $R_i \otimes F = \sum n_{ij} R_j$, where F is the two-dimensional fundamental representation by which Γ acts on $\mathbb{C}^2$.

The relation of the McKay correspondence to singularities can be seen very explicitly through the quiver construction of the resolved singularity developed in mathematics by Kronheimer and Nakajima [8] and found in D-brane physics by Douglas and Moore [9]. This even allows computing the Ricci-flat metric, as a hyperkahler quotient.

Almost all of this generalizes directly to three and higher dimensions. Now the orbifolds compatible with supersymmetry are $\mathbb{C}^N/\Gamma$ with Γ a discrete subgroup of $SU(N)$. The quotient depends on the specific action of Γ. For example, $\mathbb{Z}_n$ can act on $\mathbb{C}^3$ as $Z^i \to \exp 2\pi i/n_i Z^i$ with any (n_1, n_2, n_3) satisfying $n_1 + n_2 + n_3 = 0 \mod n$. There are also orbifolds by nonabelian groups, and the group Δ_{27} defined as a central extension of $\mathbb{Z}_3 \times \mathbb{Z}_3$ has played a significant role in string phenomenology [10, 11]. The same relation between group theory and quiver diagrams holds, now called the generalized McKay correspondence.

For $N = 3$, each of these orbifolds can be resolved to a completely smooth space, sometimes in many topologically distinct ways. There is a developed theory of this structure based on toric and birational geometry, which we will say a bit more about below.

One could use any of these resolved orbifolds as ambient spaces to construct submanifolds with $b^2 > 1$. However there is an even more general construction.

3.4. Toric varieties and gauged linear sigma models

From a physics point of view, a toric variety is the space of solutions of the D-flatness conditions of an $N = 1$ supersymmetric abelian gauge theory (defined up to gauge equivalence, of course). Thus a particularly physical construction of a Calabi-Yau manifold would be to embed it as a subvariety of a toric variety defined by F-flatness conditions which follow from a superpotential. This approach to the problem was introduced by Witten [12] and has been heavily used since.

The data of an abelian supersymmetric gauge theory can be given systematically as follows: there is the rank r of the gauge group $U(1)^r$; N chiral superfields ϕ^A; a matrix of $U(1)$ charges Q_i^A, a gauge-invariant superpotential W, and finally a set of Fayet-Iliopoulos parameters ζ_i. The

D-flatness conditions are then

$$\zeta_i = \sum_A Q_i^A |\phi^A|^2 \qquad \forall i, \tag{10}$$

while the gauge equivalences are $\phi^A \sim \exp(iQ_i^A \epsilon_i)\phi^A$.

Let us describe the quintic this way. The ambient space $\mathbb{CP}^{N-1}$ is the simplest toric variety – we take $r = 1$ and all $Q^A = 1$. Furthermore we see that the parameter ζ controls the overall scale of the metric. However, to complete the definition of a Calabi-Yau manifold in this framework, we need a superpotential such that the F-flatness conditions $\nabla W = 0$ include a defining equation such as Eq. (4). A little thought shows that this is not possible unless we introduce an additional chiral superfield P. We can then write

$$W = P f(\phi) \tag{11}$$

so that Eq. (4) is the F-flatness condition $\partial W/\partial P = 0$. Of course this modifies the D-flatness condition, and we also have to study the additional F-flatness conditions $\partial W/\partial \phi^A = 0$. Regarding the latter, one can note that, if the manifold $f = 0$ is nonsingular, their only solution will be $P = 0$ (physically, the P field pairs up with the normal direction ∇f), while if the manifold is singular, the P field could have nontrivial physics. This is the first hint that this definition knows something about singularities.

An extremely illuminating way to continue is to study the $d = 2$ sigma model with $(2, 2)$ supersymmetry defined by dimensionally reducing the $d = 4$, $N = 1$ supersymmetric gauge theory. The original paper [12] is well worth reading, but let us mention a few highlights. First, a $d = 2$ $U(1)$ gauge theory can be anomalous, and the anomaly cancellation condition is (we give it for the general case)

$$0 = \sum_A Q_i^A \,\forall i. \tag{12}$$

For the case at hand this requires $Q^P = -5$ and gauge invariance of W requires that the degree of f must be 5, corresponding to the case of zero first Chern class, to get anomaly cancellation. This is no coincidence but follows from a relation between the $U(1)$ anomaly and the beta function in these theories – in sigma model terms, we require Ricci flatness to get a conformal field theory and zero beta function.

The appearance of the first Chern class in the discussion is no coincidence either, as one can show the following. First, the second Betti number of the toric variety (so, the solutions of Eq. (10) before imposing F-flatness) is $b^2 = r$. Thus there is no limit on the b^2 which can be realized this way. Second, the $U(1)^r$ charges Q_i^A have a direct interpretation as the first Chern class of the line bundle of which ϕ^A is a section. Recall that the first Chern class is an element of H^2; thus the index i implicitly labels a basis Σ_i of H_2 and Q_i^A is the integral over the two-cycle Σ_i. Thus, a large set of natural line bundles are easy to work with in this construction.

Finally, we can explain the Kähler moduli using this construction. Their general definition is the integral of the Kähler two-form ω, which is defined in terms of the metric and complex structure, over a basis of H_2. It is not hard to show that the Kähler metric (sigma model kinetic term) restricted to the solutions of Eq. (10) (and of F-flatness as well) satisfies

$$\int_{\Sigma_i} \omega = \zeta_i. \tag{13}$$

Thus the parameters ζ_i become the Kähler moduli of the resulting toric variety, and are also Kähler moduli (not always all of them) of the Calabi-Yau manifold.

This makes the Kähler moduli concrete and easy to work with, but raises another question: what if we take some $\zeta_i \leq 0$? Consider our running example. The need to have the P field turns the D-flatness condition into

$$\zeta = |\phi^1|^2 + |\phi^2|^2 + |\phi^3|^2 + |\phi^4|^2 + |\phi^5|^2 - 5|P|^2 \tag{14}$$

and we see that, even when $\zeta \leq 0$, there is a solution, now with $P \neq 0$. What is it?

This is discussed at length in Ref. [12]. The answer is that the P field breaks $U(1)$ to a discrete $\mathbb{Z}_5$ gauge symmetry, and thus its low energy limit is an orbifold of a model with chiral superfields and a superpotential, or Landau-Ginzburg orbifold. In the limit $\zeta \to -\infty$ we find the Gepner model, an exactly solvable $d = 2$ conformal field theory. Thus, not only does this construction know about singularities, it knows about relations between seemingly different physical models, which are actually different phases of the same underlying model.

This of course implies that, although the ζ_i do parameterize the Kähler moduli space of Ricci-flat metrics on Calabi-Yau manifolds, identifying this with the Kähler moduli space used in string/M theory is simplistic and in fact false. A next step in defining the "stringy Kähler moduli space" is to note that in any string theory on a Calabi-Yau manifold, or indeed any theory which has a space-time two-form gauge potential B, there is another set of moduli arising from the integrals of B over a basis of H_2.

These combine with the geometric Kähler moduli (the ζ_i or $\int \omega$'s) to form complex variables which lead to chiral superfields in the effective field theory, complex as they must be. However this is only a first step as it turns out that physical observables which depend on Kähler moduli get instanton corrections, and these actually modify the structure of moduli space. The simplest and best definition of stringy Kähler moduli space uses mirror symmetry, as we discuss in §5.2.

Whereas in the quintic example, going from $\zeta > 0$ to $\zeta < 0$ connects a geometric sigma model to a "non-geometric" Landau-Ginzburg model, in other examples, it can connect sigma models with two distinct topologies, through a transition known as a "flop." While an example in a compact Calabi-Yau is a bit too lengthy to present here, the basic idea can be seen by considering the $U(1)$ gauge theory with four chiral superfields of charges $(+1, +1, -1, -1)$, and thus with the D-flatness condition

$$|A^1|^2 + |A^2|^2 = |B^1|^2 + |B^2|^2 + \zeta. \tag{15}$$

Suppose that $\zeta > 0$; then the right-hand side is always positive and the space of solutions is a $\mathbb{CP}^1$ parameterized by the A^i, with a vector bundle over them parameterized by the B^i (in fact it is $\mathcal{O}(-1) \oplus \mathcal{O}(-1)$ over $\mathbb{CP}^1$). This space has a single two-cycle whose volume is ζ, in keeping with our general discussion. Now suppose that we take $\zeta < 0$ instead. Clearly we will get a similar picture, but with a $\mathbb{CP}^1$ of volume $-\zeta$ parameterized by the B^i, and a fiber parameterized by the A^i.

While here the two phases are spaces of the same topology, exactly the same structure can arise as a small region in a larger, compact Calabi-Yau. In this case, the A^i and B^i could be related to coordinates on the larger space in different ways, and the two signs of ζ will lead to two different topologies for the larger space, related by a flop transition. The flop is just one example of a "birational equivalence," and the upshot is that the true stringy Kähler moduli space for a Calabi-Yau will also contain a large set of other Calabi-Yau manifolds to which it is birationally equivalent, and perhaps additional Landau-Ginzburg and other "non-geometric" phases.

While in principle all Calabi-Yau manifolds can be realized as subvarieties of toric varieties (after all they were subvarieties of projective space), it is not known how to do this in any systematic way, and even the question of whether the number of distinct Calabi-Yau threefolds is finite or infinite remains open. What has been extensively studied is the set of toric hypersurfaces, with a single defining equation. It turns out that the various conditions needed for the data Q_i^A to lead to a Calabi-Yau n-fold

can be phrased in geometric terms as the condition that they describe an
$n + 1$-dimensional "reflexive polytope." All 4-dimensional reflexive polytopes were classified some time ago by Kreuzer and Skarke [13], and it
turns out that there are $473, 800, 776$ of them, corresponding to Calabi-Yau
manifolds with Betti numbers shown in Figure 1. Many of these are different descriptions of the same manifold, or birationally equivalent manifolds,
and it is not known how many distinct toric hypersurfaces exist. Anyways,
it is easy to work out Betti numbers and intersection numbers from this
description, and there are computer databases of this information. There

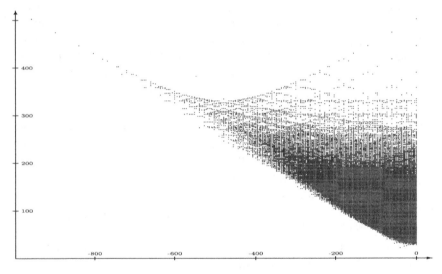

Fig. 1. $h_{11} + h_{12}$ vs. Euler number $\chi = 2(h_{11} - h_{12})$ for all pairs (h_{11}, h_{12}) with
$h_{11} \leq h_{12}$.

are other constructions of Calabi-Yau manifolds, of course. The most important is the elliptic fibration, which we will say a bit more about in §5.1.
It is thought to describe most but not all Calabi-Yau threefolds and is also
the starting point for F theory compactification. However we must move
on to other topics.

4. Complex structures and moduli

Having given a taste of the definition of Kähler moduli space, we now turn to
describe the complex structure moduli space of a Calabi-Yau manifold. We
assume that the reader has seen the elementary discussion of the complex
structure of a two-torus, with a single parameter usually called τ, and its

moduli space given by the upper half plane quotiented by $SL(2, \mathbb{Z})$. This can serve as a general paradigm for Calabi-Yau moduli space as well – it will have many complex parameters and a duality group – but in fact the global picture is not at all understood when there are more than one or two parameters.

The most elementary definition of a complex structure moduli space is in terms of the coefficients of the defining equation. Thus, in the family of quintic manifolds Eq. (5), λ is a complex structure modulus. This defining equation can be generalized to

$$0 = \sum_{I,J,K,L,M} c^{IJKLM} Z_I Z_J Z_K Z_L Z_M \tag{16}$$

which has 126 complex parameters c^{IJKLM}. On the other hand, varying the overall scale of the parameters, or performing a $SL(5)$ rotation which can be undone by a rotation of the Z_I, does not change the resulting Calabi-Yau manifold. Thus the complex structure moduli space of the quintic is at least 101 dimensional. In fact it is 101 dimensional, but in more complicated constructions it can happen that not all of the complex moduli show up in this way. This is because, as we mentioned earlier, a fully general construction might require using redundant systems of equations, and the simpler hypersurface construction might only capture part of the moduli space.

A fully general definition of complex structure can be based on periods, in the same way as for the two-torus. There we had A and B cycles, and we could use $\tau = \int_B dz / \int_A dz$ as a parameter. This was the particular case of a Calabi-Yau 1-fold and on this level the discussion generalizes directly. We consider the middle homology $H^n(M)$ for an n-complex dimensional manifold, and introduce a basis of n-cycles Σ_i. The intersection form between a pair of such cycles is symmetric for n even and antisymmetric for n odd, so the discussion branches at this point. Let us consider n odd, then we can define A and B cycles as for the two-torus, satisfying

$$(\Sigma_i^A, \Sigma_j^A) = (\Sigma_i^B, \Sigma_j^B) = 0 \, ; \tag{17}$$

$$(\Sigma_i^A, \Sigma_j^B) = -(\Sigma_i^B, \Sigma_j^A) = \delta_{ij} \, . \tag{18}$$

We then construct the holomorphic three-form Ω, which by the general discussion of §2 is unique up to an overall factor, and define its periods,

$$\Pi_i^A = \int_{\Sigma_i^A} \Omega ; \qquad \Pi_i^B = \int_{\Sigma_i^B} \Omega . \tag{19}$$

For hypersurfaces it is easy to write an explicit expression for Ω, so this is reasonably concrete.

There is then a "Torelli theorem" like that for the torus and other Riemann surfaces, to the effect that two manifolds with the same periods Π, have the same complex structure (in other words, there is a complex diffeomorphism between them). On the other hand, if the Π are different, it could be that the two manifolds are related by a duality transformation, which induces some $Sp(b^n, \mathbb{Z})$ transformation on the basis of H^n. So this is useful but not a full solution to the moduli space problem, especially as the relevant duality groups are not known.

Another point we need to settle to use the periods to define complex structure moduli space is that there are $b^3 - 1$ periods (taking out the overall scale) but only $b^3/2 - 1$ complex structure moduli. For higher Riemann surfaces, characterizing the vectors Π which are actually periods of Riemann surface is the notoriously difficult "Torelli problem," which although solvable drastically limits the value of this description. For Calabi-Yau threefolds, however, this problem is easy to solve, at least locally, through special geometry. Without entering into details, we simply recall that a special geometry is a complex space parameterized by $N + 1$ homogeneous coordinates t_i, together with a complex prepotential $\mathcal{F}$ which is a degree 2 homogenous function, $\mathcal{F}(\lambda t) = \lambda^2 \mathcal{F}(t)$. Locally, if we take the A-cycle periods as coordinates, we have

$$\Pi_i^A = t_i \, ; \qquad \Pi_i^B = \frac{\partial \mathcal{F}}{\partial t_i}. \tag{20}$$

Furthermore, there is a natural Kähler metric on this space,

$$K = \log \sum_i \Pi_i^A \bar{\Pi}_i^B - \bar{\Pi}_i^A \Pi_i^B \tag{21}$$

where the bar denotes complex conjugation. One could substitute Eq. (20) or any other explicit expression for the periods into this to get this metric, which determines a kinetic term in the effective field theory.

One can show mathematically that the complex structure moduli space of a Calabi-Yau threefold is a special geometry with dimension $N+1 = b^3/2$, so this tells us when we have found enough parameters and focuses attention on computing $\mathcal{F}$ and/or the periods. This rapidly becomes a very technical discussion as the usual way to get the periods is to derive the so-called Picard-Fuchs differential equations which they satisfy and solve them, as in the pioneering work [14]. Let us briefly outline some of the results from this paper.

The starting point is the subset of quintics defined by the equation Eq. (5). While this is only a particular one-parameter slice through the full 101-dimensional moduli space, it is also the subset of manifolds which respect an additional discrete symmetry group, namely the group $\mathbb{Z}_5^4$ generated by the rotations $Z_i \rightarrow \exp(2\pi i/5)Z_i$ (of course rotating all five coordinates is a no-op on $\mathbb{CP}^4$), and thus this restriction is natural. One can then write the periods as Eq. (19) and start differentiating with respect to λ. Since there are only 4 linearly independent periods, these derivatives should eventually become linearly dependent, and this happens by forming a combination of integrand.s which is holomorphic and single-valued, thus integrating to zero. The resulting differential equation,

$$\left(D^4 + 5z(5D + 1)(5D + 2)(5D + 3)(5D + 4)\right)\Pi(z) = 0, \qquad (22)$$

where $z = 1/\lambda^5$ and $D = z(d/dz)$, can be solved in terms of hypergeometric functions, which can be analyzed explicitly.

The basic qualitative result from this analysis is that the complex moduli space of quintics Eq. (5) is a three-punctured sphere – in other words the periods are not single-valued but have three branch points, at singularities of various types. Going around one of these three branch points induces a monodromy on the basis of periods. Geometrically, the operation of going around a branch point is a duality transformation on the basis of A and B-cycles and thus the monodromy matrices must be elements of $Sp(4, \mathbb{Z})$, which are given in Ref. [14].

The branch points are $z \rightarrow \infty$, $z = 1$ and $z = 0$. This last is actually not a singular Calabi-Yau but rather a point at which the Calabi-Yau picks up an extra $\mathbb{Z}_5$ symmetry. It corresponds to the Gepner model construction. The $z \rightarrow \infty$ limit is the "large structure limit" which plays a role in mirror symmetry, as we discuss in §5.2. Finally, the point $z = 1$ corresponds to what is called a "conifold" degeneration. This can be pictured locally in a way similar to how we pictured the flop in §Eq. (15). Now the local neighborhood of the singularity in the Calabi-Yau will be described by an "F-flatness" condition or complex equation,

$$t = Z_1^2 + Z_2^2 + Z_3^2 + Z_4^2. \qquad (23)$$

Its gradient ∇f vanishes at the point $Z_i = 0$, which is part of the space $f = 0$ only for $t = 0$, and thus this describes a region in complex structure moduli space (parameterized by t) which includes a singular point. By separating the coordinates into two real vectors, $Z_i = x_i + iy_i$, one can see that for $t > 0$ and real the general solution is the S^3 with $\sum_i (x_i)^2 = t$ and

$y_i = 0$, fibered over by y satisfying the orthogonality condition $x \cdot y = 0$. Thus this is a space with $b^3 = 1$. The volume of this three-cycle, which we will call Σ_A, is controlled by $|t|$.

While in this noncompact space, A and B-cycles need not be paired, if this geometry appeared as a region in a compact space, Σ_A would have a conjugate Σ_B (parameterized near the curve by the y^i). Going around the singularity as $t \to \exp(i\theta)t$, one can show, corresponds to the change of basis

$$\Sigma_A \to \Sigma_A \, ; \qquad \Sigma_B \to \Sigma_B + \Sigma_A. \tag{24}$$

One can show that the periods have the corresponding monodromy; in fact

$$\Pi_A \sim t \, ; \qquad \Pi_B \sim C + \frac{t}{2\pi i} \log t. \tag{25}$$

These asymptotics are also the appropriate ones for particle masses in an $N = 2$, $d = 4$ supersymmetric abelian gauge theory, and this type of relation plays an important role in string duality arguments and "geometric engineering."

While some more general theory has been developed over the years, and one can make general statements about the nature of singularities and the behavior of the periods and metric near the singularities, there is still no effective way to work with high-dimensional complex structure moduli spaces, and almost all physics work relies on one and two-parameter examples.

4.1. *Flux vacua*

One application of these results is to compactification of the type IIb superstring on a Calabi-Yau manifold with flux. The IIb string has two three-form field strengths $F^{(3)}$ and $H^{(3)}$. Turning them on produces a vacuum energy which depends on the dilaton τ and complex structure moduli t. By a combination of physical arguments first given in Ref. [15], one can show that an important part of this energy is described by the superpotential

$$W(t) = \int \Omega(t) \wedge \left(F^{(3)} + \tau H^{(3)} \right). \tag{26}$$

By Poincaré duality, the integral of a three-form wedged with one of these fluxes is equal to its integral over a dual three-cycle Σ; by flux quantization this three-cycle must live in $H^3(M, \mathbb{Z})$. Thus the superpotential can be written in terms of the periods, schematically

$$W(t) = \sum (N^F + \tau N^H) \cdot \Pi(t). \tag{27}$$

Note that the periods Π are not functions on complex moduli space, since they depend on a choice of normalization of Ω. However the superpotential in $N = 1$ supergravity is not a function either; it is a section of an line bundle, reflecting the normalization ambiguity of the covariantly constant spinor in Eq. (2), and this all fits together as it should.

To find an explicit flux vacuum with stabilized moduli, one must try different choices for the fluxes N and look for solutions of $DW = 0$. Once one sees that for each asymptotic direction in moduli space there is some period which grows in that direction, it should not be surprising that non-trivial minima exist. This is true, basically because of approximate inverse relations between the volumes of A and B-cycles in these limits. Still, given the complexity of the periods Π, explicit examples are not easy to find. Papers which do this and find vacua include [16]. One can also make a statistical treatment, in which one sums over all fluxes N and finds a density of flux vacua on complex structure moduli space, as in [17, 18].

5. Further topics

5.1. *Constructions of gauge bundles*

The original heterotic string compactifications included both $(2, 2)$ models, in which the Yang-Mills connection was set equal to the Levi-Civita connection on the tangent bundle, and $(0, 2)$ models, in which a more general Yang-Mills connection was allowed, subject to the anomaly cancellation condition that it has the same first and second Chern classes as the tangent bundle. Classifying the $(0, 2)$ models requires classifying $SU(N)$ bundles on a Calabi-Yau, where $N = 3, 4, 5$ for quasi-realistic models.

While some work has been done on this, it is far more complicated than classifying Calabi-Yau manifolds. The mathematical reason for this is that, while the moduli spaces of Calabi-Yau manifolds are themselves manifolds, with a known dimension, the moduli spaces of bundles are generally complicated singular varieties with many branches of varying dimensions. Physically, the difference is that Calabi-Yau moduli space can be studied in the context of $d = 4$, $N = 2$ supersymmetry without gauge symmetry, for which there is no superpotential. On the other hand, the heterotic string leads to $d = 4$, $N = 1$ supersymmetry, and the interesting models always have a superpotential for matter fields. Solving the F-flatness conditions generally leads to a complicated variety with many branches.

Let us explain this point a bit further as it would be attractive to have a simple expression like Eq. (27) for this superpotential. A higher-dimensional expression analogous to Eq. (26) is

$$W(t) = \int \Omega(t) \wedge \text{Tr} \left(\frac{1}{2} \bar{A} \wedge \bar{F} + \frac{1}{3} \bar{A} \wedge \bar{A} \wedge \bar{A} \right). \tag{28}$$

The parenthesized expression is the holomorphic Chern-Simons functional, whose variation leads to the condition $F^{(0,2)} = 0$ which defines a holomorphic bundle. Then, to get an analog of Eq. (27), one would need a general way to define the moduli t, and an expression for the gauge connection $A(t)$, to substitute in. Explicit expressions for $A(t)$ are very complicated, and in addition, as one moves around in the moduli space, new moduli become massless and open up new branches. For this reason, the most successful work on understanding $N = 1$ compactifications tends to work directly in the higher dimensional theory, and only reduces to effective field theory and a superpotential in the neighborhood of a solution. Thus there is no known shortcut around the problem of classifying bundles - though in §5.3 we will mention some ideas in this direction.

There are basically two general constructions of bundles which have been used in the string theory literature. The first is the so-called "monad" construction, which appears physically in $(0, 2)$ sigma models. The idea is to postulate three sums of line bundles U, V, W and to write a "chain complex"

$$0 \to U \to V \to W \to 0. \tag{29}$$

The arrows in this diagram represent linear maps, and we require that the composition of any two successive maps is zero. This is the sort of diagram used in defining cohomology, and the bundle we are constructing is exactly the cohomology, the kernel of the map $V \to W$ modulo the image of the map $U \to V$. Physically, the two maps correspond to superpotential constraints and gauge invariances, respectively.

To get a sense for this, consider a simpler example in which we drop the U term,

$$\mathcal{O}^4 \to \mathcal{O}(1)^2. \tag{30}$$

The arrow represents a 4×2 matrix whose entries are linear in the homogeneous coordinates, say

$$\begin{pmatrix} Z^1 & Z^2 & Z^3 & Z^4 \\ Z^2 & Z^3 & Z^4 & Z^1 \end{pmatrix}. \tag{31}$$

The kernel of this map will generically be two dimensional and thus might define a $U(2)$ bundle, whose Chern character is the difference $4 \operatorname{ch}(\mathcal{O}) - 2 \operatorname{ch}(\mathcal{O}(1))$. For this to be a bundle, it must be two dimensional everywhere. In our example, this fails when the two rows become colinear, but one can check that this only happens at four points. If we use this for a Calabi-Yau hypersurface which does not intersect these points, we get a bundle.

In general, checking that the objects one defines are bundles is non-trivial. But the construction makes sense without this condition – if one allows the dimensions of kernels and cokernels to vary on the manifold, one gets a more general class of objects called coherent sheaves. These include objects such as zero size instantons. However, it is not really understood which sheaves can be used in heterotic string compactification.

To claim that a bundle constructed this way leads to a solution of the DUY equation and can be used in heterotic string compactification, there is one final point to check – the bundle must be μ-stable. This condition can also be understood physically and has to do with D-flatness conditions in the effective theory.

While a fair amount of work has been done in this context, and interesting claims such as a $(0,2)$ mirror symmetry have come out, it appears to have only scratched the surface, and no general statements as compelling as the Kreuzer-Skarke diagram have emerged. Furthermore, it is known that not all bundles on a threefold can be realized this way (one would need a sequence with four terms), so it is not clear that this construction is general enough to reveal any patterns.

The second construction is the spectral cover. It only applies to a subset of Calabi-Yau manifolds, the elliptic fibrations, but it is more general, perhaps describing all bundles. We will not describe it in any detail, but the basic idea is to use T-duality on the fiber. An elliptic fibration is a manifold which locally is a product of an $n-1$-dimensional base manifold (generally not Calabi-Yau), and a fiber which is an elliptic curve, a real two-torus with a complex structure and a preferred base point. The physics intuition behind the spectral cover is to realize the bundle on Dirichlet branes which fill the Calabi-Yau, let us call them D9-branes, and then do T-duality on the fibers. This turns a $U(N)$ bundle into a collection of N D7-branes which wrap the base and sit at points on the fiber, determined by the holonomy of the original gauge connection on the fiber.

The tricky part of this construction is that (except for special cases like $K3 \times T^2$), the elliptic fibration will always have singularities at which the torus degenerates. The simplest example is the base $\mathbb{CP}^1$; to get the K3 surface one needs a fibration with 24 singularities (in F theory these would be 7-branes, but here they are just mathematical artifacts). Dealing with these singularities is the main part of the work. They are well understood for $\mathbb{CP}^1$ and largely understood for a two complex dimensional base, so this is a viable way to construct bundles. However, the need to combine information from various singularities leads one into sheaf theory and much more mathematical formalism.

5.2. *Mirror symmetry*

This was the first great discovery in the string-math interface and much of the early work on these topics was done in this context. The basic physics idea is probably familiar to the reader, that the $(2, 2)$ nonlinear sigma models with two different Calabi-Yau target spaces M and W can be shown to be equivalent by a very simple world-sheet argument. This leads to numerous predictions that mathematical objects related to M and W will be equivalent. Many of these predictions were subsequently proven by mathematicians, and the topic remains very active.

For string phenomenology, mirror symmetry is simply one type of string/M theory duality, and we can apply it in the same general way, turning hard to solve problems into duals which are easier to solve. The basic example following from our discussion so far is the treatment of the Kähler moduli space for a Calabi-Yau sigma model. These sigma models have world-sheet instantons which correct the kinetic term in a type II compactification to $d = 4$, $N = 2$ supergravity, the superpotential in type IIa with flux and heterotic compactifications, and so on. On the other hand, the analogous quantities which depend only on complex structure moduli are classically exact, and given by formulas like Eq. (21) and Eq. (27). The mirror symmetry recipe for solving all such problems is to turn them into complex structure moduli problems, which are classically exact. This includes the kinetic term on Kähler moduli space and the flux superpotential for type IIa compactifications.

As an example, we can connect the definition of Kähler moduli space made in §3.4, in terms of the Kähler form or Fayet-Iliopoulos terms, to the stringy Kähler moduli space, using mirror symmetry. The short definition is that the stringy Kähler moduli space of a Calabi-Yau M is equivalent to the complex structure moduli space of its mirror W. This requires more work to make precise as we need a map from the complexified Kähler moduli $\zeta + i \int B$ to the complex moduli t, which typically will exist only in some region (the "large structure limit") of moduli space. This can be worked out by expressing both sides in special coordinates (the t of Eq. (20)), as is done in the references.

5.3. *Dirichlet branes*

After the discovery of D-branes, many quasi-realistic models were developed in which the Standard Model and BSM matter and gauge fields come from open strings between branes. The original models involved orbifolds and branes with angles, and these can be analyzed with physics techniques, but putting branes in Calabi-Yau or special holonomy manifolds requires some mathematical help.

To give the briefest summary of this, just as supersymmetry is associated to manifolds of special holonomy, supersymmetric D-branes are associated to calibrated submanifolds. A calibration is a p-form ϕ (real or complex) which gives a lower bound on the volume: given a p-dimensional submanifold Σ, we have

$$\left| \int_\Sigma \phi \right| \leq \text{vol } (\phi). \tag{32}$$

This idea was introduced by the mathematicians Harvey and Lawlor, but turns out to be related in string theory to BPS bounds: it maps directly into the lower bound on a mass or tension M given by a central charge Z,

$$|Z| \leq M. \tag{33}$$

Furthermore, the objects which saturate the bound preserve a fraction of supersymmetry (normally $1/2$).

Given a covariantly constant spinor ϵ, one can show that

$$\phi^{IJ...K} = \epsilon^t \Gamma^{IJ...K} \epsilon \tag{34}$$

is a calibration, so all of the special holonomy manifolds have calibrations.

For manifolds of $SU(N)$ holonomy, the calibrations are ω, the Kähler form, and Ω, the holomorphic N-form. It turns out that saturating the bound for one form implies vanishing of the pullback of the other form. Thus, there are two types of supersymmetric branes.

The A-branes are 3-branes (in the Calabi-Yau; of course they can have other space-filling dimensions) which satisfy

$$0 = \omega|_\Sigma = \text{Im } e^{-i\theta} \Omega|_\Sigma , \tag{35}$$

thus the restriction of Ω to the brane has a definite phase. Now ω, being an antisymmetric tensor, is a symplectic structure. Asking that its restriction be zero defines a Lagrangian submanifold. The combination of both conditions defines a "special Lagrangian manifold." Being D-branes, they can also carry a gauge connection, but generally it must be flat.[h]

The B-type branes are even-dimensional holomorphic submanifolds. This includes the point, the entire manifold, and holomorphic curves and surfaces in the manifold. In addition, these branes can carry holomorphic gauge fields which satisfy the DUY equation Eq. (3). Many types of singular gauge fields are allowed as well, corresponding to a known subset of the coherent sheaves.

The general classification of branes on a Calabi-Yau is not a well-understood subject, but progress is being made. As little is known about special Lagrangian manifolds at present, the first step is to abandon the A-type branes, instead (by using mirror symmetry if necessary) looking at B-type branes. There is a universal construction which includes all of the B-type branes, the derived category of coherent sheaves. The physical intuition behind this construction is that we choose a finite set of branes such that sums of these branes and their antibranes reproduce all possible Chern characters (so, we need $b^0 + b^2 + b^4 + b^6$ different branes). Then, by taking bound states in all possible ways, we will get all possible branes.

Understanding the derived category of coherent sheaves appears to be a simpler problem than understanding all the bundles, for several reasons. First, it corresponds more closely to what is allowed in string theory, as zero size instantons and other coherent sheaves are allowed. Second, there are other constructions which lead to the same derived category, such as the generalized McKay correspondence for orbifolds. Another example is matrix factorization, which is a mathematical description of the branes which can be obtained from a Gepner model or orbifold theory. Because these models are connected to the original Calabi-Yau by varying Kähler moduli, one gets the same derived category of branes. The matrix factorization approach also exhibits the moduli in a more unified way and allows one to write a superpotential in terms of the moduli, analogous to Eq. (27).

To know that an object in the derived category corresponds to a physical brane and can be used in string compactification, one again has to check a stability condition, called Π-stability or Bridgeland stability. This condition is fairly well understood for orbifolds and is a subject of active research for

---

[h]There are some exceptions to this on tori, called co-isotropic branes.

compact Calabi-Yau manifolds, so one can hope for near-term progress.

5.4. *Computing the low energy Lagrangian*

Given a compactification with a quasi-realistic spectrum, one wants to compute the $d = 4$, $N = 1$ effective Lagrangian, at least for the observable sector. While the superpotential is highly constrained by holomorphy and by the more detailed mathematical structures we have been discussing, and there are many computations even including quantum effects, getting quantitative results for the spectrum and couplings requires some control over the kinetic terms (the Kähler potential) as well. This problem has only been solved in very special cases, but is not beyond hope.

One approach is to work with models with a concrete microscopic definition, such as branes in a torus or an orbifold of a torus, and use physics techniques. Examples are discussed elsewhere in this book.

We can do exact computations of kinetic terms in a $d = 4$, $N = 2$ theory, where they follow from a prepotential, using the complex structure moduli techniques outlined in §4. The current state of the art for $N = 1$ is more or less to work with cases which are close enough to an $N = 2$ theory to treat the differences as a perturbation. This could work for cases like the original CHSW $(2, 2)$ models, but there we do not have good techniques to work out the moduli potential.

In principle, the classical kinetic terms for matter and gauge fields are computable from the higher dimensional definitions. This requires knowing the Ricci-flat metric on the Calabi-Yau, and it appears that this can only be obtained using a combination of analytical and numerical techniques. Some success has been had with an approach (pioneered by Donaldson) that describes the Calabi-Yau metric as one in a family of simple metrics pulled back from the ambient space. This can be generalized to bundle metrics as in [19].

Quantum corrections to the kinetic terms remain a very difficult problem, even in simple $d = 4$ theories.

5.5. *Outlook*

The string-math interface is a very active subfield and it seems likely that all of the topics we discussed will develop substantially over the coming years. What are the most important physical questions to answer, and what progress can we expect on the 5-10 year timescale? Here are my own thoughts on this.

One clear goal would be to work out even a single quasi-realistic model to the point of general acceptance – showing that all moduli are stabilized and that supersymmetry is broken at a low scale, and estimating the Yukawa couplings to within an order of magnitude. This might be doable in a local model – one in which the Standard Model degrees of freedom are localized to branes or singularities in a small part of the Calabi-Yau. We would probably need to find a model (and duality frame) for which quantum corrections are either small or inessential at the compactification scale – although we know they are essential for the Standard Model and probably for supersymmetry breaking, we could hope to deal with them at low energies using purely $d = 4$ field theoretic techniques.

Although the basic framework for doing this may involve a good deal of mathematical input, it is hard to see how mathematics will help with the details, and hard to see how the results would lead to major physics advances. If our techniques only suffice to do this for a special class of models, without having physical arguments that these models are somehow preferred, any predictions they lead to would not be very compelling. Still, as a benchmark and foundation for further progress, this step would be very important, and probably could be achieved on the 5-10 year timescale.

Could mathematics help us get quantitative results for larger classes of compactifications? Or could it somehow suggest principles that prefer or select particular classes of compactifications? Personally, I am quite hopeful about both of these possibilities, but I believe that this will not come primarily from trying to compute the observables we could measure at LHC and/or in cosmology and astrophysics. Rather, I believe that the crucial step which would enable such progress would be to develop some sort of "big picture" which describes most or all stable four-dimensional compactifications, the possible transitions between them which could take place in early cosmology, and the possible ways that they can break supersymmetry and lead to quasi-realistic (Standard Model and similar) physics.

Why do I say this? For me, the main lessons from string phenomenology as we understand it now are that, while string theory can reproduce the physics we observe in a satisfying way, only a tiny fraction of string compactifications lead to quasi-realistic physics at all. The consistency conditions, such as anomaly and tadpole cancellation, and vacuum stability conditions, do not seem to favor the Standard Model in any way beyond what we already know from quantum field theory. They do favor low energy supersymmetry, but not necessarily at energies low enough for us to test. Other aspects of the theory, such as the ease of getting E_8 gauge symmetry and its subgroups, do favor GUTs and the Standard Model. But since they are only part of the picture, it is hard to know what to make of them.

Although superficially this might seem discouraging, I feel this is not so at all. Let us consider the analogy between solutions of string theory and solutions of the Schrödinger equation which describe atoms and molecules. Although very different, both have a large and complicated set of metastable solutions. Then, these comments are rather like saying that although the Schrödinger equation does describe specific molecules and their chemistry, nevertheless the list of all of the metastable molecules which it predicts can exist, has no apparent relation to the actual set of molecules we see around us on Earth, in the universe, or even in the chemistry lab. Nevertheless we can understand what we see, not just as a set of empirical facts, but in part from theoretical models. We have a good qualitative understanding of the possible molecules, and simple principles which drastically constrain their dynamics, such as conservation of the number of each atomic species, energetic arguments, and the basic phase structure of matter. This allows us to make models of the dynamics which created the molecules we see – on Earth, in the sun, and in other environments.

Such models can be made even for environments we cannot observe at all. We have far less observational knowledge about the center of the Earth than we do about electroweak scale particle physics, yet we feel we understand it much better. This understanding comes largely from theoretical arguments. There is no reason to think we cannot do as well some day for the physics of very high energies and the early universe, using theoretical arguments based on string/M theory.

This larger or "big picture" level is the one on which any sort of selection principle must be found. It seems very likely to me that a comparably simple overall picture of the complete set of string/M theory compactifications, and principles constraining the dynamics, will someday be developed. This will enable us to understand how certain vacua, including the one we live in, were preferentially formed in the earliest moments of cosmology. I suspect this picture and these principles will not have much in common with the details of the Standard Model, but will come out of structures in string/M theory and in mathematics which will be as surprising to us – though in retrospect foreshadowed by what we know now – as string duality was surprising to us in the 1990's. I believe we could make progress towards such a picture over the next 5-10 years, if enough people are attracted to work on it.

Acknowledgments

I thank Bobby Acharya, Sujay Ashok, Volker Braun, Ilka Brunner, Duiliu-Emanuel Diaconescu, Frederik Denef, Bartomeu Fiol, Bogdan Florea, Antonella Grassi, Brian Greene, Shamit Kachru, Robert Karp, Semyon Klevtsov, Albion Lawrence, Zhiqin Lu, Sergio Lukic, Greg Moore, Dave Morrison, Burt Ovrut, Rene Reinbacher, Christian Römelsburger, Bernard Shiffman, Wati Taylor and Steve Zelditch for collaborations on these topics over the years.

References

[1] M. Green, J. Schwarz, and E. Witten, *Superstring Theory, Vol. II*, Cambridge Monographs on Mathematical Physics.
[2] B. S. Acharya and E. Witten, hep-th/0109152.
   B. S. Acharya, hep-th/0212294.
   B. S. Acharya and S. Gukov, Phys. Rept. **392**, 121 (2004) [hep-th/0409191].
   T. Friedmann and E. Witten, Adv. Theor. Math. Phys. **7**, 577 (2003) [hep-th/0211269].
   J. A. Harvey and G. W. Moore, hep-th/9907026.
   B. S. Acharya, K. Bobkov, G. Kane, P. Kumar and D. Vaman, Phys. Rev. Lett. **97**, 191601 (2006) [hep-th/0606262].
   B. S. Acharya, K. Bobkov, G. L. Kane, P. Kumar and J. Shao, Phys. Rev. D **76**, 126010 (2007) [hep-th/0701034].
   B. S. Acharya and K. Bobkov, JHEP **1009**, 001 (2010) [arXiv:0810.3285 [hep-th]].
   B. S. Acharya, K. Bobkov, G. L. Kane, J. Shao and P. Kumar, Phys. Rev. D **78**, 065038 (2008) [arXiv:0801.0478 [hep-ph]].
[3] D.D. Joyce, Oxford University Press, 2000; A. Kovalev, math.dg/0012189.
[4] M. R. Douglas, "Statistical analysis of the supersymmetry breaking scale," hep-th/0405279.
[5] M. Dine and Z. Sun, "R symmetries in the landscape," JHEP **0601**, 129 (2006) [hep-th/0506246].
[6] M. R. Douglas and S. Kachru, "Flux compactification," Rev. Mod. Phys. **79**, 733 (2007) [hep-th/0610102].
[7] M. R. Douglas and S. Klevtsov, Commun. Math. Phys. **293**, 205 (2010) [arXiv:0808.2451 [hep-th]].
[8] P. B. Kronheimer and H. Nakajima, Math. Ann. **288**, 263-307 (1990)
[9] M. R. Douglas and G. W. Moore, hep-th/9603167.
[10] D. Berenstein, V. Jejjala and R. G. Leigh, "The Standard model on a D-brane," Phys. Rev. Lett. **88**, 071602 (2002) [hep-ph/0105042].
[11] H. Verlinde and M. Wijnholt, "Building the standard model on a D3-brane," JHEP **0701**, 106 (2007) [hep-th/0508089].

[12] E. Witten, "Phases of N=2 theories in two-dimensions," Nucl. Phys. B **403**, 159 (1993) [hep-th/9301042].

[13] M. Kreuzer and H. Skarke, Adv. Theor. Math. Phys. **4**, 1209 (2002) [hep-th/0002240].

[14] P. Candelas, X. C. De La Ossa, P. S. Green and L. Parkes, "A Pair of Calabi-Yau manifolds as an exactly soluble superconformal theory," Nucl. Phys. B **359**, 21 (1991).

[15] S. B. Giddings, S. Kachru and J. Polchinski, Phys. Rev. D **66**, 106006 (2002) [hep-th/0105097].

[16] F. Denef, M. R. Douglas, B. Florea, A. Grassi and S. Kachru, Adv. Theor. Math. Phys. **9**, 861 (2005) [hep-th/0503124].

[17] S. Ashok and M. R. Douglas, JHEP **0401**, 060 (2004) [hep-th/0307049].

[18] F. Denef and M. R. Douglas, "Distributions of nonsupersymmetric flux vacua," JHEP **0503**, 061 (2005) [hep-th/0411183].

[19] V. Braun, T. Brelidze, M. R. Douglas and B. A. Ovrut, JHEP **0805**, 080 (2008) [arXiv:0712.3563 [hep-th]].

7

The String Theory Landscape

A. N. Schellekens*

*Nikhef,
Science Park. 105, 1098 XG Amsterdam,
The Netherlands*

Perhaps the most important way string theory has affected the perspective of particle physics phenomenology is through the "string theory landscape". We discuss the evidence supporting its existence, describe the regions of the landscape that have been explored, and examine what the string theory landscape might imply for most Standard Model problems.

1. Changing Perspectives

This chapter consists of three parts. In the present section, we will give a brief historical overview of the birth of the string theory landscape and we will explain its main features. We also present the main arguments in favor of its existence, both from a top-down (string theory) and a bottom up (the Standard Model) point of view. Furthermore we underline the important changes this concept has on the perspectives for string phenomenology. In Section 2 we present an overview of the various methods of constructing string theories in four dimensions, by direct construction and by compactification. In Section 3 we discuss how far one can get towards understanding the Standard Model from the landscape perspective, in comparison to the traditional, symmetry-based approach.

Part of this chapter is based on the review article [1].

---

*Nikhef, Amsterdam; IMAPP, Radboud Unversiteit Nijmegen; IFF-CSIC, Madrid.

1.1. *The age of symmetries*

At the time when string theory started being considered as a theory of all interactions including gravity, the theoretical work on the Standard Model had reached its final form. Several decades of experimental work were still needed to establish it, but theorists started moving ahead using the concepts that had led to so much success in understanding the three non-gravitational interactions. The most important of these concepts was symmetry. It revolutionized the understanding of fundamental physics. From the seemingly hopeless chaos of nuclear and hadronic physics a very simple description of those interactions had emerged in just about two decades: a spontaneously broken gauge theory which was called the "Standard Model".

The name does not suggest much confidence in this idea, and indeed nobody saw the Standard Model as more than an intermediate stage, at best an approximate description of nature, eventually to be supplanted by something even simpler and mathematically more elegant. Indeed, by using the same group-theoretical methods that turned out to be so successful in the description of the Standard Model, new theories were found that looked more attractive. The highest achievable goal appeared to be supersymmetric Grand Unified Theories (susy-GUTs), which got even more credibility in the early nineties, when the precision results of LEP suggested that the three gauge couplings evolved to a common value at the very interesting energy scale of about 10^{16} GeV, a few orders of magnitude below the Planck scale. Even today, despite the fact that the experimental evidence has not (yet) shown up, it is hard to believe that this could all be just a coincidence.

A few years earlier string phenomenology had entered the scene. When it did, it seemed to point at even grander symmetries. In 1984, the first results obtained by compactifying the just-discovered $E_8 \times E_8$ heterotic string suggested the ultimate unification. The Standard Model appeared to emerge (almost) uniquely from the jewel of Lie algebra theory, the Lie algebra E_8.

1.2. *The birth of the landscape*

But in the remainder of the eighties there was the beginning of a slow shift away from the notions of symmetry and uniqueness that were considered almost self-understood until then. History will decide if this was the beginning of a paradigm shift or just prematurely giving up on uniqueness. But the evidence that the former is true is mounting.

Perhaps we will conclude one day that these beautiful ideas have always carried the seeds of their own destruction. A Standard Model family fits beautifully in the (5)+(10) of $SU(5)$ and even more beautifully in the (16) of $SO(10)$, but the Standard Model Higgs field does not. Furthermore, even if $SU(5)$ or $SO(10)$ exist as symmetries at short distances, there is no unique path to the Standard Model at the weak scale: in $SU(5)$ models there are two minima of the GUT Higgs potential, one leading to the Standard Model, and one to a $SU(4) \times U(1)$ gauge theory. In $SO(10)$ models the number of options increases.

The idea of low energy supersymmetry was also plagued by serious problems from the very beginning. It introduces light bosons into the spectrum that lead to rapid proton decay. By contrast, a sufficiently long life-time of the proton is automatic in the Standard Model. On general grounds, one would expect low energy supersymmetry to give rise to flavor violations that should have been observed a long time ago already. These, as well as other problems can be evaded by additional assumptions, but it is disturbing that the pieces of the puzzle do not fall into place more easily. Furthermore, although susy and GUTs are well-motivated answers to important questions, they have never led to a substantial simplification of the Standard Model.

1.2.1. *String vacua*

The uniqueness of string theory was also in doubt right from the start. The $E_8 \times E_8$ heterotic string was not unique, but part of a small set of 10- and 11-dimensional supersymmetric theories which were initially taken less seriously.

But more importantly, there was an explosion of compactifications and four-dimensional string constructions in the two years following 1984 [2–7]. Already as early as 1986 it became customary to think of the different string theories or compactifications as "vacua" or "ground states" of a fundamental theory (see for example the last line of [8] or discussion at the end of [9]; here one also finds the remark that perhaps our universe is merely a sufficiently long-lived metastable state). The proliferation of "string vacua" has not stopped since then. Here and in the following we use the word "vacuum" for the metastable state that correctly describes our Universe, and all its analogues with different gauge theories. The proper definition is itself a difficult issue, especially in de Sitter space, but if no well-defined description exists that matches our Universe, string theory would be wrong anyway.

1.2.2. Moduli

Soon it became clear that these "string compactifications" or "four-dimensional strings" had continuous deformations that can be described by vacuum expectation values of massless scalar singlets, called "moduli". Typically, there are tens or hundreds of them. All quantum field theory parameters depend on the moduli, and hence the existence of moduli is a first step towards a plethora of possibilities.

These singlets generate unobserved fifth forces and their presence is cosmologically unacceptable [10, 11], but so is the fact that supersymmetry is unbroken. For more than a decade, this left room for the possibility that the abundance of string vacua would be reduced to just a few, maybe just one, once these problems were solved. But in the beginning of this century considerable progress was made towards solving the problem of moduli stabilization and – to a lesser extent – that of supersymmetry breaking.

The large number of available ingredients (fluxes, D-branes, orientifold planes and various perturbative and non-perturbative effects) led to the nearly inevitable conclusion that if there was one solution, there were going to be many more. Almost two decades after 1984 the denial phase reached its end, marked by an influential and somewhat provocative paper by L. Susskind [12], who also gave the subject its current name, the "string landscape".

1.2.3. The cosmological constant

Remarkably, these developments were driven to a considerable extent by observation: the discovery of an accelerated expansion of the universe in 1998 [13, 14]. The most straightforward interpretation is that we live in a universe with positive vacuum energy density, which acts like a cosmological constant, and implies that we live in de Sitter (dS) rather than flat Minkowski space. Contrary to some statements in the literature, there was never any difficulty in getting positive vacuum energy in string theory. Some of the aforementioned papers from 1986 built non-supersymmetric strings, and some of those string theories have positive vacuum energy. At that time this feature was merely observed, but not yet considered to be of any interest. However, we have little computational control over non-supersymmetric strings, and at the moment the only viable path to string theory in dS space is to "up-lift" supersymmetric AdS vacua with negative vacuum energy.

The explanation of the observed accelerated expansion requires not only positive vacuum energy, but an extremely small amount of it. It is about 120 to 60 orders of magnitude less than its natural scale, depending on whether one compares to the Planck scale or the weak scale. In non-supersymmetric string theory vacuum energy comes out as a sum of positive and negative contributions of Planckian size. Everything we know suggests that this will give rise to numbers of order one in units of the natural value. If the contributions in the sum are random, the chance of finding a result near the observed value is about 1 in 10^{120}. This would imply that one needs an ensemble of at least 10^{120} vacua to have a chance of finding one like ours. With any smaller ensemble, the existence of the small observed value would be a bizarre coincidence.

1.2.4. *The Bousso-Polchinski mechanism*

It was realized decades ago [15] that anti-symmetric tensor fields $A_{\mu\nu\rho}$ might play an important rôle in solving the cosmological constant problem. Such four-index field strengths can get constant values without breaking Lorentz invariance, namely $F_{\mu\nu\rho\sigma} = c\epsilon_{\mu\nu\rho\sigma}$. If we couple the theory to gravity, it gives a contribution to the cosmological constant Λ:

$$\Lambda = \Lambda_0 - \frac{1}{48}F_{\mu\nu\rho\sigma}F^{\mu\nu\rho\sigma} = \Lambda_0 + \frac{1}{2}c^2, \tag{1}$$

where Λ_0 is the cosmological constant in the absence of anti-symmetric field strength contributions. In string theory c is not an arbitrary real number: it is quantized [16]. These quantized fields are called "fluxes". It turns out that string theory typically contains hundreds of fields $F_{\mu\nu\rho\sigma}$, which we label by $i = 1, \ldots, N$. The resulting formula for Λ is

$$\Lambda = \Lambda_0 + \frac{1}{2}\sum_{i=1}^{N} n_i^2 f_i^2. \tag{2}$$

where the f_i are numbers derived from the string theory under consideration. One would expect the values for the real numbers f_i to be different. If the values of f_i are distinct and incommensurate, then Eq. (2) defines a dense discrete set of values. Bousso and Polchinski called it a "discretuum". This realizes a dynamical neutralization of Λ first proposed by [17, 18]. See also [19] for a related realization of this idea in string theory.

The discretuum is populated by some physical process that is able to connect the different string vacua. The mechanism proposed for this is tunneling by bubble nucleation in eternal inflation, a near inevitability in most models of inflation. See [20, 21] for reviews and references. This area is – so far – less deeply connected to string theory, and therefore we will not discuss it in detail, except to mention that it leads to the very thorny issue of the multiverse measure problem. See [22–25] for various ideas about this.

1.2.5. *Existence and distribution of de Sitter vacua*

To make use of the Bousso-Polchinski neutralization of Λ a sufficiently dense discretuum of such vacua is needed. This mechanism relies on the fact that whatever the contribution of particle physics, cosmology and fundamental theory might be, it can always be canceled to 120 significant digits by flux contributions, *without making actual computations with that precision*. If in reality these distributions are severely depleted in part of the range, or have a highly complicated non-flat structure, this argument would fail. There might still exist a huge landscape, but it would be useless.

We will consider here only type-IIB (and related F-theory) compactifications where the most explicit results have been obtained. For references to work in other areas and more details see [1, 26–30].

In type-IIB theories one starts with a Calabi-Yau compactification with h_{21} complex structure ("shape") moduli and h_{11} Kähler ("size") moduli, where h_{21} and h_{11} are the Hodge numbers of the CY manifold (see Section 2.3 for more details on Calabi-Yau compactifications). One can add to this background configuration a choice of gadgets from the string theory toolbox, such as 3-form RR and NS fluxes, 5-form fluxes, denoted $F3, H3$ and $F5$ respectively, and D3 and D7 branes.

The 3-form fluxes can stabilize all complex structure moduli. This is due to a tree-level term in the superpotential that takes the form [31]

$$W_{\text{flux}} = \int (F_3 - \tau H_3) \wedge \Omega , \qquad (3)$$

where $\tau = a + ie^{-\phi}$, and a is the axion and ϕ the dilaton. The dependence on the complex structure moduli is through Ω, the holomorphic three-form of the Calabi-Yau manifold. This term also fixes the dilaton and axion. However, W_{flux} does not depend on the Kähler moduli and hence cannot fix them. Since every CY manifold has at least one Kähler modulus, this leaves therefore at least one modulus unfixed.

One may fix the size moduli with non-perturbative terms in the superpotential. These take the form $W \propto \exp(i\lambda s)$, where s is the size modulus and λ a parameter. Such terms can be generated by instantons associated with Euclidean D3-branes [32] or from gaugino condensation in gauge groups on wrapped D7 branes. If at least one of these effects is present, string vacua with all moduli stabilized can be obtained [33]. This work, usually referred to as "KKLT", builds on several earlier results, such as [34–36] and references cited therein. The solution obtained in this way has a negative vacuum energy, and is a fully stabilized supersymmetric AdS vacuum. However, the required instanton contributions may not exist in all cases. They are not generic [37] and may even be so rare that one only gets a "barren landscape" [38].

The next step is more problematic and more controversial. One must break supersymmetry and obtain a dS vacuum (this is called "up-lifting"). In KKLT this is done by adding an anti-D3 brane in a suitable location on the Calabi-Yau manifold, such that the validity of the approximations is not affected. Anti-D3 branes explicitly violate supersymmetry, and hence after introducing them one loses the control offered by supergravity. Attempts to realize the KKLT uplifting in supergravity or string theory have failed so far [39, 40], but opinions differ on the implications of that result. There exist several alternatives to D3-brane uplifting [41–45].

The result of a fully realized KKLT construction is a string vacuum that is free of tachyons, but one still has to worry about non-perturbative instability. The uplift contribution vanishes in the limit of large moduli, so there is always a supersymmetric vacuum in that limit, separated from the dS vacuum by the uplifted barrier that stabilized the AdS vacuum. One can work out the tunneling amplitude, and KKLT showed that it is generically much larger than the observed lifetime of our universe.

An alternative scenario (called the LARGE volume scenario or LVS) was presented in Ref. [46]. The starting point is type-IIB fluxes stabilizing the complex structure moduli and the dilaton and axion. By means of suitable $(\alpha')^3$ corrections these authors were able to find minima where all moduli are stabilized at exponentially large volumes in *non*-supersymmetric AdS vacua. Additional mechanisms are then needed to lift the vacuum to dS. An explicit example was presented recently [47].

The existence of the required dense distribution of vacua is still disputed, and some even question the existence of *any* such vacuum. Recent work seems to indicate that in the vast majority of cases the AdS vacua become tachyonic after uplifting [48, 49]. Another potential problem is a dramatic increase in tunneling rates as a function of the number of moduli [50]. These effects may dramatically reduce the number of dS vacua, rendering the Bousso-Polchinski argument inadequate. In [51] several criticism of the landscape are presented, including the use of effective potentials and of Coleman-de Luccia [52] tunneling between dS vacua in theories of gravity.

1.2.6. *Vacuum Counting*

The KKLT construction has been the starting point for estimates of the total number of flux vacua [53–55],

$$N_{\text{vac}} \approx \frac{(2\pi L)^{K/2}}{(K/2)!}, \tag{4}$$

where L is a number of order 1 to 100 (the "tadpole charge") and K the number of distinct fluxes. For typical manifolds this gives numbers of order 10^N, where N is of order a few hundred. A often quoted estimate is 10^{500}.

It is noteworthy that this formula turns a nuisance (a large number of moduli) into a virtue: the large number of moduli gives rise to the exponent of Eq. (4), and it is this large exponent that makes neutralization of the cosmological constant possible. All the ingredients used in the foregoing discussion are already present in string theory. Since all Standard Model parameters depend on the moduli, this results in a large distribution of options covering the environment of the Standard Model in QFT.

1.3. *A paradigm shift?*

If we were to accept that our laws of physics are picked out of a huge ensemble, and that the parameters have such special values just by coincidence, this would imply the end of science. Then the entire Standard Model could just be a random item from a huge ensemble. It is indeed remarkable that in the current state of particle physics, many of the remaining problems *could* be just environmental: the Standard Model provides an adequate description, but often with strange parameter values. Some genuine problems remain (such as dark matter and the mechanisms behind inflation and baryogenesis), but most other problems that are often discussed should really be called "worries". This means that we cannot be completely sure that there exists a solution. Perhaps these problems only exist in our minds. This includes the choice of the Standard Model gauge group, the choice of matter representations, charge quantization, the number of families, quark and lepton mass hierarchies, the smallness of neutrino masses, the gauge hierarchy problem and dark energy.

1.3.1. *Anthropic arguments*

All these worries only exist because there are minds to worry about them. It is quite plausible that this would not be the case if we allow the parameters and discrete choices of the Standard Model to vary. In an ensemble like the string landscape many such variations can occur, and it is inevitable that worrying minds will only worry about the small subset in which they can exist. This statement is one of several possible formulations of the "anthropic principle". It is a misnomer for several reasons, and in this formulation it is certainly not a principle of physics, like the equivalence principle. It is output, not input. One could even choose to ignore it, but then one would miss several potential explanations of some of the environmental problems.

Without its embedding in the string theory landscape, anthropic reasoning might also be called the end of physics, but in combination the two concepts merely are a complete change of course for traditional physics. The notion of symmetries as a fundamental concept is replaced by a combination of anthropic arguments and information about distributions of parameters in a mathematically well-defined ensemble. The lack of evidence for "new physics" may imply that we have reached the historical moment where this change is occurring. But particle physics is an experimental science, and the huge number of experimental results coming up in the next few years may revive the notion of symmetries, and postpone the emergence of a first glimpse of a landscape indefinitely.

1.3.2. *Derivability vs. Uniqueness*

One often finds criticisms of string theory like: "String theory was supposed to explain why elementary particles could only have the precise masses and forces that they do". In reality there has never been even a shred of evidence that string theory was going to lead to that. It is also nearly impossible to find quotes of this kind even in the earliest string papers. People making such statements are simply projecting their own expectations for a fundamental theory on string theory.

These expectations reflect the traditional uniqueness paradigm of particle physics: the hope that one day we will be able to derive all laws of physics, in particular the Standard Model and all its parameters. The fine structure constant α was expected to be given by some simple formula. This hope can be illustrated by famous quotes by Einstein, Feynman and others. But two concepts are often confused in this discussion, *uniqueness* and *derivability*. According to our current understanding of string theory, it is not really the uniqueness of the underlying theory that is at stake, but the uniqueness of the vacuum. If the vacuum is not unique, the Standard Model and its parameters can not be derived by purely mathematical manipulations. Additional information about the choice of vacuum, either phenomenological or anthropic, must be provided.

The fact that the Standard Model and it couplings fit nicely in an $SU(5)$ susy-GUT is often proclaimed as a strong hint in favor of uniqueness. But once again a distinction between uniqueness and derivability must be made. While GUTs may indeed point in the direction of a unique theory, SU(5) GUTs also gave the first hint *against* derivability, since they have two physically distinct vacua. The non-uniqueness of the return path towards lower energies first encountered in GUTs becomes much worse if one starts from the loftier vantage point of $E_8 \times E_8$ heterotic strings. It is like climbing a mountain: eventually one may reach a unique point, the top, but there may exist many other paths downwards, leading to other valleys.

Since string theory emerged during the height of the symmetry era, it is not surprising that it was first seen as the realization of the dream of uniqueness, interpreted as derivability. But in reality, string theory has been sending us exactly the opposite message almost from the beginning. One day, this may be recognized as its most important contribution to science.

1.3.3. *Evidence outside string theory*

From now on we will use the term "uniqueness" rather than "derivability", because that is what is commonly used. There has never been any evidence in favor of the uniqueness paradigm, the idea that the Standard Model has to be derivable. But there are several pieces of circumstantial evidence of the contrary, even without string theory.

Theories of inflation typically lead to multiple instances of new universes, a "multiverse". Even without inflation, what argument do we have to suggest that our own universe is unique, in any sense of the word? And if it is not unique, what argument do we have to tell us that the other universes must have exactly the same laws of physics as ours? The only fact that makes our universe and laws of physics unique is that they are the only ones we can observe.

The possible existence of a plethora of scalars provides another reason to question the uniqueness of the Standard Model. We have seen particles of various spins and Standard Model couplings, but only recently we may have observed the first Lorentz scalar, the Higgs boson. We have only been able to see it because it is unusually light and because it comes from a field that is not a Standard Model singlet. But experimentally we know nothing about Lorentz scalar fields that are gauge singlets. If they exist, they would appear in the Lagrangian as polynomials, modifying all dimensionless parameters. Then all parameter values are vacuum expectation values of scalar fields. One may still hope that this vacuum expectation value is somehow uniquely determined, but in almost all examples we know (including the Higgs potential of $SU(5)$ GUTs) scalar potentials tend to have more than one local minimum. One cannot say more without a more concretely defined theory, but in the only theory where such potentials can be discussed, string theory, the number of local minima appears to be astronomical.

Finally, the fact that the Standard Model is anthropically tuned provides evidence against the idea of its derivability. This is because the Standard Model stands out as a very special region in parameter space where nuclear physics and chemistry lead to complex structures we call "life". It would require an uncanny miracle for the two unrelated computations to give compatible results. Especially the last argument suggests that the ultimate fundamental theory – assuming such a notion makes sense and that we have enough intelligence and information to determine it – must have a large ensemble of physically connected vacua. This allows a process like eternal inflation to sample all these vacua, occasionally producing a universe within the anthropic domain.

1.3.4. *Uniqueness in the string landscape?*

One may still hope that the resulting scalar potential somehow has a mathematically unique local minimum, but that would be pure wishful thinking. One may even hope for a unique global minimum. However, it is not clear what to minimize, because vacuum energy is not bounded from below. However, if vacuum energy takes discrete values, there is a – presumably unique – vacuum with the lowest positive vacuum energy. Could that describe our universe? Could it be that in the sampling process of universes this particular one is somehow preferred? Another notion is the "dominant master vacuum", the state that dominates eternal inflation because it is most frequently sampled, often by a huge factor [56]. This is determined by its stability against tunneling to other states, as well as the likelihood of others tunneling into it. Both of these – the vacuum with lowest positive vacuum energy or the dominant master vacuum – have a sense of uniqueness, but the anthropic tuning argument makes it very unlikely that they happen to have the properties that allow life to exist. There is no need for

that, once there is a landscape. Then the anthropic genie is already out
of the bottle. All that is required is that there exist metastable vacua in
the anthropic zone of parameter space, and that their sampling frequency
is non-zero. There is no reason why they should dominate the multiverse
statistics. The dominant master vacuum does have an important positive
feature, because one may argue that most vacua have it in its history. Then
it can serve as a kind of eraser of initial conditions. Most observers would
find themselves in the anthropic universe most frequently reached by tunneling from the dominant master vacuum. See [57] for some speculations
regarding this idea.

1.4. Changing perspectives on string phenomenology

The existence (or non-existence) of the string theory landscape has an important impact on string phenomenology. One may distinguish at least
three different attitudes. The first is that we should simply find the exact
point in the landscape that corresponds to the Standard Model, use current
data to fix it completely, and then make an indefinite number of predictions for future experiments and observations. This includes all work on
explicit "string model building" in many corners of the landscape, reviewed
in the next section. The second is trying to extract generic predictions from
classes of models rather than individual ones. An example of work in this
category is the study of a class of M-theory models, reviewed in [58] (see
also Section 2.11). The third is to try and understand the Standard Model
by considering landscape distributions in combination with anthropic constraints. This is the point of view we take in Section 3. These three points of
view are not mutually exclusive. Their relative importance depends on how
optimistic one is about the chances of finding the exact Standard Model as
a point in the landscape.

2. The Compactification Landscape

We will present here just a brief sketch of the string compactification landscape. For further details we recommend the very complete book [59] and
references therein.

2.1. World-sheet versus space-time

In their simplest form, fermionic string theories live in ten flat dimensions.
In addition there is an 11-dimensional theory that is not described by inter-

acting strings, but closely related to string theory, known as M-theory. But in any case, to make contact with the real world we have to find theories in four dimensions.

There are essentially two ways of doing that: to choose another background space-time geometry, or to change the world-sheet theory. The geometry can be chosen as a flat four-dimensional space-time combined with a compact six-dimensional space. This is called "compactification". The world-sheet theory can be modified by choosing an appropriate two-dimensional conformal field theory. In D-dimensional flat space a string theory is described by D free two-dimensional bosons X^μ, and, if it is a fermionic string, also by D free fermions ψ^μ. Instead, one can choose another two-dimensional field theory that satisfies the same conditions of conformal invariance, called a conformal field theory (CFT). In particular one may use interacting two-dimensional theories, as long as X^μ and ψ^μ $\mu = 0, \ldots 3$ remain free fields.

The simplest compactification manifold is a six-dimensional torus. This can easily be described both from the space-time and the world-sheet point of view. The resulting theories only have non-chiral fermions in their space-time spectrum. The same is true for the more general asymmetric torus compactifications of the heterotic string with 6 left-moving and 22 right-moving "chiral" bosons found by Narain [2].

The chirality problem is easily solved by a simple generalization that yields a valid compactification manifold, namely a torus with discrete identifications. These are called orbifold compactifications [60]. These methods opened many new directions, such as orbifolds with gauge background fields ("Wilson lines") [61], and were soon generalized to *asymmetric orbifolds* [7], where "asymmetric" refers to the way left- and right-moving modes were treated. Just as torus compactifications, orbifolds can be viewed from both a space-time and a world-sheet perspective. Some orbifold compactifications can be understood as singular limits of geometric Calabi-Yau compactifications, which historically were discovered a little earlier (see Section 2.3). With more complicated compactifications, the connection between the world-sheet and space-time perspectives becomes more and more difficult to make.

2.2. General features

2.2.1. Massive and massless modes

Before introducing some of the earliest string constructions and compactifications in a little more detail, we will give a summary of the kind of spectra they generically produce. Here we will assume a supersymmetric spectrum. This already implies the prediction of a large number of particles that have not (yet) been observed. All types of particles listed below typically occur in non-supersymmetric string spectra as well, but in that case it is even less clear what their ultimate fate is, since the stability of these string theories is not understood.

Any string theory contains infinitely many additional particles: massive string excitations, Kaluza-Klein modes as in field theory compactifications, and winding modes due to strings wrapping the compact spaces.

Their masses are respectively proportional to the string scale, the inverse of the compactification radius or the compactification radius itself. In world-sheet constructions the different kinds of modes are on equal footing, and have Planckian masses. In geometric constructions one can consider large volume limits, where other mass distributions are possible. But in any case, of all the modes of the string only the massless ones are relevant for providing the Standard Model particles, which will acquire their masses from the Higgs mechanism and QCD, as usual.

Among the massless modes of string theories one may find some that match known particles, but usually there are many that do not match anything we know. This may just be an artifact of the necessarily primitive methods at our disposal. Our intuition from many years of four-dimensional string model building may well be heavily distorted by being too close to the supersymmetric limit, and by algebraically too simple constructions. Some of the additional particles may actually be a blessing, if they solve some of the remaining problems of particle physics and cosmology. The art of string phenomenology is to turn all seemingly superfluous particles into a blessing, or understand why their presence is not generic.

In addition to moduli (already introduced in Subsection 1.2.2) and axions (to be discussed in Section 3.6) string spectra generically include:

2.2.2. Chiral fermions and mirrors

All charged Standard Model fermions are chiral, and hence they can only acquire a mass after weak symmetry breaking. Therefore one can say that the weak interactions protect them from being very massive. It is very well possible that for this reason all we have seen so far at low energy is chiral fermionic matter.

In attempts at getting the Standard Model from string theory, it is therefore reasonable to require that the chiral spectra match. In general one finds additional vector-like matter, whose mass is not protected by the weak interactions. Typically, if one requires three chiral families, one gets $N + 3$ families and N mirror families. If the N families "pair off" with the N mirror families to acquire a sufficiently large mass, the low energy spectrum agrees with the data.

2.2.3. *Additional vector bosons*

Most string spectra have considerably more vector bosons than the twelve we have seen so far in nature. Even if the presence of $SU(3)$, $SU(2)$ and $U(1)$ as factors in the gauge group is imposed as a condition, one rarely finds just the Standard Model gauge group. In heterotic strings one is usually left with one of the E_8 factors. Furthermore in nearly all string constructions additional $U(1)$ factors are found. A very common one is a gauged $B - L$ symmetry.

Additional gauge groups are often needed as "hidden sectors" in model building, especially for supersymmetry breaking. Extra $U(1)$'s may be observable trough kinetic mixing [62] with the Standard Model $U(1)$, via contributions to the action proportional to $F_{\mu\nu}V^{\mu\nu}$, where F is the Y field strength, and V the one of the extra $U(1)$'s.

2.2.4. *Exotics*

One often finds particles that do not match any of the observed matter representations, nor their mirrors. Notorious examples are color singlets with fractional electric charge or higher rank tensors. These are generically called "exotics". If there are exotics that are chiral with respect to $SU(3) \times SU(2) \times U(1)$, these spectra should be rejected, because any attempt to make sense of such theories is too far-fetched to be credible. These particles may be acceptable if they are vector-like, because one may hope that they become massive under generic perturbations.

2.3. *Calabi-Yau compactifications*

The first examples of compactifications with chiral spectra and N=1 supersymmetry were found for the $E_8 \times E_8$ heterotic string in [63], even before the aforementioned mathematically simpler orbifold compactifications. The compactification manifolds are six-dimensional, Ricci-flat Kähler manifolds with $SU(3)$ holonomy, called Calabi-Yau manifolds. The $B_{\mu\nu}$ field strength $H_{\mu\nu\rho}$ was assumed to vanish, which leads to the consistency condition

$$dH = \text{Tr } R \wedge R - \frac{1}{30}\text{Tr } F \wedge F = 0. \qquad (5)$$

This implies in particular a relation between the gravitational and gauge field backgrounds. This condition can be solved by using a background gauge field that is equal to the spin connection of the manifold, embedded in an $SU(3)$ subgroup of one of the E_8 factors. In compactifications of this kind one obtains a spectrum with a gauge group $E_6 \times E_8$. The group E_6 contains the Standard Model gauge group $SU(3) \times SU(2) \times U(1)$ plus two additional $U(1)$'s. The group E_8 is superfluous but hidden (Standard Model particles do not couple to it), and may play a rôle in supersymmetry breaking.

If some dimensions of space are compactified, ten-dimensional fermion fields are split as

$$\Psi_+(x,y) = \Psi_L(x)\Psi_+(y) + \Psi_R(x)\Psi_-(y) \qquad (6)$$

where x denotes four-dimensional and y six-dimensional coordinates, $+/-$ denotes chirality in ten and six dimensions, and L, R denote chirality in four dimensions. The number of massless fermions of each chirality observed in four dimensions is determined by the number of zero-mode solutions of the six-dimensional Dirac equation in the background of interest. These numbers are equal to two topological invariants of the Calabi-Yau manifold, the Hodge numbers, h_{11} and h_{12}. As a result one obtains h_{11} chiral fermions in the representation (27) and h_{12} in the $(\overline{27})$ of E_6. The group E_6 is a known extension of the Standard Model, an example of a Grand Unified Theory, in which all three factors of the Standard Model are embedded in one simple Lie algebra. It is not the most preferred extension; a Standard Model family contains 15 or 16 (if we assume the existence of a right-handed neutrino) chiral fermions, not 27. However, since the 11 superfluous fermions are not chiral with respect to $SU(3) \times SU(2) \times U(1)$, they can acquire a mass without the help of the Higgs mechanism, in the unbroken Standard Model. Therefore these masses may be well above current experimental limits.

The number of Calabi-Yau manifolds is huge. A subset associated with four-dimensional reflexive polyhedra has been completely enumerated [64]. This list contains more than 470 million topological classes with 31,108 distinct Hodge number pairs.

In 1986 Strominger [3] considered more general geometric background geometries with torsion. This gave rise to so many possibilities that the author concluded *"all predictive power seems to have been lost"*.

2.4. Free field theory constructions

Several other methods were developed around the same time as Calabi-Yau compactifications and orbifolds. Narain's generalized torus compactifications lead to a continuous infinity of possibilities, but all without chiral fermions. Although this infinity of possibilities is not really a surprising feature for a torus compactification, Narain's paper was an eye-opener because, unlike standard six-dimensional torus compactifications, this approach allowed a complete modification of the gauge group.

More general world-sheet methods started being explored in 1986. Free field theory constructions allowed a more systematic exploration of certain classes of string theories. It became clear very quickly that also in this case there was a plethora of possibilities. But unlike Narain's constructions, these theories can have chiral fermions, and furthermore they did not seem to provide a continuum of options, but only discrete choices. With the benefit of hindsight, one can now say that all these theories *do* have continuous deformations, which can be realized by giving vacuum expectation values to certain massless scalars in the spectrum. Since these deformed theories do not have a free field theory descriptions, these deformations are not manifest in the construction. They are the world sheet construction counterparts of the geometric moduli. This does however not imply that the plethora of solutions can simply be viewed as different points in one continuous moduli space. Since many spectra are chirally distinct, it is more appropriate to view this as the discovery of a huge number of distinct moduli spaces, all leading to different physics.

An important tool in these free-field theory constructions is boson-fermion equivalence in two dimensions. In this way the artificial distinction between the two can be removed, and one can describe the heterotic string entirely in terms of free fermions [4, 6] or free bosons [5]. These constructions are closely related, and there is a huge area of overlap: constructions based on complex free fermions pairs can be written in terms of free bosons. However, one may also consider real fermions or free bosons on lattices that do not allow a straightforward realization in terms of free fermions.

2.4.1. Free fermions

Both methods have to face the problem of finding solutions to the conditions of modular invariance, a one-loop consistency condition. In the fermionic constructions this is done by allowing periodic or anti-periodic boundaries on closed cycles on the manifold for all fermions independently. Modular transformations change those boundary conditions, and hence they are constrained by the requirements of modular invariance. These constraints can be solved systematically (although in practice usually not exhaustively). Very roughly (ignoring some of the constraints), the number of modular invariant combinations is of order $2^{\frac{1}{2}n(n-1)}$ for n fermions. There are 44 right-moving and 18 left-moving fermions, so that there are potentially huge numbers of string theories. In reality there are however many degeneracies.

In-depth explorations [65] have been done of a subclass of fermionic constructions using a special set of free fermion boundary conditions that allows spectra with three families to come out. This work focuses on Pati-Salam model. Other work [66, 67] explores the variations of the "NAHE" set of free fermion boundary conditions. This is a set of fermion boundary vectors proposed in [68] that are a useful starting point for finding "realistic" spectra.

2.4.2. Free bosons: Covariant lattices

In bosonic constructions the modular invariance constraints are solved by requiring that the momenta of the bosons lie on a Lorentzian even self-dual lattice. This means that the lattice of quantized momenta is identical to the lattice defining the compactified space, and that all vectors have even norm. Both conditions are defined in terms of a metric, which is $+1$ for left-moving bosons and -1 for right-moving ones. These bosons include the ones of Narain's torus, plus eight right-moving ones representing the fermionic degrees of freedom, ψ^μ and the ghosts of superconformal invariance. These eight bosons originate from the *bosonic string map* (originally developed for ten-dimensional strings [69]) used to map the entire fermionic sector of the heterotic string to a bosonic string sector [5]. Then the Lorentzian metric has signature $((+)^{22}, (-)^{14})$, and the even self-dual lattice is denoted $\Gamma_{22,14}$. This is called a *covariant lattice* because it incorporates space-time Lorentz invariance for the fermionic string. Since the conditions for modular invariance are invariant under $SO(22, 14)$ Lorentz transformations, and since the spectrum of L_0 and $\bar{L}_0$ is changed under such transformations, their would appear to be a continuous infinity of solutions. But the right-moving modes of the lattice are strongly constrained by the requirement of two-dimensional supersymmetry, which is imposed using a non-linear realization [70] (other realizations exist, see for example [71, 72]). This leads to the so called "triplet constraint" [4]. This makes the right-moving part of the lattice rigid. The canonical linear realization of supersymmetry, relating X^μ to ψ^μ, on the other hand leads to lattices $\Gamma_{22,6} \times E_8$ with complete Lorentz rotation freedom in the first factor, which is just a Narain lattice.

2.5. An early attempt at vacuum counting

Several of the 1986 papers make attempts at getting a rough idea about the number of solutions. This is fairly straightforward for free fermions with periodic and anti-periodic boundary conditions, as explained above. However, the main problem is that not all solutions are different. Indeed, in general there are huge degeneracies among solutions that reduce the estimate by large factors. We will explain here a counting estimate used for covariant lattice constructions, because it give an interesting insight in the growth of the number of possibilities. However, this should not be confused with counting moduli stabilized points in potentials. Indeed, all these string theories have unstabilized moduli.

An interesting estimate exists for even self dual lattices, which has the advantage that it only counts distinct solutions. Unfortunately, heterotic strings are based on lorentzian lattices, for which there are no such theorems. In fact, these lattices are unique up to Lorentz transformations, but the Lorentz transformations modify the heterotic spectrum. However, covariant lattices that lead to chiral spectra have a rigid right-moving sector that forbids continuous Lorentz transformations.

The rigidity of the right-moving part of the lattice discretizes the number of solutions, which is in fact finite for a given world-sheet supersymmetry realization. A very crude attempt to estimate the number of solutions was made in [5], and works as follows. One can map the right-moving bosons to a definite set of 66 left-moving bosons, while preserving modular invariance. This brings us into the realm of even self-dual *Euclidean* lattices, for which powerful classification theorems exist.

Such lattices exist only in dimensions that are a multiple of eight, and have been enumerated for dimensions $8, 16$ and 24, with respectively $1, 2$ and 24 solutions (in 8 dimensions the solution is the root lattice of E_8, in 16 dimensions they are $E_8 \oplus E_8$ and the root lattice of D_{16} plus a spinor weight lattice, and in 24 dimensions the solutions were enumerated in [73]).

There exists a remarkable formula (the "Siegel mass formula") which gives information about the total number of distinct lattices Λ of dimension $8k$ in terms of :

$$\sum_{\Lambda} g(\Lambda)^{-1} = \frac{1}{8k} B_{4k} \prod_{j=1}^{4k-1} \frac{B_{2j}}{4j}. \tag{7}$$

Here $g(\Lambda)$ is the order of the automorphism group of the lattice Λ and B_{2j} are the Bernoulli numbers. Since the automorphisms include the reflection symmetry, $g(\Lambda) \geq 2$. If we assume that the lattice of maximal symmetry is D_{8k} (the root lattice plus a spinor, which is a canonical way to get an even self-dual lattice)) we have a plausible guess for the upper limit of $g(\Lambda)$ as well, namely the size of the Weyl group of D_{8k}, $2^{8k-1}(8k)!$. This assumption is incorrect for $k = 1$, where the only lattice is E_8, and $k = 2$, where the lattice $E_8 \times E_8$ wins against D_{16}, but for $k = 3$ and larger the Weyl group of D_{8k} is larger than the automorphism group of the lattice $(E_8)^k$. For $k = 3$ the assumption has been checked in [74] for all 24 Niemeier lattices. Making this assumption we get

$$\frac{1}{4k} B_{4k} \prod_{j=1}^{4k-1} \frac{B_{2j}}{4j} < N_{8k} < 2^{8k-1}(8k-1)! \, B_{4k} \prod_{j=1}^{4k-1} \frac{B_{2j}}{4j} \tag{8}$$

which for $k = 11$ gives $10^{930} < N_{88} < 10^{1090}$ (in [75] this number was estimated rather inaccurately as 10^{1500}; all numbers quoted here are based on an exact computation).

From a list of all N_{88} lattices one could read off all the free bosonic CFTs with the world-sheet supersymmetry realization discussed above. In particular, this shows that the total number is finite. However, there is a very restrictive subsidiary constraint due to the fact that 66 of the 88 bosons were obtained from the right moving sector. Those bosons must have their momenta on a $D_3 \times (D_7)^9$ lattice and satisfy an additional constraint inherited from world sheet supersymmetry, the triplet constraint. Perhaps a more reasonable estimate is to view this as a lattice with 32 orthogonal building blocks, $D_3 \times (D_7)^9 \times (D_1)^{22}$, which should be combinatorically similar to $(D_1)^{32}$ then the relevant number would be N_{32}, which lies between 8×10^7 and 2.4×10^{51}. But unlike N_{88}, N_{32} is not a strict limit, and furthermore is still subject to the triplet constraint.

All of this can be done explicitly for 10 dimensional strings. Then one needs the lattices of dimension 24, and eight of the 24 lattices satisfy the subsidiary constraints for ten-dimensional strings [75], namely the presence of a D_8 factor.

2.6. *Unexplored landscapes: Meromorphic CFTs*

The concept of chiral conformal field theories and even self-dual lattices can be generalized to interacting theories, the so-called *meromorphic* conformal field theories [76]. These can only exist if the central charge c (the generalization of the lattice dimension to CFT) is a multiple of 8. For $c = 8$ and $c = 16$ these meromorphic CFTs are just chiral bosons on even self-dual lattices, but for $c = 24$ there 71 CFT's are conjectured [77] to exist including the 24 Niemeier lattices (most of them have indeed been constructed). Gauge symmetries in the vast majority of the heterotic strings in the literature (for exceptions see for example [78]) are mathematically described in terms of affine Lie algebras, a kind of string generalization of simple Lie-algebras, whose representations are characterized by a Lie-algebra highest weight and an additional integer parameter k called the *level*. In the free boson theories the only representations one encounters have $k = 1$, and the total rank equals the number of compactified bosons in the left-moving sector, 22 for four-dimensional strings, and 24 for Niemeier lattices. All even self-dual lattices are direct sums of level 1 affine algebras plus a number of abelian factors (U(1)'s), which we will call the gauge group of the theory. In meromorphic CFT's the restriction to level one is removed. The list of 71 meromorphic CFTs contains 70 cases with a gauge group whose total central charge is 24, plus one that has no gauge group at all, the "monster module". Just one of these yields an additional ten-dimensional string theory with tachyons and an E_8 realized as an affine Lie algebra at level 2. This solution was already known [79], and was obtained using free fermions.

The importance of the meromorphic CFT approach is that it gives a complete classification of all solutions without assuming a particular construction method. In four dimensions the same method can be used. For example, from a list of meromorphic CFTs with $c = 88$ all four-dimensional string theories with a given realization of world-sheet supersymmetry (namely the same one used above) can be obtained, independent of the construction method. Unfortunately next to nothing is known about meromorphic CFTs for $c \geq 32$. It is not known if, like lattices, they are finite in number. Their gauge groups can have central charges that are not necessarily 0 or the total central charge of the meromorphic CFT. It is not known if the gauge groups are typically large or small. There is an entire landscape here that is totally unexplored, but hard to access.

So far this method of mapping a heterotic theory to a meromorphic CFT has only been applied to a world-sheet supersymmetry realization using the triplet constraint. But this can be generalized to other realizations of world-sheet supersymmetry, including perhaps the ones discussed in the next section.

The point we are trying to make here is that despite many decades of work, we are probably still only able to see the tip of a huge iceberg.

2.7. Gepner models

In 1987 world-sheet constructions were extended further by the use of interacting rather than free two-dimensional conformal field theories [80]. The "building blocks" of this construction are two-dimensional conformal field theories with $N = 2$ world-sheet supersymmetry. These building blocks are combined ("tensored") in such a way that they contribute in the same way to the energy momentum tensor as six free bosons and fermions. This is measured in terms of the central charge of the Virasoro algebra, which must have a value $c = 9$. In principle the number of such building blocks is huge, but in practice only a very limited set is available, namely the "minimal models" with central charge $c = 3k/(k + 2)$, for $k = 1 \ldots \infty$. There are 168 distinct ways of adding these numbers to 9, so that only a few members of the infinite set are actually used.

With the constraints of superconformal invariance solved, one now has to deal with modular invariance. In exact CFT constructions the partition function takes the form

$$P(\tau, \bar{\tau}) = \sum_{ij} \chi_i(\tau) M_{ij} \bar{\chi}_j(\bar{\tau}) \tag{9}$$

where χ_i are characters of the Virasoro algebra, traces over the entire Hilbert space built on the ground state labeled i by the action of the Virasoro generators L_n:

$$\chi_i(\tau) = \text{Tr} e^{2\pi i \tau (L_0 - c/24)}. \tag{10}$$

The multiplicity matrix M indicates how often the ground states $|i\rangle|j\rangle$ occurs in the spectrum. Its entries are non-negative integers, and it is severely constrained by modular invariance. Note that in (9) we allowed for the possibility that the left- and right-moving modes have a different symmetry (a different extension of superconformal symmetry) with different sets of characters χ and $\bar{\chi}$. But then the conditions for modular invariance are very hard to solve. They can be trivially solved if the left and right algebras are the same. Then modular invariance demands that M must commute with

the matrices S and T that represent the action of the modular transformations $\tau \to -1/\tau$ and $\tau \to \tau + 1$ on the characters. This has always at least
one solution, $M_{ij} = \delta_{ij}$.

However, assuming identical left and right algebras is contrary to the
basic idea of the heterotic string. Instead Gepner model building focuses
on a subset, namely those spectra that can be obtained from a symmetric
type-II spectrum by mapping one of the fermionic sectors to a bosonic
sector. For this purpose we can use the same bosonic string map discussed
above. This results in a very special and very limited subset of the possible
bosonic sectors.

Using the discrete symmetries of the building blocks, for each of the
168 tensor combinations, a number of distinct modular invariant partition
functions can be constructed, for a grand total of about five thousand [81].
Each of them gives a string spectrum with a gauge group $E_6 \times E_8$ (or
occasionally an extension of E_6 to E_7 or E_8) with massless chiral matter
in the representations (27) and ($\overline{27}$) of E_6, exactly like the Calabi-Yau
compactifications discussed above.

Indeed, it was understood not long thereafter that there is a close relationship between these "Gepner models" and geometric compactifications
on Calabi-Yau manifolds. Exact correspondences between their spectra
were found, including the number of singlets. This led to the conjecture that
Gepner Models are Calabi-Yau compactifications in a special point of moduli space. Evidence was provided by a conjectured relation between $N = 2$
minimal models and critical points of Landau-Ginzburg models [82, 83].

Getting the right number of families in this class of models has been
challenging, since this number turns out to be quantized in units of six or
four in nearly all cases that were studied initially. The only exception is a
class studied in [84].

2.8. *Dualities, M-theory and F-theory*

In general, four-dimensional string theories are related to others by maps
like S-duality [85] (strong-weak dualities due to inversion of coupling constants), T-duality (transformations involving inversion of compactification
radii) and combinations thereof. This suggests a connected "landscape" of
four-dimensional strings. However, this is still largely based on anecdotal
evidence. A complete picture of the four-dimensional string landscape is
still very far away.

In ten (and eleven) dimensions, the picture is better understood. Under

S-duality, Type-IIA string theory is mapped to an 11-dimensional theory compactified on a circle [86, 87]. The 11-dimensional theory is not a string theory. It is called "M-theory". Its field theory limit is $D = 11$ supergravity.

A similar relation holds for the $E_8 \times E_8$ heterotic string. Its strong coupling limit can be formulated in terms of 11-dimensional M-theory compactified on a line-segment [88], the circle with two halves identified. This is sometimes called "heterotic M-theory".

Strong coupling duality maps type-IIB strings to themselves [89]. Furthermore the self-duality can be extended from an action just on the string coupling, and hence the dilaton, to an action on the entire dilaton-axion multiplet. This action is mathematically identical to the action of modular transformations on the two moduli of the torus, and corresponds to the group $SL(2, \mathbb{Z})$. This isomorphism suggests a geometric understanding of the self-duality in terms of a compactification torus T_2, whose degrees of freedom correspond to the dilaton and axion field. An obvious guess would be that the type-IIB string may be viewed as a torus compactification of some twelve-dimensional theory [90]. But there is no such theory. The first attempts to develop this idea led instead to a new piece of the landscape called "F-theory", consisting only of compactifications and related to $E_8 \times E_8$ heterotic strings and M-theory by chains of dualities.

2.9. New directions in heterotic strings

2.9.1. New embeddings

The discovery of heterotic M-theory opened many new directions. Instead of the canonical embedding of the $SU(3)$ valued spin-connection of a Calabi-Yau manifold, some of these manifolds admit other bundles that can be embedded in the gauge group. In general, condition (5) is then not automatically satisfied, but in heterotic M-theory one may get extra contributions from heterotic five branes [91, 92].

In this way one can avoid getting the Standard Model via the complicated route of E_6 Grand Unification. Some examples that have been studied are $SU(4)$ bundles [93], $U(1)^4$ bundles [94] and $SU(N) \times U(1)$ bundles [95], which break E_8 to the more appealing $SO(10)$ GUTs, to $SU(5)$ GUTs, or even directly to the Standard Model. Extensive and systematic searches are underway that have resulted in hundreds of distinct examples [96] with the exact supersymmetric Standard Model spectrum, without even any vector-like matter (but with extra gauge groups and the usual large numbers of singlets). However, the gauge group contains extra $U(1)$'s and an E_8 factor, and large numbers of gauge singlets, including unstabilized moduli. There can be several Higgs multiplets. To break the GUT groups down to the Standard Model, background gauge fields on suitable Wilson lines are used. For this purpose one needs a manifold with a freely acting (*i.e.* no point on the manifold is fixed by the action) discrete symmetry. One then identifies points on the manifold related by this symmetry and adds a background gauge field on a closed cycle on the quotient manifold (a Wilson line).

2.9.2. *The heterotic mini-landscape*

The Heterotic Mini-landscape is a class of orbifold compactifications on a torus $T^6/\mathbf{Z}_6$ cleverly constructed so that the heterotic gauge group $E_8 \times E_8$ is broken down to different subgroups in different fixed points, such as $SO(10)$, $SU(4)^2$ and $SU(6) \times SU(2)$. This leads to the notion of *local unification* [97–99]. The Standard Model gauge group is the intersection of the various "local" gauge realized at the fixed points. Fields that are localized near the fixed points must respect its symmetry, and hence be in complete multiplets of that group. Unlike field theory GUTs, these models have no limit where $SO(10)$ is an exact global symmetry. In this way one can make sure that matter families are in complete spinor representations of $SO(10)$, while Higgs bosons need not be in complete representations of $SO(10)$, avoiding the notorious doublet splitting problem of GUTs. The number of 3-family models in this part of the landscape is of order a few hundred, and there is an extensive body of work on their phenomenological successes and problems, see for example [100, 101] and references therein.

2.9.3. *Heterotic Gepner models*

As explained above, the original Gepner models are limited in scope by the requirement that the left and right algebras should be the same. There is no such limitation in free CFT constructions, but they are limited in being non-interacting in two dimensions. What we would like to have is asymmetric, interacting CFT constructions. Examples in this class have been obtained using a method called "heterotic weight lifting" [102]. In the left-moving sector one of the superconformal building blocks (combined with one of the E_8 factors) is replaced by another CFT that has no superconformal symmetry, but is isomorphic to the original building block as a modular group representation. This opens up an entirely new area of the heterotic string landscape. It turns out that the difficulty in getting three families now disappears.

2.10. *Orientifolds and intersecting branes*

The Standard Model comes out remarkably easily from the simplest heterotic strings. But that is by no means the only way. One may also get gauge groups in string theory from stacks of membranes. If open strings end on a D-brane that does not fill all of space-time, a distinction must be made between their fluctuations away from the branes, and the fluctuations of their endpoints on the branes. The former are standard string vibrations leading to gravity (as well as a dilaton, and other vibrational modes of closed strings), whereas fluctuations of the endpoints are only observable on the brane, and give rise to fermions and gauge interactions.

2.10.1. *Chan-Paton groups*

The possibility of getting gauge theories and matter from branes sparked another direction of research with the goal of getting the Standard Model from open string theories. To get towards the Standard Model, one starts with type-II string theory, and compactifies six dimensions on a manifold. This compactified manifold may have a large radius, as in the brane world scenario, but this is optional. In these theories one finds suitable D-branes coinciding with four-dimensional Minkowski space, and intersecting each other in the compactified directions. These can be D5, D7 or D9 branes in type-IIB and D6 branes in type-IIA. Each such brane can give rise to a gauge group, called a Chan-Paton gauge group, which can be $U(N)$, $Sp(N)$ or $O(N)$ [103]. By having several different branes one can obtain a gauge group consisting of several factors, like the one of the Standard Model. The brane intersections can give rise to massless string excitations of open strings with their ends on the two intersecting branes. These excitations can be fermions, and they can be chiral. Each open string end endows the fermion with a fundamental representation of one of the two Chan-Paton groups, so that the matter is in a bi-fundamental representation of those gauge groups.

Remarkably, a Standard Model family has precisely the right structure to be realized in this manner. The first example is the so-called "Madrid model" [104]. It consists of four stacks of branes, a $U(3)$ stack giving the strong interactions, a $U(2)$ or $Sp(2)$ stack for the weak interactions, plus two $U(1)$ stacks. The Standard Model Y charge is a linear combination of the unitary phase factors of the first, third and fourth stack (the stacks are labeled **a** ... **d**)

$$Y = \frac{1}{6}Q_\mathbf{a} + \frac{1}{2}Q_\mathbf{c} - \frac{1}{2}Q_\mathbf{d}.$$

This configuration is depicted in Fig. 1(a).

2.10.2. *The three main classes*

There are other ways of getting the Standard Model. They fall into three broad classes, labeled by a real number x. The Standard Model generator is in general some linear combination of all four brane charges (assuming stack **b** is $U(2)$ and not $Sp(2)$), and takes the form [105]

$$Y = (x - \frac{1}{3})Q_\mathbf{a} + (x - \frac{1}{2})Q_\mathbf{b} + xQ_\mathbf{c} + (x - 1)Q_\mathbf{d}. \tag{11}$$

Two values of x are special. The case $x = \frac{1}{2}$ leads to a large class containing among others the Madrid model, Pati-Salam models [106] and flipped $SU(5)$ [107] models. The value $x = 0$ gives rise to classic $SU(5)$ GUTs [108]. To get Standard Model families in this case one needs chiral anti-symmetric rank-2 tensors, which originate from open strings with both their endpoints on the same brane. The simplest example is shown in Fig. 1(b). It has one $U(5)$ stack giving rise to the GUT gauge group, but needs at least one other brane in order to get matter in the $(\bar{5})$ representation of $SU(5)$.

Other values of x can only occur for oriented strings, which means that there is a definite orientation distinguishing one end of the string from the other end. An interesting possibility in this class is the trinification model, depicted in Fig. 1(c).

Note that it was assumed here that there are at most four branes participating in the Standard Model. If one relaxes that condition, the number of possibilities is unlimited.

2.10.3. *Orientifolds*

An important issue in open string model building is the cancellation of tadpoles of the disk diagram. These lead to divergences and can lead to chiral anomalies. These tadpoles can sometimes be canceled by adding another object to the theory, called an orientifold plane. In fact, the usual procedure is to start with an oriented type-II string, and consider an involution

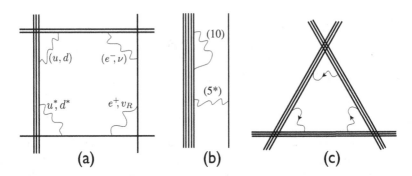

Fig. 1.   Brane configurations: (a) the Madrid model, (b) SU(5) GUTs and (c) Trinifi-cation.

of the world-sheet that reverses its orientation. Then one allows strings to close up to that involution. In terms of world-sheet topology, this amounts to adding surfaces with the topology of a Klein bottle. The combination of torus and Klein-bottle diagram acts like a projection on the closed string theory, removing some of its states. In most cases, removing states from string theory comes at a price: other states must be added to compensate what was removed. In this case, this rôle is played by open strings. These ideas were pioneered in [109, 110].

Orientifold model building has been very actively pursued during the first decade of this century (see [111] for a review).

2.10.4. *Anomalies, axions and massive abelian vector bosons*

Canceling all tadpoles between the disk and crosscap diagram removes most anomalies, but some factorized anomalies remain which can then be canceled by the Green-Schwarz mechanism [112] involving tree-level diagrams with exchange of axions. In contrast to perturbative heterotic strings the anomaly factorizes in terms of several factors. These anomalies are then canceled by a Green-Schwarz mechanism involving multiple axions, which are available in the Ramond-Ramond sector of the closed theory.

In four dimensions, a factorized anomaly always involves a $U(1)$. The corresponding $U(1)$ vector bosons acquire a mass by "eating" the axion, which provides the missing longitudinal mode. String theory will always remove anomalous symmetries in this manner, but it turns out that this can happen for non-anomalous $U(1)'s$ as well. This can be traced back to anomalies in six dimensions [113].

2.10.5. *Boundary RCFT constructions*

Just as in the heterotic string, one can construct orientifold spectra using purely geometric methods, orbifold methods or world-sheet constructions. Most work in the literature uses the second approach.

World-sheet approaches use boundary CFT: conformal field theory on surfaces with boundaries and crosscaps. This requires an extension of the closed string Hilbert space with "states" (in fact not normalizable, and hence not in the closed string Hilbert space) that describe closed strings near a boundary, or in the presence of orientation reversal. An extensive formalism for computing boundary and crosscap states in (rational) CFT was developed in the last decade of last century, starting with [114], developed further by several groups [115–119], culminating in a simple and general formula [120]. For an extensive review of this field see [121]. This was applied to orientifolds of Gepner models [122], and led to a huge (of order 200.000) number of distinct string spectra that match the chiral Standard Model. This set provides an extensive scan over the orientifold landscape.

These spectra are exact in perturbative string theory and not only the massless but also all massive states are known explicitly. There are no chiral exotics, but in general there are large numbers of the ubiquitous vector-like states that plague almost all exact string spectra. All tadpoles are canceled, but in most cases this requires hidden sectors. However, there are a few cases where all tadpoles cancel entirely among the Standard Model branes (hence no hidden sector is present) and furthermore the superfluous $B - L$ vector bosons acquires a mass from axion mixing. These spectra have a gauge group which is exactly $SU(3) \times SU(2) \times U(1)$ (there are a few additional vector bosons from the closed sector, but the perturbative spectrum contains no matter that is charged under these bosons; this is the same as in the type IIA string, which contains a vector boson that only couples to non-perturbative states, D0-branes).

2.11. *Decoupling limits*

Brane model building led to an interesting change in strategy. Whereas string theory constructions were originally "top-down" (one constructs a string theory and then compares with the Standard Model), using branes one can to some extent work in the opposite direction, "bottom-up". The idea is to start with the Standard Model and construct a brane configuration to match it, using branes localized at (orbifold) singularities. Then this brane configuration may be embedded in string theory at a later stage.

This point of view was pioneered in [123], who found examples with $\mathbb{Z}_3$ singularities. See *e.g.* [124, 125] for other kinds of singularities.

One extreme possibility is to decouple gravity by sending the compactification radius to infinity. In heterotic string models both gravity and gauge interactions originate from closed string exchange, and such a decoupling limit would not make sense.

The other extreme is to take the details of the Standard Model for granted and focus on issues like moduli, supersymmetry breaking and hierarchies. In this case one has to assume that once the latter are solved, the Standard Model can be added.

Both points of view are to some extent a return to the "old days" of quantum field theory. On the one hand, the techniques of branes and higher dimensions are used to enrich old ideas in GUT model building; on the other hand, string theory is treated as a "framework", analogous to quantum field theory, where gauge groups, representations and couplings are input rather than output.

Decoupling of gravity is an important element in recent work on F-theory GUTs [126–128] obtained by compactifying F-theory on elliptically fibered Calabi-Yau fourfolds. This allows the construction of models that may be thought of as non-perturbative realizations of the orientifold $SU(5)$ GUT models depicted in Fig. 1(b), solving some of their problems, especially the absence of the top-Yukawa coupling, which is perturbatively forbidden. This has led to a revival of Grand Unified Theories, invigorated with features of higher dimensional theories. See the reviews [129–132] for further details.

An example in the second category is recent work in the area of M-theory compactifications [58]. Getting chiral $N=1$ supersymmetric spectra in M-theory requires compactification on a seven dimensional manifold with G_2 holonomy [133], also known as a Joyce manifold. Much less is known about M-theory than about string theory, and much less is known about Joyce manifolds than about Calabi-Yau manifolds, since the powerful tool of complex geometry is not available. For this reason the Standard Model is treated as input rather than output, in the spirit of QFT.

Another kind of compactification that allows splitting the problem into decoupled parts is the LARGE Volume Scenario [46], originally invented for the purpose of moduli stabilization (see Section 1.2.5). Here both kinds of decoupling limits have been discussed, and there have also been steps towards putting both parts together [134]. This illustrates that focusing on decoupling limits does not mean that the original goal of a complete theory is forgotten. Indeed, there also exist *global* F-theory constructions [135, 136].

2.12. *Non-supersymmetric strings*

Although the vast majority of the literature on string constructions concerns space-time supersymmetric spectra, in world-sheet based methods – free bosons and fermions, Gepner models, and certain orbifolds – it is as easy to construct non-supersymmetric ones. In fact, it is easier, because space-time supersymmetry is an additional constraint. These spectra are generally plagued by tachyons, but by systematic searches one can find examples where no tachyons occur. This was first done in ten dimensions in [137, 138]. These authors found a heterotic string theory with a $SO(16) \times SO(16)$ gauge group, the only tachyon-free non-supersymmetric theory in ten dimensions, out of a total of seven. Four-dimensional non-supersymmetric strings were already constructed shortly thereafter [5, 79]. Non-supersymmetric strings can also be constructed using orientifold methods, see for example [139–143].

Non-supersymmetric strings can have a vacuum energy Λ of either sign. See for example [144] for a distribution of values of the vacuum energy for a class of heterotic strings. There also exist examples where Λ vanishes exactly to all orders in perturbation theory [145] but probably this feature does not hold beyond perturbation theory [146].

One might think that in the absence of any evidence for low energy supersymmetry, and because of the evidence in favor of an accelerated expansion of the universe, non-supersymmetric strings with a positive cosmological constant are a better candidate for describing our universe than the much more frequently studied supersymmetric ones. But the absence of supersymmetry is a serious threat for the stability of these theories, even in the absence of tachyons in the perturbative spectrum.

3. The Standard Model in the Landscape

In this chapter we will discuss how the main features of the Standard Model fit in the String Theory Landscape, taking into account anthropic restrictions and analytical and numerical work on landscape distributions. We focus on questions related to susy-GUTs, where the stress between symmetry and landscape anarchy has been building up in the last few years.

3.1. The gauge group

It is by now abundantly clear that string theory can reproduce the discrete structure of the Standard Model: the gauge group $SU(3) \times SU(2) \times U(1)$ with chiral fermion representations. Indeed, the gauge group is easy to get in many construction methods: Heterotic Calabi-Yau and orbifold compactifications with GUT symmetries broken by Wilson lines, orientifold models with various kinds of intersection branes, strings at singularities, free fermion and free boson constructions, heterotic Gepner models and Gepner orientifolds, higher level heterotic strings with symmetry breaking by the standard adjoint Higgs, F-theory with Y-flux, and others. See section 2 for references to all this work.

But this work also demonstrates very clearly that there is nothing special about the Standard Model from the top down perspective, except that it is rather simple. Many other gauge theories and representations are possible as well, although both are limited in size. Unlike quantum field theory, string theory only allows small representations. Furthermore, the size of the gauge group tends to be limited by the conformal anomaly in closed strings, or dilaton tadpole cancellation in open strings. This is not a theorem: there are remarkable exceptions with very large gauge groups, but it seems plausible that these are far out in the tail of landscape distributions, and hence statistically very rare. The Standard Model gauge group does have one remarkable feature, namely that it fits beautifully in a Grand Unified gauge theory. We will discuss that below in section 3.2.3.

From the landscape perspective, one might hope that the gauge group can be understood using string theory plus anthropic constraints. The anthropic constraints are hard to determine, but all three factors of the gauge group are needed for *our* kind of life. Electromagnetism is so essential that it is impossible to imagine life without it. One can imagine life without $SU(3)_{color}$ and only electromagnetism, but it is by no means obvious that such universes will really come to life. Rigorous evidence of such statements is unlikely to emerge soon, as it requires to work out the full nuclear and atomic spectrum as well as astrophysics, nucleosynthesis and baryogenesis for alternatives of the Standard Model. But given what we know, the presence of electromagnetic and strong interactions are well-motivated anthropic assumptions.

The weak interactions also play a crucial rôle in our universe, but perhaps not in every habitable one. See [147] for a detailed discussion of a "weakless" universe that may yield acceptable nuclear and atomic physics even though the weak scale is pushed towards the Planck scale. Perhaps the main rôle of the weak interactions is to provide chirality to fermions, protecting them from getting a large mass. For this to be true, one has to be able to argue that in the string theory landscape it is easier to have a single light scalar than several light fermions. See [148] for a discussion along these lines.

3.2. Family structure and charge quantization

3.2.1. *Quantum field theory*

A Standard Model family is described by the following reducible $SU(3) \times SU(2) \times U(1)$ representation:

$$(3, 2, \frac{1}{6}) + (3^*, 1, \frac{1}{3}) + (3^*, 1, -\frac{2}{3}) + (1, 2, -\frac{1}{2}) + (1, 1, 1) \tag{12}$$

where we ignore singlets. This occurs three times, and in addition to this there is a Higgs field in the representation $(1, 2, -\frac{1}{2})$. At first, this looks arbitrary and unintuitive, but on closer examination some structure becomes apparent. For example, the three entries of each irreducible term multiply to an integer. This fact implies that all color singlet bound states of the broken $SU(3) \times U(1)$ spectrum have charges that are integer multiples of the electron charge. This fact is not explained in the Standard Model. An arbitrary representation has the form (R_3, R_2, q) where R_n is an $SU(n)$ representation and q a real number. But the observed representations satisfy the rule

$$\frac{t_3}{3} + \frac{t_2}{2} + \frac{1}{6} = 0 \bmod 1, \tag{13}$$

where t_3 is the triality of the $SU(3)$ representation and t_2 the duality of $SU(2)$, twice the spin modulo integers. Group-theoretically this means that all observed representations are in fact representations of the group $S(U(3) \times U(2))$, which has the same Lie algebra as $SU(3) \times SU(2) \times U(1)$.

It seems clear that the family structure is more than just an environmental fact. Some of it is explained by the consistency conditions imposed by anomaly cancellation. This implies that four cubic traces and a linear one must vanish. There is also a global $SU(2)$ anomaly [149] and perhaps one should impose a string-inspired non-abelian $SU(2)$ anomaly [150]. These anomaly cancellation conditions are sufficient to explain charge quantization if one assumes that there is just a single family with the observed $SU(3) \times SU(2)$ content. But quantum field theory offers no reason for these assumptions, and the fact that there are three families ruins the argument anyway.

3.2.2. String theory

String theory makes important contributions towards understanding the family structure. First of all, it limits the choice of representations to only a handful of options. Secondly, as far as anyone knows, string theory always implies absolute charge quantization. By this we mean that the charges are rational numbers, though not necessarily the right ones. And thirdly, string theory provides a rationale for anomaly cancellation that is somewhat more deeply rooted than the ad-hoc rules of quantum field theory.

3.2.3. Grand unification

Usually Grand Unification is invoked to explain Eq. (12). In the context of quantum field theory, this would offer a plausible explanation for the fact that particles fit in $S(U(3) \times U(2))$ representations. There is no good motivation in QFT to consider just $S(U(3) \times U(2))$. From the traditional symmetry-based perspective, assuming that what we see is a broken $SU(5)$ (or larger) gauge theory looks like a natural idea. But it is by no means perfect. It does not explain the family structure, but it just limits the allowed combinations of $SU(3) \times SU(2) \times U(1)$ representations. It does not explain why a breaking to $SU(3) \times SU(2) \times U(1)$ is preferred, and it has difficulties accommodating the Higgs field. In $SU(5)$ the Standard Model Higgs has an $SU(3)$ triplet partner which must remain heavy and cannot have a vacuum expectation value: the doublet triplet splitting problem. The GUT hypothesis would get a lot more credibility if a second nontrivial coincidence is established: the renormalization group convergence of the coupling constants to a single value at the GUT scale. This does not hold if one extrapolates the current low-energy couplings, but it would work if more or less standard supersymmetric partners of all Standard Model particles are discovered at a future run of LHC.

Since many Standard Model realizations in string theory look superficially like GUT theories, one might have expected that all facts mentioned in the foregoing paragraphs are naturally combined to get a satisfactory explanation of family structure and charge quantization. But this has never worked as easily as it should.

3.2.4. *Grand unification in heterotic strings*

The oldest examples studied are compactifications of the heterotic string. There are two equivalent ways of understanding why Grand Unification emerges so easily in $E_8 \times E_8$ heterotic strings. In Calabi-Yau compactification this comes from the embedding of the $SU(3)$ holonomy group of the manifold in one of the E_8 factors, breaking it to E_6, which contains $SO(10)$ (and hence $SU(5)$ as a subgroup). In world-sheet constructions this is a consequence of the "bosonic string map" [5] used to map the fermionic (right-moving) sector of the theory into a bosonic one, in order to be able to combine it in a modular invariant way with the left-moving sector. The bosonic string map takes the fermionic sector of a heterotic or type-II string, and maps it to a bosonic sector. The world-sheet fermions ψ^μ transform under the D-dimensional Lorentz group $SO(D-1,1)$. The bosonic string map replaces this by an $SO(D+6) \times E_8$ affine Lie algebra, which manifests itself as a gauge group in space-time. In [5] this trick was used to map the problem of finding modular invariants to the already solved problem of characterizing even self-dual lattices. This automatically gives rise to a four-dimensional theory with an $SO(10) \times E_8$ gauge group and chiral fermions in the spinor representation of the first factor.

With only slight exaggeration one can state that this ideal GUT group, $SO(10)$, emerges uniquely from the heterotic string. All we had to do is specify the space-time dimension, $D = 4$, and apply the bosonic string map, and we get $SO(10)$ for free.

3.2.5. *Fractional charges in heterotic spectra*

A mechanism to break $SO(10)$ to $SU(3) \times SU(2) \times U(1)$ can also be found, but it does not come out automatically. Furthermore, it does not work as nicely as in field theory GUTs, because the heterotic string spectrum does not contain the Higgs representation used in field theory. The breaking can instead be achieved by adding background fields (Wilson lines).

But in that case the full spectrum of these heterotic strings will never satisfy (13), and it is precisely the deep underlying structure of string theory that is the culprit. In a string spectrum every state is relevant, as is fairly obvious from the modular invariance condition. Removing one state destroys modular invariance. In this case, what one would like to remove are the extra gauge bosons in $SU(5)$ in comparison to $SU(3) \times SU(2) \times U(1)$. To do this one has to add something else to the spectrum, and it turns out that the only possibility is to add something that violates (13) and hence is fractionally charged [151]. The possible presence of unconfined fractional charges in string spectra was first pointed out in [152] and the implications were discussed further in [153].

The occurrence of fractional charges in heterotic string spectra has been studied systematically for free fermion constructions and for heterotic Gepner models. All these models realize the gauge group in the canonical heterotic way, as a subgroup of $SO(10)$ (which may be further extended to E_6). There is a total of four distinct subgroups that one may encounter within $SO(10)$. These subgroups are further subdivided into several classes, distinguished by the minimal electric charge quantum that occurs in their spectra. These charge quanta are *not* determined by group theory in quantum field theory, but by affine Lie algebras in string theory. This gives a total of eight possibilities, with charge quanta given in curly brackets:

$$SU(3) \times SU(2) \times U(1) \times U(1) \quad \{\tfrac{1}{6}, \tfrac{1}{3}, \tfrac{1}{2}\}$$
$$SU(3) \times SU(2)_L \times SU(2)_R \times U(1) \quad \{\tfrac{1}{6}, \tfrac{1}{3}\}$$
$$SU(4) \times SU(2)_L \times SU(2)_R \quad \{\tfrac{1}{2}\}$$

plus $SU(5) \times U(1)$ and $SO(10)$, which automatically yield integer charges. This classification applies to all constructions in the literature where the Standard Model is realized with level 1 affine Lie algebras, with a standard Y charge normalization, embedded via an $SO(10)$ group. The minimal electric charge *must* be realized in the spectrum, but it is in principle possible that fractionally charged particles are vector-like (so that they might become massive under deformations of the theory), have Planck-scale masses or are coupled to an additional interaction that confines them into integer charges, just as QCD does with quarks.

Fractional charges can be avoided by looking for spectra where all these particles have Planckian masses. In [65] a large class of free fermionic theories with Pati-Salam spectra was investigated. These authors did find examples with three families where all fractionally charged particles are at the Planck mass, but only in about 10^{-5} of the chiral spectra. In [102, 154–156] a similar small fraction was seen, but examples were only found for even numbers of families. These authors also compared the total number of spectra with chiral and vector-like fractional charges, and found that about in 5% to 20% of the chiral, non-GUT spectra the fractional charges are massless, but vector-like. They also found some examples of confined fractional charges.

In a substantial fraction of explicitly constructed string vacua the fractionally charged particles are vector-like. If one assumes that in genuine string vacua vector-like particles will always be very massive, this provides a way out. There is a more attractive possibility. In orbifold models $SO(10)$ is broken using background gauge fields on Wilson lines. In this process fractional charges must appear, and therefore they must be in the twisted sector of the orbifold model. If the Wilson lines correspond to freely acting discrete symmetries of the manifold (see [157]), the twisted sector fields are massive, and hence all fractionally charge particles are heavy. This method is commonly used in Calabi-Yau based constructions, *e.g.* [158], but is chosen for phenomenological reasons, and hence this does not answer the question why nature would have chosen this option. Also in the heterotic mini-landscape an example was found [159]. These authors suggested another rationale for using freely acting symmetries, namely that otherwise the Standard Model Y charge breaks if the orbifold singularities are "blown up".

But even though there are ways out, it is disappointing that charge quantization comes out less easily than it does in field theoretic $SU(5)$ GUTs, without string theory.

3.2.6. *GUT unification in brane models*

There is another important region in the landscape where $SU(5)$ GUTs can be obtained, namely intersecting brane models. The simplest possibility is to intersect a stack of $U(3)$ branes with a stack of $U(2)$ branes. The entire Standard Model group can be embedded in these two groups, but to get the matter representations one needs not only bi-fundamentals (from strings stretching between the two stacks) and rank-2 anti-symmetric tensors, but also $U(3)$ and $U(2)$ vectors. They would come from endpoints of an open string, but then additional neutral branes are needed for the other endpoint to end on. The resulting configuration is exactly as shown in fig. 5b, but with the $U(5)$ stack split in $U(3)$ and $U(2)$.

It has been known for a long time already that $SU(5)$ GUTs can be obtained from configurations like 5b [160]. These authors noticed however that solutions to the tadpole conditions do not generically lead to the expected anomaly free representation $(\bar{5}) + (10)$, but to more complicated solutions involving the symmetric tensor (15). One can also start with a split stack, but then more input would seem to be required. This includes not only the brane configuration, but also the exact embedding of the $U(1)$ group in $U(3) \times U(2)$ and the particle assignment. In other words, a set of allowed massless open string states is hand-picked to match the Standard Model spectrum. With all these assumptions, one can indeed find numerous solutions [105].

Interestingly, these $U(3) \times U(2)$ intersecting brane models do provide a convincing rationale for the gauge group $U(3) \times U(2)$, which is not easily found in field theory models. But how do we get the restriction to $S(U(3) \times U(2))$ to give the correct charge conjugation, and how do we justify the choice of representations in a Standard Model family? For example, if even in the $U(5)$ limit symmetric tensors are hard to avoid, with split stacks there are even more possibilities. Note that by "split stacks" we do not necessarily mean a $U(5)$ stack with branes move apart. There are more general possibilities, where the $U(3)$ and $U(2)$ stacks occupy unrelated cycles on a compactification manifold.

It turns out that there is an extremely simple answer to this question if one allows a mild anthropic condition [148]. It turns out that for all anomaly free matter configurations, and for all possible $U(1)$ embeddings the electromagnetic $U(1)$ is chiral after Higgs symmetry breaking, or there remain massless charged leptons in the spectrum, with one exception: the Standard Model, with a number of families of the form (12). The motivation for these conditions is that a chiral $U(1)$ will be broken by the color group, and massless charged leptons can be pair-produced without limit, so that the entire universe becomes an opaque plasma of lepton-antilepton pairs [161]. Although we cannot prove that life is impossible without a massless photon or in an opaque plasma, the circumstances for our kind of life – indeed, any kind of life based on atomic physics – are so adverse that one can certainly defend this as a well-motivated anthropic assumption.

One can say that the assumption of symmetry at high energies has been traded for these anthropic assumptions, and remarkably, the latter are more powerful. Even the Higgs choice does not have to be put in, but is determined. Indeed, unless we see evidence for gauge coupling unification because of new matter bending the coupling constant curves, full $SU(5)$ unification has nothing useful to offer.

The argument can be extended to more general $U(M) \times U(N)$ stacks with a few additional assumptions. The group $SU(M)$ is assumed not to be broken by the Higgs, and to be a strong interaction group that is asymptotically free, and dominates over the other gauge interactions. The Higgs is assumed to give mass to all charge fermions, not just the leptons (above we just required the quarks to become non-chiral, not necessarily massive). These simple conditions have a few solutions: the Standard Model, a series of models without leptons, a few cases with $SU(2)$ "color" and no conserved baryon number, and a few models with just electromagnetism and no strong interactions. Finally, there is an $SU(4) \times U(1)$ model, broken by a Higgs boson to $SU(3)_{\text{color}} \times U(1)_{\text{em}}$. This uses the alternative breaking pattern of $SU(5)$ to $SU(4) \times U(1)$ instead of the Standard Model. However, it has baryon-number violating weak interactions that are probably fatal. All other alternatives appear to have fatal problems for life as well, and the Standard Model really stands out as the optimal, and probably unique anthropic solution within this class of brane models.

In this class, $SU(5)$ symmetry is not needed to explain charge quantization, nor the structure of a family. In the heterotic case, it does not work as an explanation of charge quantization. Similar remarks may apply to F-theory GUTs, where the GUT group is present by choice, and not because it is required. There are some corners in the landscape where $SU(5)$ really works as in field theory. The only example we know are heterotic string theories with GUT group with affine level larger than 1 [162].

3.3. *The number of families*

The string theory landscape does not offer, according to our current understanding, an answer to the question why we observe three families. Although early constructions, for example the first Gepner models, had some difficulties getting three families (the number was predominantly a multiple of four or six [81, 163]), further work showed that the number of families in heterotic strings has a slow exponential fall-off, with the number three appearing not much less frequently than 2 (see *e.g.* [154, 164]). In orientifold models the fall off seems to be much faster [122].

There is no convincing anthropic argument for three families. We are built out of just one family. The most often mentioned feature is that three families are needed for CP violation in the CKM matrix, which in its turn might be required for baryogenesis, which is obviously anthropically relevant. But CP violation in the CKM matrix is not believed to be sufficient. The top quark plays an interesting rôle in the running of couplings, and the stability of our vacuum under tunneling depends in a remarkable way on both the top and the Higgs mass. Perhaps this points to an important rôle for the third family, but then why does the second family exist? The s-quark is not completely irrelevant in QCD, and the muon affects biological mutations, but neither of these arguments provides a convincing reason.

3.4. Quark and lepton mass hierarchies

An area where there is an interesting rivalry between symmetry-based ideas
and landscape anarchy is the understanding of quark and lepton mass hierarchies. Already long ago Grand Unified theories predicted interesting
mass relations for the second and third family fermions. Certain string approaches, such as the heterotic mini landscape point to top Yukawa-gauge
unification. Using F-theory compactifications, part of the structure of the
observed masses and mixing angles can be nicely explained.

But on the other hand, one can also get a long way towards the correct
distribution of quark and lepton masses by assuming statistical distributions of Yukawa couplings. Clearly, flat distributions will not work, because
the quark and lepton masses have an unmistakable hierarchical structure,
and the mixing angles are small, and seem to get smaller as the hierarchies
get larger. However, scale invariant distributions (with a cut-off fit to the
data) [165, 166] or Gaussian overlaps [167, 168] work rather well. They
even lead generically to small mixing angles, but it is not automatic in all
cases that the mass eigenvalues of the up and down quarks are ordered
correctly. Then in an alternative universe the three charge $\frac{2}{3}$ quarks would
predominantly couple to a different permutation of the three charge $-\frac{1}{3}$
quarks, and only in one-sixth of all universes with $SU(3) \times SU(2) \times U(1)$
the Standard Model ordering would be observed. One may view this either
as a minor statistical problem, or an indication that something essential is
missing. For a more detailed discussion and references see [1].

This issue is far from settled. Without prior knowledge of the answer,
none of the aforementioned ideas would have given an accurate description
of the quark and lepton spectrum, even if the existence of three families is
provided as information, and even if the anthropic constraints on the light
fermions are used. Furthermore it is plausible that the top quark mass, in
combination with weak symmetry breaking and the Higgs mass, plays an
important rôle that remains to be elucidated.

The smallness of neutrino masses has a well-known natural explanation,
the see-saw mechanism. This is so natural and requires so few changes to
the Standard Model that it is generally seen as the most plausible kind
of beyond the Standard Model physics. Indeed, all that is required are
additional singlets (right-handed neutrinos) having their natural mass. This
mass is not proportional the the Higgs vev, and hence one would expect
it to be large. How large, and how it is distributed depends on several
assumptions, and there are also several anthropic issues related to neutrino

masses. See [1] for more details.

3.5. *The scales of the Standard Model*

The Standard Model has two scales, the strong and the weak scale. To first approximation the strong scale, Λ_{QCD}, determines the proton mass, and the weak scale M_{weak} determines the masses of the quarks and leptons. The proton mass owes less than 1% of its mass to the up and down quarks. However, the smallness of both scales with respect to the Planck scale has an important environmental impact.

In quantum field theory, the strong scale is said to be determined by "dimensional transmutation", which turns a dimensionless coupling constant into a scale. The appearance of a scale from a dimensionless theory is due to quantum loop effects. The relation is:

$$\Lambda_{QCD} = Q \, e^{-1/(2\beta_0 \alpha_s(Q^2))}, \tag{14}$$

where α_s is the strong coupling constant, Q the scale where it is defined and β_0 the leading coefficient of the β-function. This relation does not determine the scale, since $\alpha_s(Q^2)$ is input, but from a landscape perspective it affects the distribution of scales in such a way that a large hierarchy is easy to obtain. How easy depends on the distribution of α_s in a fundamental theory, but what we know about the landscape suggests that this argument is valid.

This then leaves the weak scale M_{weak} to worry about. In contrast to Λ_{QCD}, the μ^2 parameter in the Standard Model Higgs potential receives quadratic quantum corrections from higher scales. This worry has been the focus of decades of work on natural solutions to the hierarchy problem. All of these solutions lead to predictions of new particles near the weak scale, although some can be pushed a few orders of magnitude higher. So far no such particles have been found. One can take the point of view that we simply have to be more patient. After all, the top quark and the Higgs boson also required a lot of patience. But the current situation clearly demands a reassessment of the arguments. During the year 2013, after the existence of the Higgs boson was convincingly established, there has been a lot of discussion about this. The different lines of argument are roughly as follows.

- There is no hierarchy. Some people argue that the hierarchy should be viewed as a misconception in quantum field theory. They point to the fact that unlike logarithmic corrections, quadratic corrections are not present in every regularization scheme, and argue that they may not really be physical; see *e.g.* [169–172]. However, these papers deal with just the Standard Model in quantum field theory, and ignore potential BSM physics (such as GUTs) and certain BSM physics (gravity). There is general agreement that the existence of new massive particles beyond the weak scale implies a quadratic hierarchy for μ^2 [173]. One can try to get around this by following a minimalistic approach in which new physics beyond the weak scale is completely avoided, as in [174]. But then one still has to deal with gravity. It is possible that the naive notion that since there is a Planck scale, there must be new physics at that scale is wrong. Perhaps nature is fundamentally scale invariant (as suggested for example in [175–177]; see however [178] for criticism). However, this escape route is closed off in string theory, which clearly predicts not just new physics but concrete new particles at the Planck scale. This still leaves the next option:

- The Planck scale is at the weak scale. This possibility exists if there are large extra dimensions at a length scale far larger than typical length scale of particle physics, even as large as 0.1 mm. Such a scenario was proposed in [179], but it predicts observable gravitational physics, perhaps even black holes, at the weak scale. This idea is put under severe stress by the latest LHC results.
- The weak/GUT or weak/Planck scale hierarchy is real and is generated by new dynamics (technicolor, compositeness) at the weak scale. In this approach the weak scale is generated analogously to the strong scale by dimensional transmutation. Also in this case absence of new physics at the LHC is a serious problem. Furthermore it is hard to build credible examples where the entire Standard Model (including quark and lepton masses) is reproduced.
- The weak/GUT or weak/Planck scale hierarchy is real and is "protected" by low energy supersymmetry. Since low energy supersymmetry does not determine the weak scale, this still requires an additional mechanism to generate it, and often this is related to dimensional transmutation (*e.g.* gaugino condensation). This idea is facing similar observational problems as the previous two.
- The hierarchy is mainly anthropic (see below).

From the string theory perspective, the second option is very attractive since it would give direct access to gravitational physics; from the landscape perspective the question is how often large extra dimensions occur in conjunction with a Standard Model with a weak scale as we observe. The third option is unattractive from the landscape point of view since it would imply that our vision of fundamental physics would be blurred by an additional "onion shell" which we have to peel off first (since technicolor gained some renewed interest due to an interpretation as a dual of a Randal-Sundrum model [180], some people view this as an appealing prospect).

At this moment, low energy supersymmetry is usually considered to be the most attractive of options 2, 3 and 4. It is a well-motivated idea because it not only controls the hierarchy, but also provides additional particles needed for the convergence of the running coupling constants as well as dark matter candidates. Furthermore string theory may come with supersymmetry built in. The latter statement holds for all string theories with a controlled perturbative expansion. Without supersymmetry, quantum corrections diverge beyond one loop. Supersymmetry may be required for a fundamental understanding of quantum gravity, or it may just be a calculational tool we need because our current understanding is too primitive.

Since, despite good arguments in its favor, no sign of low energy supersymmetry has been seen so far, we have to ask which mistakes may have been made in arriving at the overly optimistic expectations. We will mention three here, labeled as "anthropic", "landscape" and "string theory".

Anthropic. The idea that the hierarchy might be anthropic was not even mentioned during three of the four decades of discussion of naturalness, and during the past decade was mentioned only to ridicule it. And yet it is true. It is true in the sense that in an ensemble of theories with a range of gauge hierarchies, (intelligent) life can exist only for theories with a large hierarchy. The simplest argument is based on the weakness of gravity. The largest structures that can exist without being crushed into a black hole have N building blocks, where $N = (M_{\text{Planck}}/m)^3$ and m is the mass of the building block (the proton mass in our universe). This already requires a hierarchy of nine orders of magnitude for something with the complexity of a human brain to exist. More detailed arguments with stronger (and more debatable) assumptions pin the weak scale down with a precision of less than ten per cent (see [1] for more details and references). But even if the entire hierarchy of 18 orders of magnitude is fully understood anthropically, this still does not imply that the hierarchy is explained by anthropic arguments alone, as suggested in [181] and [182].

Landscape.   One may still ask the question if the underlying physics helps in getting the anthropically required hierarchy. In a landscape of variants of the Standard Model with a fixed Planck scale and a full range of values of μ^2, flatly distributed, the chance of getting an anthropically acceptable theory is about 10^{-34}. That is the fraction of theories with a μ^2 parameter that is small enough, if the parameter μ^2 can be though of as a sum of uncorrelated terms of order $(M_{\text{Planck}})^2$. So with a large enough landscape, one could consider the problem solved. But this is not true if the landscape contains other vacua where the statistical penalty of 10^{-34} does not have to be paid. Technically natural theories escape this penalty. However, then the question arises what the cost is in getting technical naturalness.

These questions can only be addressed in a context where the relative abundance of theories can be compared. In particular, it makes no sense in quantum field theory. Indeed, the very concept of technical naturalness is at best a poorly defined intuitive notion without the context of an ensemble of theories, a landscape. The string theory landscape certainly contains supersymmetric theories; indeed, these are the only ones under computational control. To predict if they dominate, one would have to estimate the ratio of the number of supersymmetric and non-supersymmetric theories, given the hierarchy. Supersymmetric theories start with a huge advantage of a factor 10^{34}, but they may still lose by being far less abundant, or because they have to satisfy additional anthropic constraints to avoid fast, catastrophic proton decay. Unfortunately, answering this question is not feasible at present. What has been tried is comparing supersymmetric theories with different supersymmetry breaking scales. Very roughly (for many more details see [27]) one would like to compute

$$P\left(\frac{M_{\text{weak}}}{M_{\text{susy}}}\right) P\left(\frac{M_{\text{susy}}}{M_{\text{Planck}}}\right) , \tag{15}$$

the probability for getting a weak scale given a supersymmetry breaking scale times the probability for getting a certain supersymmetry breaking scale given the Planck scale. The first factor is $(M_{\text{weak}}/M_{\text{susy}})^2$ by the usual naturalness argument. It varies between 1 and 10^{-34} if we move M_{susy} from the weak scale to the Planck scale, and this is the basis for the prediction $M_{\text{susy}} \approx M_{\text{weak}}$. But all QFT-based, bottom up naturalness arguments completely ignore the second factor, which is not even defined in QFT. The first attempts to compute it in the string landscape produced the result that it was proportional to $M_{\text{susy}}/M_{\text{Planck}}$ to a power larger than 2, so that large susy scales dominate over small ones by a large factor [183, 184]. Meanwhile that conclusion has been shown to be too simplistic [27, 185], and furthermore an important contributing factor was underestimated in earlier work, namely that vacua with broken susy are less likely to be stable. This can lead to a huge suppression [48, 49]. But one conclusion remains: even if the second factor in Eq. (15) is difficult to compute in a well-defined setting, this does not imply that it can be ignored in approaches where it cannot even be defined.

String Theory. An obvious weakness of the "MSSM" hypothesis is the first M, which stands for "minimal". There is no good fundamental reason to expect minimality, but dropping this restriction implies a substantial loss of predictive power. If the supersymmetric Standard Model is realized in string theory, the result is rarely minimal, see Section 2.2. But while most of the additional particles can at least be avoided in special constructions, one kind is essentially inevitable: moduli.

It has been known for a long time that moduli can lead to cosmological problems [10, 11, 186]. If they decay during or after BBN they will produce additional baryonic matter and destroy the successful BBN predictions. Bosonic moduli have potentials, and will in general be displaced from their minima. Their time evolution is governed by the equation

$$\ddot{\phi} + 3H\dot{\phi} + \frac{\partial V}{\partial \phi} = 0, \tag{16}$$

where H is the Hubble constant. If $V = \frac{1}{2}m^2\phi^2 +$ higher order terms, and $H \gg m$, then the second term dominates over the third, and ϕ gets frozen at some constant value ("Hubble friction"). This lasts until H drops below m. Then the field starts oscillating in its potential, and releases its energy. The requirement that this does not alter BBN predictions leads to a lower bound on the scalar moduli mass of a few tens of TeV (30 TeV, for definiteness). For higher masses moduli decay can reheat the universe sufficiently to restart BBN from electroweak equilibrium.

Furthermore one can argue [187] that the mass of the lightest modulus is of the same order of magnitude as the gravitino mass, $m_{3/2}$. The latter mass is generically of the same order as the soft susy breaking scalar masses: the squarks and sleptons searched for at the LHC. Gaugino masses can be one or two orders of magnitude less. This chain of arguments leads to the prediction that the sparticle masses will be a few tens of TeV, out of reach for the LHC, probably even after its upgrade, but there would still be a good chance to observe gauginos. Circumstantial evidence in favor of this scenario is that it prefers a Higgs mass near the observed value [188], whereas bottom-up supersymmetric models, ignoring moduli, suggested an upper limit of at most 120 GeV.

But there is one worrisome point. The 30 TeV bound on moduli masses is not an anthropic bound. Observers in universes with moduli masses below that bound would be deeply puzzled that their attempts at computing BBN abundances gave incorrect answers. They might see a helium abundance of only 19% while their computations predicted 24%, but nothing we know suggests that this has adverse effects on life. Now if supersymmetry prefers a lower scale because of naturalness, this would imply that universes with deeply puzzled observers should dominate universes with observers enjoying successful BBN predictions, such as ourselves.

None of these three mistakes is obviously fatal, but taking into account the "anthropic landscape of string theory" and all its implications definitely lowers the confidence level of predictions of low energy supersymmetry.

3.6. Axions and the strong CP problem

The Standard Model Lagrangian contains a term

$$\theta \frac{g_3^2}{32\pi^2} \sum_{a=1}^{8} F_{\mu\nu}^a F_{\rho\sigma}^a \epsilon^{\mu\nu\rho\sigma} \tag{17}$$

where the sum is over the eight generators of $SU(3)$. The parameter θ, an angle with values between 0 and 2π, is not an observable by itself. By making suitable phase rotations of the fermions its value can be changed, but then these phase rotations end up in the mass-matrices of the quarks. In the end, this leads to one new physical parameter, $\bar{\theta} = \theta - \arg \det (M_u M_d)$, where M_u and M_d are the quark mass matrices. A non-zero value for this parameter would produce a non-zero dipole moment for the neutron and certain nuclei, which so far has not been observed. This puts an upper limit on $\bar{\theta}$ of about 10^{-10}. Since there is no anthropic argument in favor of such a small value – the smallest one can argue for anthropically [189] is 10^{-3} – this is the one of the most serious naturalness problems in the Standard Model. No one would argue that we observe such a small value just by chance.

This problem has a rather simple solution, the Peccei-Quinn [190] mechanism. It works by postulating an additional pseudoscalar boson with a dimension 5 interaction with the QCD vector bosons

$$\Delta \mathcal{L} = \frac{1}{2} \partial_\mu a \partial^\mu a + \frac{a}{32 \pi^2 f_a} \sum_a F_{\mu\nu}^a F_{\rho\sigma}^a \epsilon^{\mu\nu\rho\sigma}, \tag{18}$$

where f_a is the "axion decay constant". Since $F\tilde{F}$ (where $\tilde{F}_{\mu\nu} = \frac{1}{2} \epsilon_{\mu\nu\rho\sigma} F^{\rho\sigma}$) is a total derivative, after integration by parts the second term is proportional to $\partial_\mu a$. Hence there is a shift symmetry $a \to a + \epsilon$. This allows us to shift a by a constant $-\bar{\theta} f_a$ so that the $F\tilde{F}$ term (17) is removed from the action. However, the shift symmetry is anomalous with respect to QCD because the $F\tilde{F}$ term is a derivative of a gauge non-invariant operator. Through non-perturbative effects the anomaly generates a potential with a minimum at $a = 0$ of the form

$$V(a) \propto \Lambda_{\text{QCD}}^4 \left(1 - \cos(a/f_a)\right). \tag{19}$$

Note that $\bar{\theta}$ is periodic with period 2π, so that the shift symmetry is globally a $U(1)$ symmetry. It was pointed out in [191, 192] that this breaking of the $U(1)$ symmetry leads to a pseudo-scalar pseudo-Goldstone boson, which was called "axion".

The mass of this particle is roughly $\Lambda_{\text{QCD}}^2 / f_a$, but if we take into account the proportionality factors in (19) the correct answer is

$$m_a = \frac{m_\pi f_\pi}{f_a} F(m_q), \tag{20}$$

where f_π is the pion decay constant and $F(m_q)$ a function of the (light) quark masses that is proportional to their product. The scale f_a was originally assumed to be that of the weak interactions, leading to a mass prediction of order 100 KeV, that is now ruled out. But soon it was realized that f_a could be chosen freely, and in particular much higher, making the axion "harmless" or "invisible" (see [193] and references therein). This works if the coupling f_a is within a narrow window. For small f_a the constraint is due to the fact that supernovae or white dwarfs would cool too fast by axion emission. This gives a lower limit $f_a > 10^9$ GeV. There is an upper limit of $f_a < 10^{12}$ GeV because if f_a were larger the contribution of axions to dark matter would be too large. This results in a small allowed window for the axion mass: $6~\mu\text{eV} < m_a < 6$ meV.

The upper limit of f_a is interesting since on the one hand it is large, but on the other hand it does not quite reach the string scale. The non-renormalizable interaction in the axion Lagrangian points to new physics at that scale, and from a string theory perspective the natural candidate for such new physics would be string theory itself. But then the scale seems uncomfortably low in comparison to typical string scales. Indeed, in [194] the difficulties are examined and possible ways out are discussed.

One way out is suggested by the fact that the amount of axion dark matter is proportional to $\sin^2\theta_0$, where θ_0 is the initial misalignment angle of the axion potential. In deriving the upper bound, one assumes that our universe emerges from a configuration with random alignments, resulting in an average of $\sin^2\theta_0$. This would give a value of $\frac{1}{2}$ for $\langle\sin^2\theta_0\rangle$.

The fact that the parameter is an angle and that axions are not strongly coupled to the rest of the landscape makes it an ideal arena for anthropic reasoning [195]. It is possible that our universe comes from a single inflated region with a small value of θ_0. For a larger value of θ_0 (given an axion decay constant at the string scale) too much dark matter would be produced. One has to argue that, even though the likelihood of living in a region with small θ_0 is small, this is compensated by the fact that more observers will find themselves in such regions, because larger dark matter densities are detrimental for the existence of life. The most likely reason for that is galaxy formation and the density of matter in galaxies, both of which depend on the dark matter fraction. See [196–199] for further discussion. It has even been argued that finding a high scale axion would provide evidence for the multiverse and the string theory landscape [200]. The upper bound on the axion decay constant can also be raised if there is a non-thermal cosmological history, for example caused by decay of heavy moduli [58].

Candidate axions occur abundantly in string theory, but their survival as light particles is affected by the moduli stabilization mechanism. They may be thought of as phase factors of complex fields. The real parts of those fields must be stabilized. They would otherwise give rise to fifth forces or affect BBN predictions. But thanks to their derivative couplings, axions are far less constrained. However, not every mechanism to stabilize moduli is capable of giving mass to the real part and leave the imaginary part unaffected. For example, stabilization by fluxes or by instanton induced terms in the superpotential gives mass to both the moduli and the corresponding axions.

Axions that survive moduli stabilization may in principle play the role of PQ axions that solve the strong CP problem, provided they do not acquire masses by other mechanisms. The usual folklore that gravity does not allow exact global symmetries suggests that all axions will eventually get a mass. Otherwise their presence as massless particles would imply the existence of an exact, but spontaneously broken global symmetry, with axions as Goldstone bosons. Just as the QCD instanton get a mass from non-perturbative QCD effects, all other axions should get a mass as well. The Peccei-Quinn mechanism works as long as there is an axion coupling to QCD with a mass contribution from any other sources that is at least ten orders of magnitude smaller than the QCD contribution.

It is clear from the previous paragraph that if string theory produces a PQ axion, it is likely that it produces many other axions in addition to this. Since their masses are generated by non-perturbative effects, it is natural to expect them to be distributed in a scale invariant way, spanning many orders of magnitude. This plenitude of axions has been called the "axiverse" [200]. Since the masses of the additional axions (not involved in the PQ mechanism) are not limited to the QCD window, this provides ample opportunities for observations in many mass regions.

Realizations of an axiverse have been discussed in fluxless M-theory compactifications [201] and in type-IIB models in the LARGE Volume Scenario [202]. Both papers consider compactifications with many Kähler moduli that are stabilized by a single non-perturbative contribution rather than a separate contribution for each modulus. Then all Kähler moduli can be stabilized, but just one "common phase" axion acquires a large mass. For supersymmetric moduli stabilization (such as the KKLT scenario, but unlike LVS) a no-go theorem was proved in [203]. Axions in the heterotic mini-landscape were discussed in [204]. They consider discrete symmetries that restrict the superpotential, so that the lowest order terms have accidental $U(1)$ symmetries that may include a PQ symmetry.

There are numerous possibilities for experiments and observations that may shed light on the rôle of axions in our universe, and thereby provide information on the string theory landscape. The observation of tensor modes in the CMB might falsify the axiverse [201, 205]. See [200, 206, 207] for a variety of possible signatures, ongoing experiments and references.

4. Conclusions

The past four decades have been a magnificent golden age for particle physics, but also for the leading approach to understanding it in terms of a fundamental theory of quantum gravity, string theory. Now we appear to be approaching an interesting and perhaps decisive moment in history. The Standard Model seems complete, and string theory has led to a mathematical challenge of monstrous proportions: the string theory landscape.

Whether one likes it or not, successful string phenomenology requires taming this monster. This will almost certainly involve a reconsideration of the questions we wish to answer. In the future, the current time may be remembered as the transition from the era of symmetries to a new era with different ways of thinking about fundamental problems.

In experimental physics, we are faced with a situation where positive results may emerge but are not guaranteed, and where negative results tell us fairly little. Apart from the traditional "new physics", examples of positive results that would have an impact on the landscape are variations of constants of nature, the observation of axions, dark matter of any kind, neutrino Majorana masses, sterile neutrinos, massive vector bosons, and many serendipitous discoveries ranging from desirable (*e.g.* proton decay or magnetic monopoles) to totally unexpected (*e.g.* something like faster-than-light neutrinos). With the large number of experiments and astrophysical observations still underway, it seems unthinkable that the Higgs particle will turn out to be the last discovery in particle physics. But if it is, this has to be viewed as circumstantial evidence in favor of a landscape. Any of the aforementioned positive results can be good or bad for the landscape idea, but there is no gold-plated experiment that verifies or falsifies it.

Further supportive or damaging evidence must come from pure theory. The (non)-existence of a broadly spread distribution of de Sitter vacua in string theory is a decidable issue, but it has turned out to be very difficult to reach a conclusion. The fundamental principle behind string theory still eludes us. The quantization of vibrating strings is not an acceptable fundamental starting point, and does not describe everything we call string theory anyway. The best hope for acceptance of the landscape idea is that it is derived from a fundamental theory of gravity which in its turn is derived from a plausible principle of nature. Meanwhile, we can try to bridge the gap between the Standard Model and the string landscape. We must convince ourselves that our Universe is indeed contained in the string theory landscape. We can explore our environment in the landscape, to see if we can understand why we observe the Standard Model and all of its features, especially the puzzling ones. This requires determining landscape distributions in various regions, and using anthropic arguments where possible.

This may all be postponed until the indefinite future if new physics still emerges during the next few years. If that new physics is due to large extra dimensions and a low higher-dimensional Planck scale, we can explore quantum gravity directly. This has always seemed too good to be true, and probably it is. If the new physics is low energy supersymmetry, which is still a well-motivated option, perhaps the winding road to the string landscape becomes a broad avenue all the way to the Planck scale. Most other options may provide more decades of exciting particle physics, but the historic moment we might be witnessing now would have passed and may never return.

Acknowledgments

It is a pleasure to thank Beatriz Gato-Rivera for reading the manuscript and many useful comments. This work has been partially supported by funding of the Spanish Ministerio de Economía y Competitividad, Research Project FIS2012-38816, and by the Project CONSOLIDER-INGENIO 2010, Programme CPAN (CSD2007-00042).

References

[1] A. Schellekens, Life at the Interface of Particle Physics and String Theory, *Rev.Mod.Phys.* **85**, 1491 (2013). doi: 10.1103/RevModPhys.85.1491.

[2] K. Narain, New Heterotic String Theories in Uncompactified Dimensions, *Phys.Lett.* **B169**, 41 (1986). doi: 10.1016/0370-2693(86)90682-9.

[3] A. Strominger, Superstrings with Torsion, *Nucl.Phys.* **B274**, 253 (1986). doi: 10.1016/0550-3213(86)90286-5.

[4] H. Kawai, D. C. Lewellen, and S. H. Tye, Construction of Four-Dimensional Fermionic String Models, *Phys.Rev.Lett.* **57**, 1832 (1986). doi: 10.1103/PhysRevLett.57.1832.

[5] W. Lerche, D. Lust, and A. N. Schellekens, Chiral Four-Dimensional Heterotic Strings from Selfdual Lattices, *Nucl.Phys.* **B287**, 477 (1987). doi: 10.1016/0550-3213(87)90115-5.

[6] I. Antoniadis, C. Bachas, and C. Kounnas, Four-Dimensional Superstrings, *Nucl.Phys.* **B289**, 87 (1987). doi: 10.1016/0550-3213(87)90372-5.

[7] K. Narain, M. Sarmadi, and C. Vafa, Asymmetric Orbifolds, *Nucl.Phys.* **B288**, 551 (1987). doi: 10.1016/0550-3213(87)90228-8.

[8] L. J. Dixon, J. A. Harvey, C. Vafa, and E. Witten, Strings on Orbifolds. 2., *Nucl.Phys.* **B274**, 285–314 (1986). doi: 10.1016/0550-3213(86)90287-7.

[9] H. Kawai, D. C. Lewellen, and S. H. Tye, Construction of Fermionic String Models in Four-Dimensions, *Nucl.Phys.* **B288**, 1 (1987). doi: 10.1016/ 0550-3213(87)90208-2.

[10] B. de Carlos, J. Casas, F. Quevedo, and E. Roulet, Model independent properties and cosmological implications of the dilaton and moduli sectors of 4-d strings, *Phys.Lett.* **B318**, 447–456 (1993). doi: 10.1016/ 0370-2693(93)91538-X.

[11] T. Banks, D. B. Kaplan, and A. E. Nelson, Cosmological implications of dynamical supersymmetry breaking, *Phys.Rev.* **D49**, 779–787 (1994). doi: 10.1103/PhysRevD.49.779.

[12] L. Susskind. The Anthropic landscape of string theory. In ed. B. J. Carr, *Universe or Multiverse?*, pp. 247–266. Cambridge University Press (2003).

[13] A. G. Riess et al., Observational evidence from supernovae for an accelerating universe and a cosmological constant, *Astron.J.* **116**, 1009–1038 (1998). doi: 10.1086/300499.

[14] S. Perlmutter et al., Measurements of Omega and Lambda from 42 high redshift supernovae, *Astrophys.J.* **517**, 565–586 (1999). doi: 10.1086/307221.

[15] A. D. Linde, The Inflationary Universe, *Rept.Prog.Phys.* **47**, 925–986 (1984). doi: 10.1088/0034-4885/47/8/002.

[16] R. Bousso and J. Polchinski, Quantization of four form fluxes and dynamical neutralization of the cosmological constant, *JHEP.* **0006**, 006 (2000).

[17] J. D. Brown and C. Teitelboim, Dynamical Neutralization Of The Cosmological Constant, *Phys.Lett.* **B195**, 177–182 (1987). doi: 10.1016/ 0370-2693(87)91190-7.

[18] J. D. Brown and C. Teitelboim, Neutralization of the Cosmological Constant by Membrane Creation, *Nucl.Phys.* **B297**, 787–836 (1988). doi: 10.1016/0550-3213(88)90559-7.

[19] J. L. Feng, J. March-Russell, S. Sethi, and F. Wilczek, Saltatory relaxation of the cosmological constant, *Nucl.Phys.* **B602**, 307–328 (2001). doi: 10. 1016/S0550-3213(01)00097-9.

[20] A. H. Guth, Inflation and eternal inflation, *Phys.Rept.* **333**, 555–574 (2000). doi: 10.1016/S0370-1573(00)00037-5.

[21] A. D. Linde. Inflation, quantum cosmology and the anthropic principle. In ed. J. Barrow, *Science and Ultimate Reality*, pp. 426–458 (2002).

[22] A. Vilenkin, Probabilities in the landscape (2006).

[23] A. H. Guth, Eternal inflation and its implications, *J.Phys.A.* **A40**, 6811–6826 (2007). doi: 10.1088/1751-8113/40/25/S25.

[24] B. Freivogel, Making predictions in the multiverse, *Class.Quant.Grav.* **28**, 204007 (2011). doi: 10.1088/0264-9381/28/20/204007.

[25] Y. Nomura, Quantum Mechanics, Gravity, and the Multiverse, *Astron.Rev.* **7**, 36 (2012).

[26] M. Graña, Flux compactifications in string theory: A Comprehensive review, *Phys.Rept.* **423**, 91–158 (2006). doi: 10.1016/j.physrep.2005.10.008.

[27] M. R. Douglas and S. Kachru, Flux compactification, *Rev.Mod.Phys.* **79**, 733–796 (2007). doi: 10.1103/RevModPhys.79.733.

[28] R. Blumenhagen, B. Kors, D. Lust, and S. Stieberger, Four-dimensional

String Compactifications with D-Branes, Orientifolds and Fluxes, *Phys.Rept.* **445**, 1–193 (2007). doi: 10.1016/j.physrep.2007.04.003.

[29] F. Denef. Les Houches Lectures on Constructing String Vacua. In *String theory and the real world: From particle physics to astrophysics*, Proceedings, Summer School in Theoretical Physics, Les Houches, France, July 2-27, 2007, pp. 483–610, North Holland (2008).

[30] F. Denef, M. R. Douglas, and S. Kachru, Physics of String Flux Compactifications, *Ann.Rev.Nucl.Part.Sci.* **57**, 119–144 (2007). doi: 10.1146/annurev.nucl.57.090506.123042.

[31] S. Gukov, C. Vafa, and E. Witten, CFT's from Calabi-Yau four folds, *Nucl.Phys.* **B584**, 69–108 (2000). doi: 10.1016/S0550-3213(00)00373-4.

[32] E. Witten, Nonperturbative superpotentials in string theory, *Nucl.Phys.* **B474**, 343–360 (1996). doi: 10.1016/0550-3213(96)00283-0.

[33] S. Kachru, R. Kallosh, A. D. Linde, and S. P. Trivedi, De Sitter vacua in string theory, *Phys.Rev.* **D68**, 046005 (2003). doi: 10.1103/PhysRevD.68.046005.

[34] K. Dasgupta, G. Rajesh, and S. Sethi, M theory, orientifolds and G - flux, *JHEP.* **9908**, 023 (1999).

[35] I. R. Klebanov and M. J. Strassler, Supergravity and a confining gauge theory: Duality cascades and chi SB resolution of naked singularities, *JHEP.* **0008**, 052 (2000).

[36] S. B. Giddings, S. Kachru, and J. Polchinski, Hierarchies from fluxes in string compactifications, *Phys.Rev.* **D66**, 106006 (2002). doi: 10.1103/PhysRevD.66.106006.

[37] F. Denef, M. R. Douglas, and B. Florea, Building a better racetrack, *JHEP.* **0406**, 034 (2004). doi: 10.1088/1126-6708/2004/06/034.

[38] D. Robbins and S. Sethi, A Barren landscape?, *Phys.Rev.* **D71**, 046008 (2005). doi: 10.1103/PhysRevD.71.046008.

[39] I. Bena, M. Grana, S. Kuperstein, and S. Massai, Anti-D3's - Singular to the Bitter End, *Phys.Rev.* **D87**, 106010 (2013). doi: 10.1103/PhysRevD.87.106010.

[40] I. Bena, M. Graña, S. Kuperstein, and S. Massai, Polchinski-Strassler does not uplift Klebanov-Strassler (2012).

[41] C. Burgess, R. Kallosh, and F. Quevedo, De Sitter string vacua from supersymmetric D terms, *JHEP.* **0310**, 056 (2003).

[42] A. Saltman and E. Silverstein, The Scaling of the no scale potential and de Sitter model building, *JHEP.* **0411**, 066 (2004). doi: 10.1088/1126-6708/2004/11/066.

[43] O. Lebedev, H. P. Nilles, and M. Ratz, De Sitter vacua from matter superpotentials, *Phys.Lett.* **B636**, 126–131 (2006). doi: 10.1016/j.physletb.2006.03.046.

[44] A. Westphal, Lifetime of Stringy de Sitter Vacua, *JHEP.* **0801**, 012 (2008). doi: 10.1088/1126-6708/2008/01/012.

[45] L. Covi, M. Gomez-Reino, C. Gross, J. Louis, G. A. Palma, et al., de Sitter vacua in no-scale supergravities and Calabi-Yau string models, *JHEP.* **0806**, 057 (2008). doi: 10.1088/1126-6708/2008/06/057.

[46] V. Balasubramanian, P. Berglund, J. P. Conlon, and F. Quevedo, Systematics of moduli stabilisation in Calabi-Yau flux compactifications, *JHEP.* **0503**, 007 (2005). doi: 10.1088/1126-6708/2005/03/007.

[47] J. Louis, M. Rummel, R. Valandro, and A. Westphal, Building an explicit de Sitter, *JHEP.* **1210**, 163 (2012). doi: 10.1007/JHEP10(2012)163.

[48] X. Chen, G. Shiu, Y. Sumitomo, and S. H. Tye, A Global View on The Search for de-Sitter Vacua in (type IIA) String Theory, *JHEP.* **1204**, 026 (2012). doi: 10.1007/JHEP04(2012)026.

[49] D. Marsh, L. McAllister, and T. Wrase, The Wasteland of Random Supergravities, *JHEP.* **1203**, 102 (2012). doi: 10.1007/JHEP03(2012)102.

[50] B. Greene, D. Kagan, A. Masoumi, E. Weinberg, and X. Xiao, Tumbling through a landscape: Evidence of instabilities in high-dimensional moduli spaces (2013).

[51] T. Banks, The Top 10^{500} Reasons Not to Believe in the Landscape (2012).

[52] S. R. Coleman and F. De Luccia, Gravitational Effects on and of Vacuum Decay, *Phys.Rev.* **D21**, 3305 (1980). doi: 10.1103/PhysRevD.21.3305.

[53] M. R. Douglas, Basic results in vacuum statistics, *Comptes Rendus Physique.* **5**, 965–977 (2004). doi: 10.1016/j.crhy.2004.09.008.

[54] S. Ashok and M. R. Douglas, Counting flux vacua, *JHEP.* **0401**, 060 (2004). doi: 10.1088/1126-6708/2004/01/060.

[55] F. Denef and M. R. Douglas, Distributions of flux vacua, *JHEP.* **0405**, 072 (2004). doi: 10.1088/1126-6708/2004/05/072.

[56] J. Garriga, D. Schwartz-Perlov, A. Vilenkin, and S. Winitzki, Probabilities in the inflationary multiverse, *JCAP.* **0601**, 017 (2006). doi: 10.1088/1475-7516/2006/01/017.

[57] M. R. Douglas. The string landscape and low energy supersymmetry. In *Strings, Gauge Fields and the Geometry Behind. The Legacy of Max Kreuzer.*, pp. 261–288, World Scientific (2012).

[58] B. S. Acharya, G. Kane, and P. Kumar, Compactified String Theories – Generic Predictions for Particle Physics, *Int.J.Mod.Phys.* **A27**, 1230012 (2012).

[59] L. E. Ibañez and A. M. Uranga, *String theory and particle physics: An introduction to string phenomenology.* Cambridge University Press (2012).

[60] L. J. Dixon, J. A. Harvey, C. Vafa, and E. Witten, Strings on Orbifolds, *Nucl.Phys.* **B261**, 678–686 (1985). doi: 10.1016/0550-3213(85)90593-0.

[61] L. E. Ibañez, H. P. Nilles, and F. Quevedo, Orbifolds and Wilson Lines, *Phys.Lett.* **B187**, 25–32 (1987). doi: 10.1016/0370-2693(87)90066-9.

[62] M. Goodsell and A. Ringwald, Light Hidden-Sector U(1)s in String Compactifications, *Fortsch.Phys.* **58**, 716–720 (2010). doi: 10.1002/prop. 201000026.

[63] P. Candelas, G. T. Horowitz, A. Strominger, and E. Witten, Vacuum Configurations for Superstrings, *Nucl.Phys.* **B258**, 46–74 (1985). doi: 10.1016/0550-3213(85)90602-9.

[64] M. Kreuzer and H. Skarke, Complete classification of reflexive polyhedra in four-dimensions, *Adv.Theor.Math.Phys.* **4**, 1209–1230 (2002).

[65] B. Assel, K. Christodoulides, A. E. Faraggi, C. Kounnas, and J. Rizos,

Classification of Heterotic Pati-Salam Models, *Nucl.Phys.* **B844**, 365–396 (2011). doi: 10.1016/j.nuclphysb.2010.11.011.

[66] T. Renner, J. Greenwald, D. Moore, and G. Cleaver, Initial Systematic Investigations of the Landscape of Low Layer NAHE Extensions, *Eur.Phys.J.* **C72**, 2167 (2012). doi: 10.1140/epjc/s10052-012-2167-y.

[67] T. Renner, J. Greenwald, D. Moore, and G. Cleaver, Initial Systematic Investigations of the Landscape of Low Layer NAHE Variation Extensions (2011).

[68] I. Antoniadis, J. R. Ellis, J. Hagelin, and D. V. Nanopoulos, The Flipped SU(5) x U(1) String Model Revamped, *Phys.Lett.* **B231**, 65 (1989). doi: 10.1016/0370-2693(89)90115-9.

[69] F. Englert, H. Nicolai, and A. Schellekens, Superstrings From Twentysix-Dimensions, *Nucl.Phys.* **B274**, 315–348 (1986). doi: 10.1016/0550-3213(86)90288-9.

[70] I. Antoniadis, C. Bachas, C. Kounnas, and P. Windey, Supersymmetry Among Free Fermions and Superstrings, *Phys.Lett.* **B171**, 51 (1986). doi: 10.1016/0370-2693(86)90996-2.

[71] G. Waterson, Bosonic Construction Of An N=2 Extended Superconformal Theory In Two-Dimensions, *Phys.Lett.* **B171**, 77 (1986). doi: 10.1016/0370-2693(86)91002-6.

[72] A. N. Schellekens and N. Warner, Weyl Groups, Supercurrents And Covariant Lattices, *Nucl.Phys.* **B308**, 397 (1988). doi: 10.1016/0550-3213(88)90570-6.

[73] H. Niemeier, Definite quadratische formen der dimension 24 und diskriminate 1, *J. Number Theory.* (5), 142 (1973).

[74] J. Conway and N. Sloane, On the enumeration of lattices of determinant one, *Journal of Number Theory.* **14**, 83–94 (1982).

[75] W. Lerche, D. Lust, and A. N. Schellekens, Ten-Dimensional Heterotic Strings From Niemeier Lattices, *Phys.Lett.* **B181**, 71 (1986). doi: 10.1016/0370-2693(86)91257-8.

[76] P. Goddard, Meromorphic Conformal Field Theory (1989).

[77] A. Schellekens, Meromorphic C = 24 conformal field theories, *Commun.Math.Phys.* **153**, 159–186 (1993). doi: 10.1007/BF02099044.

[78] P. Candelas, E. Perevalov, and G. Rajesh, Toric geometry and enhanced gauge symmetry of F theory / heterotic vacua, *Nucl.Phys.* **B507**, 445–474 (1997). doi: 10.1016/S0550-3213(97)00563-4.

[79] H. Kawai, D. Lewellen, and S. Tye, Classification of Closed Fermionic String Models, *Phys.Rev.* **D34**, 3794 (1986). doi: 10.1103/PhysRevD.34.3794.

[80] D. Gepner, Space-Time Supersymmetry in Compactified String Theory and Superconformal Models, *Nucl.Phys.* **B296**, 757 (1988). doi: 10.1016/0550-3213(88)90397-5.

[81] A. N. Schellekens and S. Yankielowicz, New Modular Invariants For N=2 Tensor Products And Four-Dimensional Strings, *Nucl.Phys.* **B330**, 103 (1990). doi: 10.1016/0550-3213(90)90303-U.

[82] C. Vafa and N. P. Warner, Catastrophes and the Classification of Conformal Theories, *Phys.Lett.* **B218**, 51 (1989). doi: 10.1016/0370-2693(89)90473-5.

[83] W. Lerche, C. Vafa, and N. P. Warner, Chiral Rings in N=2 Superconfor-

mal Theories, *Nucl.Phys.* **B324**, 427 (1989). doi: 10.1016/0550-3213(89) 90474-4.

[84] D. Gepner, String Theory On Calabi-Yau Manifolds: The Three Generations Case (1987).

[85] A. Font, L. E. Ibañez, D. Lust, and F. Quevedo, Strong - weak coupling duality and nonperturbative effects in string theory, *Phys.Lett.* **B249**, 35–43 (1990). doi: 10.1016/0370-2693(90)90523-9.

[86] E. Witten, String theory dynamics in various dimensions, *Nucl.Phys.* **B443**, 85–126 (1995). doi: 10.1016/0550-3213(95)00158-O.

[87] P. Townsend, The eleven-dimensional supermembrane revisited, *Phys.Lett.* **B350**, 184–187 (1995). doi: 10.1016/0370-2693(95)00397-4.

[88] P. Horava and E. Witten, Heterotic and type I string dynamics from elevendimensions, *Nucl.Phys.* **B460**, 506–524 (1996). doi: 10.1016/0550-3213(95) 00621-4.

[89] C. Hull and P. Townsend, Unity of superstring dualities, *Nucl.Phys.* **B438**, 109–137 (1995). doi: 10.1016/0550-3213(94)00559-W.

[90] C. Vafa, Evidence for F theory, *Nucl.Phys.* **B469**, 403–418 (1996). doi: 10.1016/0550-3213(96)00172-1.

[91] A. Lukas, B. A. Ovrut, and D. Waldram, Nonstandard embedding and five-branes in heterotic M theory, *Phys.Rev.* **D59**, 106005 (1999). doi: 10.1103/PhysRevD.59.106005.

[92] Z. Lalak, S. Pokorski, and S. Thomas, Beyond the standard embedding in M theory on S**1 / Z(2), *Nucl.Phys.* **B549**, 63–97 (1999). doi: 10.1016/ S0550-3213(99)00136-4.

[93] V. Braun, Y.-H. He, B. A. Ovrut, and T. Pantev, The Exact MSSM spectrum from string theory, *JHEP.* **0605**, 043 (2006). doi: 10.1088/1126-6708/ 2006/05/043.

[94] L. B. Anderson, J. Gray, A. Lukas, and E. Palti, Heterotic Line Bundle Standard Models, *JHEP.* **1206**, 113 (2012). doi: 10.1007/JHEP06(2012) 113.

[95] R. Blumenhagen, S. Moster, and T. Weigand, Heterotic GUT and standard model vacua from simply connected Calabi-Yau manifolds, *Nucl.Phys.* **B751**, 186–221 (2006). doi: 10.1016/j.nuclphysb.2006.06.005.

[96] L. B. Anderson, J. Gray, A. Lukas, and E. Palti, Two Hundred Heterotic Standard Models on Smooth Calabi-Yau Threefolds, *Phys.Rev.* **D84**, 106005 (2011). doi: 10.1103/PhysRevD.84.106005.

[97] S. Forste, H. P. Nilles, P. K. Vaudrevange, and A. Wingerter, Heterotic brane world, *Phys.Rev.* **D70**, 106008 (2004). doi: 10.1103/PhysRevD.70. 106008.

[98] W. Buchmuller, K. Hamaguchi, O. Lebedev, and M. Ratz, Dual models of gauge unification in various dimensions, *Nucl.Phys.* **B712**, 139–156 (2005). doi: 10.1016/j.nuclphysb.2005.01.038.

[99] W. Buchmuller, K. Hamaguchi, O. Lebedev, and M. Ratz, Supersymmetric standard model from the heterotic string, *Phys.Rev.Lett.* **96**, 121602 (2006). doi: 10.1103/PhysRevLett.96.121602.

[100] O. Lebedev, H. P. Nilles, S. Raby, S. Ramos-Sanchez, M. Ratz, et al., A

Mini-landscape of exact MSSM spectra in heterotic orbifolds, *Phys.Lett.* **B645**, 88–94 (2007). doi: 10.1016/j.physletb.2006.12.012.

[101] H. P. Nilles, S. Ramos-Sanchez, M. Ratz, and P. K. Vaudrevange, From strings to the MSSM, *Eur.Phys.J.* **C59**, 249–267 (2009). doi: 10.1140/epjc/s10052-008-0740-1.

[102] B. Gato-Rivera and A. N. Schellekens, Asymmetric Gepner Models II. Heterotic Weight Lifting, *Nucl.Phys.* **B846**, 429–468 (2011). doi: 10.1016/j.nuclphysb.2011.01.011.

[103] N. Marcus and A. Sagnotti, Group Theory from Quarks at the Ends of Strings, *Phys.Lett.* **B188**, 58 (1987). doi: 10.1016/0370-2693(87)90705-2.

[104] L. E. Ibañez, F. Marchesano, and R. Rabadan, Getting just the standard model at intersecting branes, *JHEP.* **11**, 002 (2001).

[105] P. Anastasopoulos, T. Dijkstra, E. Kiritsis, and A. N. Schellekens, Orientifolds, hypercharge embeddings and the Standard Model, *Nucl.Phys.* **B759**, 83–146 (2006). doi: 10.1016/j.nuclphysb.2006.10.013.

[106] J. C. Pati and A. Salam, Lepton Number as the Fourth Color, *Phys.Rev.* **D10**, 275–289 (1974). doi: 10.1103/PhysRevD.10.275,10.1103/PhysRevD.11.703.2.

[107] S. M. Barr, A New Symmetry Breaking Pattern for SO(10) and Proton Decay, *Phys.Lett.* **B112**, 219 (1982). doi: 10.1016/0370-2693(82)90966-2.

[108] H. Georgi and S. Glashow, Unity of All Elementary Particle Forces, *Phys.Rev.Lett.* **32**, 438–441 (1974). doi: 10.1103/PhysRevLett.32.438.

[109] A. Sagnotti, Closed Strings And Their Open String Descendants, *Phys.Rept.* **184**, 167–175 (1989). doi: 10.1016/0370-1573(89)90036-7.

[110] P. Horava, Strings on World Sheet Orbifolds, *Nucl.Phys.* **B327**, 461 (1989). doi: 10.1016/0550-3213(89)90279-4.

[111] R. Blumenhagen, M. Cvetic, P. Langacker, and G. Shiu, Toward realistic intersecting D-brane models, *Ann.Rev.Nucl.Part.Sci.* **55**, 71–139 (2005). doi: 10.1146/annurev.nucl.55.090704.151541.

[112] M. B. Green and J. H. Schwarz, Anomaly Cancellation in Supersymmetric D=10 Gauge Theory and Superstring Theory, *Phys.Lett.* **B149**, 117–122 (1984). doi: 10.1016/0370-2693(84)91565-X.

[113] E. Kiritsis, D-branes in standard model building, gravity and cosmology, *Phys.Rept.* **421**, 105–190 (2005). doi: 10.1016/j.physrep.2005.09.001,10.1002/prop.200310120.

[114] J. L. Cardy, Boundary Conditions, Fusion Rules and the Verlinde Formula, *Nucl.Phys.* **B324**, 581 (1989). doi: 10.1016/0550-3213(89)90521-X.

[115] M. Bianchi and A. Sagnotti, On the systematics of open string theories, *Phys.Lett.* **B247**, 517–524 (1990). doi: 10.1016/0370-2693(90)91894-H.

[116] G. Pradisi, A. Sagnotti, and Y. Stanev, Completeness conditions for boundary operators in 2-D conformal field theory, *Phys.Lett.* **B381**, 97–104 (1996). doi: 10.1016/0370-2693(96)00578-3.

[117] J. Fuchs and C. Schweigert, Branes: From free fields to general backgrounds, *Nucl.Phys.* **B530**, 99–136 (1998). doi: 10.1016/S0550-3213(98)00352-6.

[118] R. E. Behrend, P. A. Pearce, V. B. Petkova, and J.-B. Zuber, Boundary

conditions in rational conformal field theories, *Nucl.Phys.* **B570**, 525–589 (2000). doi: 10.1016/S0550-3213(99)00592-1.

[119] L. Huiszoon, A. Schellekens, and N. Sousa, Klein bottles and simple currents, *Phys.Lett.* **B470**, 95–102 (1999). doi: 10.1016/S0370-2693(99) 01241-1.

[120] J. Fuchs, L. Huiszoon, A. Schellekens, C. Schweigert, and J. Walcher, Boundaries, crosscaps and simple currents, *Phys.Lett.* **B495**, 427–434 (2000). doi: 10.1016/S0370-2693(00)01271-5.

[121] C. Angelantonj and A. Sagnotti, Open strings, *Phys.Rept.* **371**, 1–150 (2002). doi: 10.1016/S0370-1573(02)00273-9.

[122] T. Dijkstra, L. Huiszoon, and A. N. Schellekens, Supersymmetric standard model spectra from RCFT orientifolds, *Nucl.Phys.* **B710**, 3–57 (2005). doi: 10.1016/j.nuclphysb.2004.12.032.

[123] G. Aldazabal, L. E. Ibañez, F. Quevedo, and A. Uranga, D-branes at singularities: A Bottom up approach to the string embedding of the standard model, *JHEP.* **0008**, 002 (2000).

[124] D. Berenstein, V. Jejjala, and R. G. Leigh, The Standard model on a D-brane, *Phys.Rev.Lett.* **88**, 071602 (2002). doi: 10.1103/PhysRevLett.88. 071602.

[125] H. Verlinde and M. Wijnholt, Building the standard model on a D3-brane, *JHEP.* **0701**, 106 (2007). doi: 10.1088/1126-6708/2007/01/106.

[126] C. Beasley, J. J. Heckman, and C. Vafa, GUTs and Exceptional Branes in F-theory - I, *JHEP.* **0901**, 058 (2009). doi: 10.1088/1126-6708/2009/01/058.

[127] C. Beasley, J. J. Heckman, and C. Vafa, GUTs and Exceptional Branes in F-theory - II: Experimental Predictions, *JHEP.* **0901**, 059 (2009). doi: 10.1088/1126-6708/2009/01/059.

[128] R. Donagi and M. Wijnholt, Model Building with F-Theory, *Adv. Theor. Math .Phys.* **15**, 1237–1318 (2011).

[129] T. Weigand, Lectures on F-theory compactifications and model building, *Class.Quant.Grav.* **27**, 214004 (2010). doi: 10.1088/0264-9381/27/21/ 214004.

[130] J. J. Heckman, Particle Physics Implications of F-theory, *Ann. Rev. Nucl. Part. Sci.* **60**, 237–265 (2010). doi: 10.1146/annurev.nucl.012809.104532.

[131] G. K. Leontaris, Aspects of F-Theory GUTs, *PoS.* **CORFU2011**, 095 (2011).

[132] A. Maharana and E. Palti, Models of Particle Physics from Type IIB String Theory and F-theory: A Review, *Int.J.Mod.Phys.* **A28**, 1330005 (2013). doi: 10.1142/S0217751X13300056.

[133] B. S. Acharya and E. Witten, Chiral fermions from manifolds of G(2) holonomy (2001).

[134] J. P. Conlon, A. Maharana, and F. Quevedo, Towards Realistic String Vacua, *JHEP.* **0905**, 109 (2009). doi: 10.1088/1126-6708/2009/05/109.

[135] R. Blumenhagen, T. W. Grimm, B. Jurke, and T. Weigand, Global F-theory GUTs, *Nucl.Phys.* **B829**, 325–369 (2010). doi: 10.1016/j.nuclphysb. 2009.12.013.

[136] J. Marsano, H. Clemens, T. Pantev, S. Raby, and H.-H. Tseng, A Global

SU(5) F-theory model with Wilson line breaking, *JHEP*. **1301**, 150 (2013). doi: 10.1007/JHEP01(2013)150.

[137] L. J. Dixon and J. A. Harvey, String Theories in Ten-Dimensions Without Space-Time Supersymmetry, *Nucl.Phys.* **B274**, 93–105 (1986). doi: 10.1016/0550-3213(86)90619-X.

[138] L. Alvarez-Gaumé, P. H. Ginsparg, G. W. Moore, and C. Vafa, An O(16) x O(16) Heterotic String, *Phys.Lett.* **B171**, 155 (1986). doi: 10.1016/0370-2693(86)91524-8.

[139] A. Sagnotti. Some properties of open string theories. In eds. I. Antoniadis and H. Videau, *Supersymmetry and unification of fundamental interactions, Palaiseau, Susy 95*, pp. 473–484, Ed. Frontieres (1995).

[140] C. Angelantonj, Nontachyonic open descendants of the 0B string theory, *Phys.Lett.* **B444**, 309–317 (1998). doi: 10.1016/S0370-2693(98)01430-0.

[141] S. Sugimoto, Anomaly cancellations in type I D-9 - anti-D-9 system and the USp(32) string theory, *Prog.Theor.Phys.* **102**, 685–699 (1999). doi: 10.1143/PTP.102.685.

[142] B. Gato-Rivera and A. Schellekens, Non-supersymmetric Orientifolds of Gepner Models, *Phys.Lett.* **B671**, 105–110 (2009). doi: 10.1016/j.physletb.2008.11.039.

[143] I. Antoniadis, E. Dudas, and A. Sagnotti, Brane supersymmetry breaking, *Phys.Lett.* **B464**, 38–45 (1999). doi: 10.1016/S0370-2693(99)01023-0.

[144] K. R. Dienes, Statistics on the heterotic landscape: Gauge groups and cosmological constants of four-dimensional heterotic strings, *Phys.Rev.* **D73**, 106010 (2006). doi: 10.1103/PhysRevD.73.106010.

[145] S. Kachru, J. Kumar, and E. Silverstein, Vacuum energy cancellation in a nonsupersymmetric string, *Phys.Rev.* **D59**, 106004 (1999). doi: 10.1103/PhysRevD.59.106004.

[146] J. A. Harvey, String duality and nonsupersymmetric strings, *Phys.Rev.* **D59**, 026002 (1999). doi: 10.1103/PhysRevD.59.026002.

[147] R. Harnik, G. D. Kribs, and G. Perez, A Universe without weak interactions, *Phys.Rev.* **D74**, 035006 (2006). doi: 10.1103/PhysRevD.74.035006.

[148] B. Gato-Rivera and A. N. Schellekens, GUTs without guts (2014).

[149] E. Witten, An SU(2) Anomaly, *Phys.Lett.* **B117**, 324–328 (1982). doi: 10.1016/0370-2693(82)90728-6.

[150] J. Halverson, Anomaly Nucleation Constrains SU(2) Gauge Theories, *Phys. Rev. Lett.* **111**, 261601 (2013).

[151] A. N. Schellekens, Electric Charge Quantization In String Theory, *Phys.Lett.* **B237**, 363 (1990). doi: 10.1016/0370-2693(90)91190-M.

[152] X.-G. Wen and E. Witten, Electric and Magnetic Charges in Superstring Models, *Nucl.Phys.* **B261**, 651 (1985). doi: 10.1016/0550-3213(85)90592-9.

[153] G. G. Athanasiu, J. J. Atick, M. Dine, and W. Fischler, Remarks On Wilson Lines, Modular Invariance And Possible String Relics In Calabi-Yau Compactifications, *Phys.Lett.* **B214**, 55 (1988). doi: 10.1016/0370-2693(88)90451-0.

[154] B. Gato-Rivera and A. N. Schellekens, Asymmetric Gepner Models: Revisited, *Nucl.Phys.* **B841**, 100–129 (2010). doi: 10.1016/j.nuclphysb.2010.07.

020.

[155] B. Gato-Rivera and A. Schellekens, Asymmetric Gepner Models III. B-L Lifting, *Nucl.Phys.* **B847**, 532–548 (2011). doi: 10.1016/j.nuclphysb.2011. 02.004.

[156] M. Maio and A. Schellekens, Permutation orbifolds of heterotic Gepner models, *Nucl.Phys.* **B848**, 594–628 (2011). doi: 10.1016/j.nuclphysb.2011. 03.012.

[157] E. Witten, Symmetry Breaking Patterns in Superstring Models, *Nucl.Phys.* **B258**, 75 (1985). doi: 10.1016/0550-3213(85)90603-0.

[158] L. B. Anderson, J. Gray, Y.-H. He, and A. Lukas, Exploring Positive Monad Bundles And A New Heterotic Standard Model, *JHEP.* **1002**, 054 (2010). doi: 10.1007/JHEP02(2010)054.

[159] M. Blaszczyk, S. Nibbelink Groot, M. Ratz, F. Ruehle, M. Trapletti, et al., A $Z_2 \times Z_2$ standard model, *Phys.Lett.* **B683**, 340–348 (2010). doi: 10.1016/ j.physletb.2009.12.036.

[160] M. Cvetic, I. Papadimitriou, and G. Shiu, Supersymmetric three family SU(5) grand unified models from type IIA orientifolds with intersecting D6-branes, *Nucl.Phys.* **B659**, 193–223 (2003). doi: 10.1016/j.nuclphysb. 2004.06.041.

[161] C. Quigg and R. Shrock, Gedanken Worlds without Higgs: QCD-Induced Electroweak Symmetry Breaking, *Phys.Rev.* **D79**, 096002 (2009). doi: 10. 1103/PhysRevD.79.096002.

[162] Z. Kakushadze and S. H. Tye, Three family SU(5) grand unification in string theory, *Phys.Lett.* **B392**, 335–342 (1997). doi: 10.1016/ S0370-2693(96)01543-2.

[163] J. Fuchs, A. Klemm, C. Scheich, and M. G. Schmidt, Spectra And Symmetries Of Gepner Models Compared To Calabi-Yau Compactifications, *Annals Phys.* **204**, 1–51 (1990). doi: 10.1016/0003-4916(90)90119-9.

[164] A. E. Faraggi, C. Kounnas, and J. Rizos, Chiral family classification of fermionic Z(2) x Z(2) heterotic orbifold models, *Phys.Lett.* **B648**, 84–89 (2007). doi: 10.1016/j.physletb.2006.09.071.

[165] J. F. Donoghue, The Weight for random quark masses, *Phys.Rev.* **D57**, 5499–5508 (1998). doi: 10.1103/PhysRevD.57.5499.

[166] J. F. Donoghue, K. Dutta, and A. Ross, Quark and lepton masses and mixing in the landscape, *Phys.Rev.* **D73**, 113002 (2006). doi: 10.1103/ PhysRevD.73.113002.

[167] L. J. Hall, M. P. Salem, and T. Watari, Quark and Lepton Masses from Gaussian Landscapes, *Phys.Rev.Lett.* **100**, 141801 (2008). doi: 10.1103/ PhysRevLett.100.141801.

[168] L. J. Hall, M. P. Salem, and T. Watari, Statistical Understanding of Quark and Lepton Masses in Gaussian Landscapes, *Phys.Rev.* **D76**, 093001 (2007). doi: 10.1103/PhysRevD.76.093001.

[169] W. A. Bardeen, On naturalness in the standard model (1995).

[170] H. Aoki and S. Iso, Revisiting the Naturalness Problem – Who is afraid of quadratic divergences? –, *Phys.Rev.* **D86**, 013001 (2012). doi: 10.1103/ PhysRevD.86.013001.

[171] B. W. Lynn, Spontaneously broken Standard Model (SM) symmetries and the Goldstone theorem protect the Higgs mass and ensure that it has no Higgs Fine Tuning Problem (HFTP) (2011).

[172] F. Jegerlehner, The hierarchy problem of the electroweak Standard Model revisited (2013).

[173] S. P. Martin, A Supersymmetry primer (1997).

[174] T. Asaka and M. Shaposhnikov, The nuMSM, dark matter and baryon asymmetry of the universe, *Phys.Lett.* **B620**, 17–26 (2005). doi: 10.1016/j.physletb.2005.06.020.

[175] K. A. Meissner and H. Nicolai, Conformal Symmetry and the Standard Model, *Phys.Lett.* **B648**, 312–317 (2007). doi: 10.1016/j.physletb.2007.03.023.

[176] M. Shaposhnikov and D. Zenhausern, Quantum scale invariance, cosmological constant and hierarchy problem, *Phys.Lett.* **B671**, 162–166 (2009). doi: 10.1016/j.physletb.2008.11.041.

[177] M. Heikinheimo, A. Racioppi, M. Raidal, C. Spethmann, and K. Tuominen, Physical Naturalness and Dynamical Breaking of Classical Scale Invariance (2013).

[178] G. Marques Tavares, M. Schmaltz, and W. Skiba, Higgs mass naturalness and scale invariance in the UV, *Phys.Rev.* **D89**, 015009 (2014). doi: 10.1103/PhysRevD.89.015009.

[179] N. Arkani-Hamed, S. Dimopoulos, and G. Dvali, The Hierarchy problem and new dimensions at a millimeter, *Phys.Lett.* **B429**, 263–272 (1998). doi: 10.1016/S0370-2693(98)00466-3.

[180] L. Randall and R. Sundrum, A Large mass hierarchy from a small extra dimension, *Phys.Rev.Lett.* **83**, 3370–3373 (1999). doi: 10.1103/PhysRevLett.83.3370.

[181] V. Agrawal, S. M. Barr, J. F. Donoghue, and D. Seckel, The Anthropic principle and the mass scale of the standard model, *Phys.Rev.* **D57**, 5480–5492 (1998). doi: 10.1103/PhysRevD.57.5480.

[182] S. Weinberg. Living in the multiverse. In ed. B. Carr, *Universe or Multiverse?*, pp. 29–42. Cambridge University Press (2005).

[183] M. R. Douglas, Statistical analysis of the supersymmetry breaking scale (2004).

[184] L. Susskind. Supersymmetry breaking in the anthropic landscape. In ed. M. Shifman, *From fields to strings*, vol. 3, pp. 1745–1749. World Scientific (2004).

[185] F. Denef and M. R. Douglas, Distributions of nonsupersymmetric flux vacua, *JHEP.* **0503**, 061 (2005). doi: 10.1088/1126-6708/2005/03/061.

[186] G. Coughlan, W. Fischler, E. W. Kolb, S. Raby, and G. G. Ross, Cosmological Problems for the Polonyi Potential, *Phys.Lett.* **B131**, 59 (1983). doi: 10.1016/0370-2693(83)91091-2.

[187] B. S. Acharya, G. Kane, and E. Kuflik, String Theories with Moduli Stabilization Imply Non-Thermal Cosmological History, and Particular Dark Matter (2010).

[188] G. Kane, P. Kumar, R. Lu, and B. Zheng, Higgs Mass Prediction for

Realistic String/M Theory Vacua, *Phys.Rev.* **D85**, 075026 (2012). doi: 10.1103/PhysRevD.85.075026.

[189] L. Ubaldi, Effects of theta on the deuteron binding energy and the triple-alpha process, *Phys.Rev.* **D81**, 025011 (2010). doi: 10.1103/PhysRevD.81. 025011.

[190] R. Peccei and H. R. Quinn, CP Conservation in the Presence of Instantons, *Phys.Rev.Lett.* **38**, 1440–1443 (1977). doi: 10.1103/PhysRevLett.38.1440.

[191] S. Weinberg, A New Light Boson?, *Phys.Rev.Lett.* **40**, 223–226 (1978). doi: 10.1103/PhysRevLett.40.223.

[192] F. Wilczek, Problem of Strong p and t Invariance in the Presence of Instantons, *Phys.Rev.Lett.* **40**, 279–282 (1978). doi: 10.1103/PhysRevLett.40.279.

[193] J. E. Kim, Light Pseudoscalars, Particle Physics and Cosmology, *Phys.Rept.* **150**, 1–177 (1987). doi: 10.1016/0370-1573(87)90017-2.

[194] P. Svrcek and E. Witten, Axions In String Theory, *JHEP.* **0606**, 051 (2006). doi: 10.1088/1126-6708/2006/06/051.

[195] F. Wilczek. A Model of anthropic reasoning, addressing the dark to ordinary matter coincidence. In ed. B. Carr, *Universe or Multiverse?*, pp. 151–162. Cambridge University Press (2004).

[196] A. D. Linde, Axions in inflationary cosmology, *Phys.Lett.* **B259**, 38–47 (1991). doi: 10.1016/0370-2693(91)90130-I.

[197] S. Hellerman and J. Walcher, Dark matter and the anthropic principle, *Phys.Rev.* **D72**, 123520 (2005). doi: 10.1103/PhysRevD.72.123520.

[198] M. Tegmark, A. Aguirre, M. Rees, and F. Wilczek, Dimensionless constants, cosmology and other dark matters, *Phys.Rev.* **D73**, 023505 (2006). doi: 10.1103/PhysRevD.73.023505.

[199] B. Freivogel, Anthropic Explanation of the Dark Matter Abundance, *JCAP.* **1003**, 021 (2010). doi: 10.1088/1475-7516/2010/03/021.

[200] A. Arvanitaki, S. Dimopoulos, S. Dubovsky, N. Kaloper, and J. March-Russell, String Axiverse, *Phys.Rev.* **D81**, 123530 (2010). doi: 10.1103/ PhysRevD.81.123530.

[201] B. S. Acharya, K. Bobkov, and P. Kumar, An M Theory Solution to the Strong CP Problem and Constraints on the Axiverse, *JHEP.* **1011**, 105 (2010). doi: 10.1007/JHEP11(2010)105.

[202] M. Cicoli, M. Goodsell, A. Ringwald, M. Goodsell, and A. Ringwald, The type IIB string axiverse and its low-energy phenomenology, *JHEP.* **1210**, 146 (2012). doi: 10.1007/JHEP10(2012)146.

[203] J. P. Conlon, The QCD axion and moduli stabilisation, *JHEP.* **0605**, 078 (2006). doi: 10.1088/1126-6708/2006/05/078.

[204] K.-S. Choi, H. P. Nilles, S. Ramos-Sanchez, and P. K. Vaudrevange, Accions, *Phys.Lett.* **B675**, 381–386 (2009). doi: 10.1016/j.physletb.2009.04. 028.

[205] P. Fox, A. Pierce, and S. D. Thomas, Probing a QCD string axion with precision cosmological measurements (2004).

[206] D. J. Marsh, E. Macaulay, M. Trebitsch, and P. G. Ferreira, Ultra-light Axions: Degeneracies with Massive Neutrinos and Forecasts for Future Cosmological Observations, *Phys.Rev.* **D85**, 103514 (2012).

[207] A. Ringwald, Exploring the Role of Axions and Other WISPs in the Dark Universe, *Phys.Dark Univ.* **1**, 116–135 (2012). doi: 10.1016/j.dark.2012.10. 008.

8

Local String Models and Moduli Stabilisation

Fernando Quevedo

ICTP, Abdus Salam International Centre for Theoretical Physics,
Strada Costiera 11, Trieste, Italy
DAMTP, Department of Applied Mathematics and Theoretical Physics,
University of Cambridge, Wilberforce Road, Cambridge, UK
f.quevedo@damtp.cam.ac.uk

A brief overview is presented of the progress made during the past few years on the general structure of local models of particle physics from string theory including: moduli stabilisation, supersymmetry breaking, global embedding in compact Calabi-Yau compactifications and potential cosmological implications. Type IIB D-brane constructions and the Large Volume Scenario (LVS) are discussed in some detail emphasising the recent achievements and the main open questions.

1. Introduction

The aim of string phenomenology is well defined and very ambitious: to uncover string theory scenarios that satisfy all particle physics and cosmological observations and hopefully lead to measurable predictions (for a comprehensive treatment of the field with a very complete set of references, see Ref. 1).

This defines a list of concrete challenges for string constructions that have been addressed over the years:

(1) *Gauge and matter structure of the Standard Model (SM).*
(2) *Hierarchy of scales and quark/lepton masses (including a proper account of neutrino masses).*
(3) *Realistic flavour structure with right quark (CKM), lepton (PMNS) mixings and right amount of CP violation, avoiding flavour changing neutral currents (FCNC).*
(4) *Hierarchy of gauge couplings at low energies potentially unified at high*

energy.

(5) *Almost stable proton but with a realistic quantitative account of baryogenesis.*

(6) *Inflation or alternative early universe scenarios that can explain the CMB fluctuations.*

(7) *Dark matter (but avoid overclosing).*

(8) *Dark radiation ($4 \geq N_{eff} \geq 3.04$).*

(9) *Dark energy (with equation of state $w = \rho/p \sim -1$).*

Addressing all of these issues has kept the string phenomenology community busy for several decades now. Partial success has been achieved on each of the points but not for all of them at the same time in particular classes of models. This list can be seen as a guideline for the present overview. Notice that in string theory, contrary to field theoretical model building, if a model fails with one of the above requirements it has to be ruled out.

The prospect of obtaining a proper ultraviolet complete extension of the Standard Model (SM) not only justifies efforts in this direction but provides for the first time a well defined alternative to the traditional bottom-up approach to model building beyond the SM that has very limited guidelines beyond experiment and that is currently being under pressure by the LHC results so far.

In order to address these issues, several approaches have been followed. Ideally the first attempts are to try as much as possible to extract 'generic model independent implications of string theory. Regarding the general string predictions relevant for our universe (see for instance the discussion in[2]) we can mention only very few:

- Gravity + dilaton + antisymmetric tensors + gauge fields + matter.
- Supersymmetry (SUSY) (with 32, 16 or less supercharges, but breaking scale not fixed).
- Extra dimensions (6 or 7) (flat, small, large, warped?).
- No (massless) continuous spin representations (CSR) in perturbative string theory.

These have to be compared with general model independent field theoretical predictions which are also very few (identity of particles, existence of antiparticles, relation between spin statistics, the CPT theorem and the running of physical couplings with energy, following the renormalisation group (RG) equations). The 4th prediction above is relatively less known,[3] it essentially states the fact that in perturbative string theory massive and massless representations of the Poincare group are linked to each other and since the massive representations are finite dimensional the same should be true for the massless representations, forbidding then the continuous spin representations.[a] Notice that being massless they could have been relevant at low energies and perturbative string theory indicates that they should not exist at least as perturbative string states, an statement consistent with all observations that has no clear explanation otherwise.

Having stated the general properties of string models we may concentrate on their general 4-dimensional implications. The most promising compactifications have $\mathcal{N} = 1$ SUSY that guarantees stability and chirality in the spectrum. The generic properties of 4D string compactifications are:

(1) Moduli: gravitationally-coupled scalar fields that usually measure size and shape of extra dimensions. They are massless as long as supersymmetry is unbroken.

(2) Antisymmetric tensors of different ranks implying the generic existence of axions, the possibility of turning on their fluxes in the extra dimensions and the generic appearance of branes that couple to these antisymmetric tensors and may even host the Standard Model.

(3) Matter appears on low dimensional group representations: (bifundamentals, symmetric, antisymmetric, adjoints).

(4) If the 4D theory has SUSY broken at the TeV scale the moduli tend to receive a mass of an order similar to the soft terms, implying the Cosmological Moduli Problem (CMP) (they overclose the universe or ruin nucleosynthesis upon late decay unless the mass of all the moduli can be made $m > 10$ TeV).

These generic predictions are very powerful and can be used to study general 'string inspired' scenarios to try to make contact with observations but are clearly not enough for a proper phenomenology and we have to consider concrete ways to explicitly build models. Over the years, two general classes of models have been studied.

• *Global string models*: 10D string theory compactified on 6D manifold. Gauge and matter fields in 4D come from gauge multiplets in 10D.

---

[a]Continuous spin representations (CSR) are representations of the little group for massless states of the Poincare group which is the Euclidean group in two dimensions. Eliminating these infinite dimensional unitary representations (arguing for instance that these particles have not been observed in nature[4]) limits to the subgroup $SO(2)$ with the standard helicity quantum number.

- *Local string models*: Standard model lives on a D-brane localised in some point in the extra dimensions.

The global models are essentially the heterotic models in which the gauge symmetry is already present in the 10-dimensional theory and upon compactification it may lead to chiral string models. Local models are essentially the type II string models in which the gauge and matter fields are localised on D-branes. The extreme case could be the Standard Model localised on a D3 brane which is just a point in the extra dimensions. This allows to separate the physics questions between the questions that can be addressed only by how the Standard Model fits inside the D-brane and the global questions that do depend on the full structure of the six or seven extra dimensions. This is known in the string literature as the 'bottom-up' approach to string model building.

Local Questions	Global Questions
Gauge Group	Moduli Stabilisation
Chiral Spectrum	Cosmological Constant
Yukawa Couplings	Supersymmetry Breaking
Gauge Couplings	Physical scales (unification, SUSY breaking, axions)
Proton Stability	Inflation or alternative, Reheating
Flavour issues (CKM. PMNS)	Cosmological Moduli Problem

The bottom-up approach is simply a systematic way to organise the challenge of realistic string model building. It is midway between traditional field theoretical model building and fully-fledged string constructions. Since the challenges are so big, it makes the search for realistic models more manageable asking a set of questions at the time and follow a modular approach to model building.[b] In contrast in global models such as the heterotic string all the physics questions have to be addressed at once. Heterotic models have other advantages such as starting already with a unified group like E_6. Local models at the end have to be fully embedded into a complete compactification that is not straightforward. Both approaches have been followed with different amounts of success.

Over the years the main obstacles for realistic string model building have been, more than getting precisely the Standard Model spectrum at low energies, the stabilisation of moduli (which otherwise would source

---

[b]At this stage this is only a convenient computational strategy rather than physically motivated. That will come later in this article.

unobserved long-range interactions) and supersymmetry breaking. These are the questions that we will address next.

2. Moduli Stabilisation

This model independent sector for string compactifications is usually composed of the following fields: The axio-dilaton S, the complex structure moduli or size of the non-trivial 3-cycles U_a and the Kähler moduli measuring the size of the non-trivial 4 and 2 cycles: T_i. Being scalar fields, their vevs have to be specified dynamically. The simplest supersymmetric compactifications leave them unspecified. The effective field theory depending on the Kähler potential K, superpotential W and gauge kinetic function f are such that the corresponding scalar potential is flat and here the SUSY non-renormalisation theorems protect the flatness of their potential perturbatively. Therefore there are limited sources of their scalar potential:

(1) Fluxes of Antisymmetric tensors to generate a non-vanishing tree-level superpotential.
(2) Non-perturbative corrections to the superpotential W.
(3) Perturbative (and non-perturbative) corrections to the Kähler potential K in both α' and string-loop expansions.
(4) Induced D-terms.

We will use all of them.[c] Fluxes are usually complicated to deal with and it took many years before people learned to manage them.[6,7] The main reason is that they tend to change the structure of the compact manifold in such a way that there is no much mathematical understanding on the compactification manifold. Fortunately the case of IIB compactifications is such that the manifold after flux compactifications is conformal to the well studied Calabi-Yau manifolds and this is one of the main reasons these compactifications have attracted much attention in the past decade. This is just a computational rather than conceptual advantage and from string dualities we know all other string compactifications should lead to similar physics once the technical aspects are sorted out. Since the spectrum of IIB supergravity has two 3-index antisymmetric tensors, fluxes on 3-cycles are able to fix all the U_a fields and the axio-dilaton through a flux superpotential

---

[c]Notice that the simplest constructions in which all these effects are neglected are not viable since neglecting all these effects is either inconsistent or very non-generic (setting by hand all fluxes to zero for instance). It is good news that including all available sources lead to more realistic physics.

$W_0(S, U)$. Naively the fluxes of a three form field strength H_3 tend to fix the size U of a three-cycle γ by the quantisation condition $\int_\gamma H_3 = 2\pi n$, this effect is captured in the EFT by the flux superpotential $W_0(U, S)$.[d] The perturbative superpotential cannot depend on the T fields since their imaginary components are axion-like fields having a perturbative Peccei-Quinn shift symmetry: $\mathrm{Im}T_i \to \mathrm{Im}T_i + c_i$ and the holomorphicity of W would then not allow dependence on the full superfield T_i. Therefore they can only appear in W through non-perturbative effects.

$$W_{np} = \sum_i A_i e^{-a_i T_i} \tag{1}$$

in which the A_i may be functions of other moduli or even matter fields.

Combining this with the flux superpotential gives the full $W = W_0 + W_{np}$ which combined with the corrections to K are able to fix all moduli. This has been done in practice for only a handful of models.

The scalar potential derived from the general $\mathcal{N} = 1$ supergravity expression $V = V_F + V_D$, with:

$$V_F = e^K \left[K^{I\bar{J}} D_I W D_{\bar{J}} \bar{W} - 3|W|^2 \right] \tag{2}$$

where $K^{I\bar{J}}$ is the inverse of the Kähler metric $K_{I\bar{J}} = \partial_I \partial_{\bar{J}} K$ and $D_I W = \partial_I W + W \partial_I K$ is the Kähler covariant derivative. The D-term part of the salar potential is:

$$V_D = \frac{1}{\mathrm{Re}f} \left(\xi_{FI}(T) + K_\Phi T \Phi \right)^2 \tag{3}$$

where $\xi_{FI} \sim \partial K / \partial T$ are the (misnamed) field-dependent Fayet-Iliopoulos terms, only present for abelian groups, Φ a matter field transforming under the corresponding gauge group and T are the corresponding generators (charges in the case of a $U(1)$). Gauge indices suppressed.

Concentrating on the moduli dependence, the typical shape of the moduli scalar potential takes the form:

$$V_F \propto \left(\frac{K^{S\bar{S}}|D_S W|^2 + K^{a\bar{b}} D_a W \bar{D}_{\bar{b}} \bar{W}}{\mathcal{V}^2} \right) + \left(\frac{A e^{-2a\tau}}{\mathcal{V}} - \frac{B e^{-a\tau} W_0}{\mathcal{V}^2} + \frac{C|W_0|^2}{\mathcal{V}^3} \right) \tag{4}$$

Here $\tau = \mathrm{Re}T$ represents a typical T modulus, with $\mathcal{V}$ the overall volume (function of the T fields) and the potential is meant to be seen as

---

[d]More explicitly the flux superpotential takes the form $\int G_3 \wedge \Omega$ where $G_3 = H_3 + iSF_3$ with H_3, F_3 the two 3-form field strengths of the two stringy 2-form potentials. Here Ω is the unique $(3, 0)$ form that exists for every CY manifold. Expanding Ω in a basis of three-forms generates a superpotential dependence on the U_a fields.

an expansion in large volume, where the effective field theory treatment is justified. In this case the first terms in parentheses are of order $1/\mathcal{V}^2$ and being positive definite they have to vanish at the minimum, imposing $D_S W = D_a W = 0$ and therefore fixing S and U_a generically. This in turn fix the values of W_0 at the minima which is a huge distribution of values but mostly fitting in the range $0.1 \leq |W_0| \leq 100$. The second parentheses is not positive definite and depending on the signs of the coefficients A, B, C it gives a minimum for the Kähler moduli T. In particular the sign of C depends on the sign of the Euler number of the Calabi-Yau manifold, by mirror symmetry half of them have negative Euler number and then positive C implying a minimum at volumes of order

$$\mathcal{V} \sim e^{a\tau} \quad \text{with} \quad \tau \sim \text{Re } S \sim 1/g_s > 1. \tag{5}$$

Implying an exponentially large volume. This gives rise to the LARGE volume scenario or LVS.[8] For very particular values of W_0, the large number of solutions allows for a few of them to satisfy $|W_0| \ll 1$. In this case W_0 can be tuned so that $W_0 \sim W_{np} \sim e^{-a\tau}$ so the term proportional to C can be neglected and a minimum can be found for τ. This is essentially the KKLT scenario.[9] But for generic values of W_0 only the LVS works. It has been shown that this holds as long as the number of 3-cycles is larger than the number of 4-cycles and both greater than one ($h_{12} > h_{11} > 1$) which is satisfied for half of the CY manifolds by mirror symmetry. The second condition is the existence of at least one collapsible 4-cycle which is the generic case.

In both KKLT and LVS the position of the minimum is at negative values of V_F so leading to AdS vacua. The main difference is that in KKLT this minimum is supersymmetric ($D_T W = 0$) but in LVS supersymmetry is broken. They both have small parameters in which to base the approximate effective theory, $1/\mathcal{V}$ for LVS and W_0 for KKLT. Notice that we have not yet used the D-term part of the scalar potential. This is more model dependent since it depends on charged matter fields for which we need to specify the concrete model. Being positive definite it will tend to uplift the minimum found from purely V_F.[10] However for KKLT there is a strong restriction, since the minimum is supersymmetric, it means all F-terms vanish which in turn imply that the D-terms have to vanish. In LVS D-terms can lift the minimum opening the possibility of leading to de Sitter space. But this is very model dependent and there is a need to have a full global compactification with all matter fields to make a proper study. This will be addressed later on in this article.

Another way to uplift the minimum to de Sitter was proposed in KKLT by introducing anti D3 branes at the tip of a warped region in the compact manifold. This provides a positive contribution to the vacuum energy given by the warped brane tension and an explicit supersymmetry breaking source. The effective field theory is more complicated to handle since it leaves the regime of validity of $\mathcal{N} = 1$ supergravity. For LVS this is also an option but D-terms (and matter F-terms) provide a more promising avenue to obtain de Sitter within a purely supersymmetric effective action.[e]

In LVS there is a clear hierarchy of scales shown in the table below.

Table 1.   Relevant physical scales in LVS.

Physical scale	Volume dependence
Planck mass	M_P
String scale	$M_s = \frac{M_P}{\mathcal{V}^{1/2}}$
Kaluza-Klein scale	$M_{KK} = \frac{M_P}{\mathcal{V}^{2/3}}$
Gravitino mass	$m_{3/2} = \frac{M_P W_0}{\mathcal{V}}$
Volume modulus mass	$m_V = \frac{M_P W_0}{\mathcal{V}^{3/2}}$

Notice that a clear bound for W_0 is $|W_0| \ll \mathcal{V}^{1/3}$ in order to have a proper hierarchy ($M_{KK} \gg m_{3/2}$) and guarantee the consistent use of an effective field theory to describe the physical implications of the scenario (an even stronger bound is $|W_0|^{1/6}$ guarantees the effective potential being smaller than M_{KK}^4). Also even though the gravitino mass is supposed to set the scale of all particles that receive a mass after SUSY breaking, all moduli S, U_a and most of T_i receive a mass of the order of the gravitino mass, however the overall volume modulus has a mass much lighter and remains small after quantum effects even though it is not protected by supersymmetry.[34] Furthermore, in some Calabi-Yau manifolds which happen to be fibrations of 4D manifolds such as K3, the corresponding modulus does not receive a mass until loop effects are taken into account and therefore their mass is even smaller than that of the volume modulus ($m \sim W_0/\mathcal{V}^{5/3}$).

---

[e]For other interesting proposals to obtain de Sitter from purely supersymmetric EFTs see Ref. 11.

Some general properties of LVS:

- *Stability.* Even though the overall minimum is locally stable the fact that even the AdS vacuum is not supersymmetric makes it subject to non-perturbative instabilities, such as bubble of nothing decay. This was studied in.[12] As long as the effective field theory is valid the AdS minimum is stable and no indication to a bubble of nothing decay. This leads to the possibility of having a CFT dual and therefore a proper non-perturbative description of these vacua despite being non-supersymmetric. The dS minima are clearly metastable and the decay rate goes like $\Gamma \sim e^{-\mathcal{V}^3}$. The probability to decay to an AdS minimum is preferred over a dS as a ratio $P_{dS}/P_{ads} \sim e^{-\mathcal{V}}$ whereas its decay towards the 10D decompactification vacuum ($\mathcal{V} \to \infty$) is further suppressed $P_{dec}/P_{dS} \sim e^{-\mathcal{V}^2}$. Clearly the larger the volume the more stable the vacuum.

- *Bounds on the volume.* However the volume cannot be arbitrarily large since for values $\mathcal{V} \sim 10^{30}$ the string scale becomes smaller than the TeV scale, also beyond $\mathcal{V} \sim 10^{15}$ the gravitino mass (and usually soft terms) will be smaller than the TeV scale. Finally for volumes $\mathcal{V} \geq 10^9$ the volume modulus becomes lighter than 10 TeV which would lead to the cosmological moduli problem (CMP). Smaller volumes $10^3 < \mathcal{V} \geq 10^8$ are consistent and survive overclosing (with the larger volumes being the more stable from the previous item) but still imply a special cosmological role for the volume modulus (or any lighter one in particular cases). This modulus is the latest to decay and its decay would be the source of reheating of the observable universe leading to interesting post-inflationary cosmology (see for instance Ref. 16).

- *Inflation.* The three terms in the second parentheses for V_F hint at a concrete realisation of inflation. Assuming the volume is already at its minimum value, the potential for τ is precisely of the form $A - Be^{-x}$ for large values of τ which is one of the preferred inflationary potentials for a canonically normalised inflaton field x. In order to achieve this concretely at least three T_i fields are required which is very generic in string compactifications. Loop corrections may destabilise the flatness of the potential during inflation. A more elaborated and stable under quantum corrections model of inflation has been proposed in which the inflaton is a fibre modulus. For this scenario the spectral index and tensor to scalar ratio $r \sim 10^{-3}$ falls just in the preferred Planck regime (see Ref. 14 for a recent overview). However, if the recent results from

BICEP are confirmed $r \sim \mathcal{O}(0.1)$ then these scenarios are ruled out by experiment. An example on how string scenarios can be predictive and contrasted with experiment. The string scenarios consistent with BICEP: N-flation, axion monodromy and Wilson line inflation[15] can be embedded in the LVS. More work in this direction is needed.

- *Axions.* There are plenty of axions in string compactifications, many can survive at low energies but some do not. In LVS it is clear that the axion partners of the dilaton and Kähler moduli stabilised by non-perturbative effects acquire a mass of order the gravitino mass. Other axions are eaten by anomalous $U(1)$s by the Stuckelberg mechanism. But some survive at very low energies, in particular the axion partner of the volume modulus is essentially massless after moduli stabilisation and may have some implications for late time cosmology. In particular contributong to dark radiation. Also axions coming from phases of matter fields may survive low energies but are more model dependent (see Ref. 17 for a recent overall review on stringy axions).

Before finishing this section let us also mention the progress made in other string compactifications. In heterotic strings despite many efforts there is no yet a compelling scenario for stabilisation of all moduli. Substantial progress has been made in the past few years to stabilise most of them. Contrary to IIB strings that have two 3-index antisymmetric tensors to turn on, in the heterotic there is only one and so fluxes are not as efficient as they are in IIB strings. The number of equations and unknowns is similar leaving no room for a landscape and no mechanism to solve the cosmological constant problem. Furthermore they move the model away from the Calabi-Yau spaces. Yet, since heterotic strings carry a large gauge group already before compactification, this introduces new moduli (called 'bundle moduli') that can actually help to fix the complex structure moduli by consistency gauge conditions that have to be satisfied. Nonperturbative effects still can fix the Kähler moduli, similar to the IIB case. Clearly further progress is expected in this direction since for realistic model building heterotic models are probably the most developed.

G2-holonomy manifolds compactifications of the 11D supergravity limit of M-theory have been studied also. There is no explicit model of particle physics from these compactifications which need much more mathematical developments although they are also clearly local models also. For moduli stabilisation there are interesting properties that can be extracted without entering into details. In particular all moduli are similar to the T moduli of type IIB, making the moduli stabilisation issue easier to define. Fluxes are not an option in this formalism and therefore the full superpotential is non-perturbative. A superpotential of the form $W = \sum_{ij} A_i e^{-a_{ij} T_j}$ has been proposed with the potential to fix all T fields. The general properties of this scenario have been summarised in Ref. 25.

3. Supersymmetry Breaking

The breaking of supersymmetry is intimately related to moduli stabilisation. This explains that only after a well defined framework for moduli stabilisation it was possible to extract information about supersymmetry breaking in string theory. Progress in moduli stabilisation eventually extends to progress in supersymmetry breaking. Contrary to moduli stabilisation in which to large extent the location of the standard model can be ignored, here it is fundamental. In IIB local models the standard model can be inside a D3 brane at a singular point in the extra dimensions or at a D7 brane wrapping a 4-cycle of the extra dimensions. Other odd dimensional branes are dual to these and even dimensional branes would appear in the IIA case.

What we can see from both LVS and KKLT[f] is that both the S and U_a fields do not break supersymmetry at leading order since they are fixed by the condition $D_S W = D_a W = 0$.

An important information is regarding the contribution to supersymmetry breaking of the cycle where the standard model lives on the D7 brane case. On the D3 brane case this may also be thought as a collapsed 4-cycle. The point is the following. In any brane sector where there is chiral matter, the corresponding T modulus, measuring the size of the 4-cycle that the D7 brane is wrapping, acquires a charge under an anomalous $U(1)$. Therefore it is not possible to have a term of the form $W_{np} = A e^{-a T_{SM}}$ in the superpotential in which A is a constant, since this term would not be gauge invariant. Therefore whenever a dependence on T appears it has to come together with a dependence of A on charged matter fields that compensate the gauge variation of T_{SM}. This makes it very difficult to stabilise T_{SM} by the LVS or KKLT methods since the SM fields are supposed to have zero vev at the high scales (otherwise they may induce colour breaking for instance). Let us call this the BMP constraint.[26] As long as one of the

---

[f]For KKLT even though the source of SUSY breaking is the uplift by anti D3 branes a very interesting scenario called mirage mediation has emerged with interesting phenomenological implications.[18]

moduli T_h from a hidden sector (in which chirality is not required) appears in $W_{np} = Ae^{-aT_h}$ with constant (or only moduli dependent) A the LVS minimum is obtained. The T_{SM} cycle may be fixed by D-terms or even by loop corrections to V.

Some general properties of SUSY breaking in LVS are

(1) The source of SUSY breaking is well identified coming from generic values of the 3-form fluxes for which $D_S W = D_a W = 0$ but $W_0 \neq 0$ and the F- term of the Kähler moduli is non-vanishing. This is major difference as compared to KKLT in which before uplifting SUSY is not broken and its breaking is fully determined by the anti-D3 brane that performs the uplift. In LVS the uplift can be done in different ways but even if the anti-brane is added, its contribution to soft terms is usually negligible.

(2) The existence of the landscape allows for the first time to address simultaneously the cosmological constant and the hierarchy problems. This justifies the standard strong assumption that has been made over the years regarding the use of low-energy SUSY to address the hierarchy problem: that something else takes care of the cosmological constant problem and has no direct influence on the calculation of soft SUSY breaking terms (*good*). But, by the nature of the landscape, it prevents us to find new physical phenomena at low energies determined by the cosmological constant (*bad*). It also opens the possibility to use anthropic arguments to address the hierarchy problem (*ugly*).

(3) SUSY is broken by the Kähler moduli which do not enter in the tree-level matter superpotential, therefore as long as Kähler and complex structure moduli do not mix (true at tree-level Kähler potential) then flavour problems, generic for gravity mediation, are ameliorated.[19,27] The correct estimate on how much flavour violation is induced by quantum corrections mixing the moduli is an open question.

(4) The dominant source of SUSY breaking in the EFT is the F-term of the volume modulus. But this gives rise, to leading order, a no-scale model with vanishing soft terms. Therefore next order corrections are relevant. In order to explicitly compute the soft terms requires knowledge of the F-term of the cycle that hosts the SM. If its F-term vanishes (to avoid the BMP obstruction) then soft terms can be very much suppressed $(M_{1/2} \sim \mathcal{O}(m_{3/2}/\mathcal{V}))$. Otherwise they are proportional to the gravitino mass up to a small loop factor. We list the different scenarios in the table below. Notice that if we use a TeV gravitino mass, it would

select one of the first two scenarios. The second one needs a strong fine-tuning in W_0 in order to simultaneously obtain the unification and SUSY breaking scales at the preferred values. The first one does not need the tuning at the cost of lowering the unification scale. Both suffer from the cosmological moduli problem (CMP). If we use the avoidance of this problem as the selection criterion then the last three scenarios are preferred. The first one of those gives up a natural explanation of the TeV scale. The last two are sequestered scenarios in which the F term of the SM modulus vanishes to avoid the BMP obstruction. Sequestered scenarios may be subject to modifications due to quantum corrections and at the moment only models in which the SM is at D3 branes at singularities seem to remain truly sequestered. If so then both TeV SUSY breaking and the preferred GUT scale can be obtained without the CMP.

Table 2.   SUSY Breaking Scenarios in LVS.

Name	String Scale	W_0	$m_{3/2}$	Soft masses	CMP
Intermediate Scale	10^{11} GeV	$\mathcal{O}(1)$	1 TeV	$M_{soft} \sim 1$ TeV	Yes
Tuned GUT Scale	10^{15} GeV	10^{-10}	1 TeV	$M_{soft} \sim 1$ TeV	Yes
Generic GUT Scale	10^{15} GeV	$\mathcal{O}(1)$	10^{10} GeV	$M_{soft} \sim 10^{10}$ GeV	No
Sequestered Unsplit	10^{15} GeV	$\mathcal{O}(1)$	10^{10} GeV	$M_{soft} \sim \frac{m_{3/2}}{\mathcal{V}} \sim 1$ TeV	No
Sequestered Split	10^{15} GeV	$\mathcal{O}(1)$	10^{10} GeV	$M_{1/2} \sim \frac{m_0}{\mathcal{V}^{1/2}} \sim \frac{m_{3/2}}{\mathcal{V}} \sim 1$ TeV	No

The intermediate scale scenario and the tuned GUT scenario have been studied in the past with some detail.[19] Despite the fact that they both have the CMP the soft terms can be calculated in a more explicit way since the dominant contribution comes from the F-term of the SM cycle. The generic GUT scale scenario has not been studied in detail since the superpartners are much heavier than the TeV scale and no hope to be detected not even in the long term. This scenario however realises explicitly the large SUSY scale proposals recently discussed[20,21] in which the hierarchy problem is not solved by low-energy supersymmetry but fits with the measured value of the Higgs mass. The two sequestered scenarios proposed in Ref. 22 are very interesting because they both have the preferred GUT scale while at the same time superpartners have the TeV scale that solve the hierarchy problem. However since the SM cycle does not break SUSY then the explicit expressions for the soft terms are difficult to compute since they are small compared to the gravitino mass and they are more model dependent. In particular they depend on the uplifting mechanism which gives negligible contributions in the other scenarios. Therefore these scenarios have not been studied in detail. See however Ref. 23.

4. Local and Global Model Building

One of the implications of the LVS is that the standard model has to be localised. The reason is that if it lives on a D7 brane wrapping a four-cycle, this cycle cannot be the one dominating the volume, since the volume is exponentially large and the gauge coupling of the gauge theory living on the brane is inversely proportional to the size of the cycle $g_{SM}^{-2} \sim \mathrm{Re}T_{SM}$ and would generically be too small to fit realistic values $\mathcal{O}(20)$ expected at the GUT scale. Therefore either the SM lives on a D7 brane wrapping a small cycle or at a D3 at a singularity. In both cases it is localised. This provides an independent argument to consider local string models in IIB compactifications. Notice that local F-theory models can be seen as strong-coupling generalisations of magnetised D7 brane models and in principle also fit with this analysis.

We will restrict here to local models at singularities (for a more comprehensive discussion see the nice review[28] in which both local F-theory models and branes at singularities are reviewed). An argument to justify this selection is the following. Since the SM is chiral the corresponding modulus T_{SM} is usually charged under anomalous $U(1)$ groups on the brane. This has two important implications, first as mentioned before there is the BMP obstruction to fix T_{SM} from nonperturbative effects. Second the corresponding $U(1)$ group has a Fayet-Iliopoulos (FI) term $\xi \propto \mathrm{Re}T$. The corresponding D-term potential

$$V_D \propto \left(\xi - q_i|\phi_i|^2\right)^2 \tag{6}$$

combined with soft mass terms induced for matter fields $m_i^2|\phi_i|^2$ tend to prefer $\xi = 0$ and then towards a collapsed cycle $\mathrm{Re}T_{SM} \to 0$. This implies that the effective field theory (EFT) valid for large values of moduli (compared with the string scale) is not valid and a different EFT has to be used valid as an expansion around the singularity. Fortunately this is also known for orbifold-like singularities in which the FI term is also proportional to the size of the blow-up mode $\xi \propto \rho_{SM}$ and again the D-term minimisation tends to prefer the collapsed cycle $\rho_{SM} \to 0$.[g] Quantum corrections may

---

[g]Notice that EFTs can be written in the two regimes in which the cycle is either much larger than the string scale or very close to zero. It is a complicated open question to

blow-up the singularity to non-vanishing values of the blow-up mode but in general keeping it within the singularity regime (*i.e.* $\rho_{SM} \le l_s^4$) with l_s the string length scale.

This argument is however not a proof that the SM has to live at a singularity since there are two ways out: the FI term is model dependent and usually is a linear combination of moduli fields. There may be a way to engineer the models so that these combinations vanish with non-vanishing fields (see for instance Ref. 37). Furthermore the soft terms contributions to the matter fields ϕ_i may be tachyonic and then $\phi_i \ne 0$ at the minimum.

Local models of D-branes at singularities have been studied over the years.[30-32] But it is only until very recently that they have been systematically embedded in compact Calabi-Yau compactifications including moduli stabilisation[35] (see also Ref. 36).

Let us start with the fully local constructions first and discuss the global embedding later. The gauge theory of branes at singularities can be described by quiver diagrams with nodes and arrows, node i represents n_i D-branes implying a group $U(n_i)$ and arrows going from the i node to the j node represents a bifundamental $(n_i, \bar{n}_j)$. Usually a closed loop in a quiver represents a gauge invariant superpotential term although the precise structure of the superpotential needs further techniques based on dimer diagrams that we will not discuss here.[32] A simple example is provided by the $\mathbb{Z}_3$ singularity with a triangular quiver and three arrows connecting the nodes. Choosing $n_i = n_j$ guarantees an anomaly free model (except for an anomalous $U(1)$ that become massive from the Stuckelberg mechanism) which can be easily evaluated by counting the number of arrows coming in and out each node which should match for anomalies to be canceled. We show in the figure the simplest of these cases including the SM which corresponds to $n_i = 3$ and is precisely the trinification model $SU(3)^3$ with three families.

$$3\left[(3, \bar{3}, 1) + (1, 3, \bar{3}) + (\bar{3}, 1, 3)\right] \tag{7}$$

which is actually three families of 27s of E_6.

This simple orbifold singularity is actually the simplest of a very special class of singularities called del Pezzo n or dP_n singularities. These singularities are actually collapsed del Pezzo surfaces which are four-dimensional surfaces (which in our case are four-cycles inside a Calabi-Yau space). These are defined as the complex 2-dimensional projective space $\mathbb{P}_2$ blow up at n points with $n = 0, 1 \cdots 8$. The $\mathbb{Z}_3$ singularity corresponds to dP_0. The

find the matching between the two EFTs going through the domain of string scale cycle.

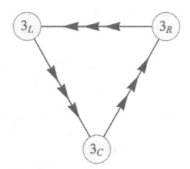

Fig. 1. The simplest realisation of a chiral model at D3 brane singularities containing the SM (the trinification model at dP_0 in this case). The three gauge groups are identical, $SU(3)^3$. The three arrows imply three families of bi-fundamental fields. The equal number of ingoing and outgoing arrows at each node guarantees anomaly cancellation.

special property of the del Pezzo surfaces is that they are the simplest 4-cycles that can collapse to a single point in a Calabi-Yau manifold. The quiver diagrams and superpotential couplings have been uncovered on the past years. Leading to phenomenological considerations. The corresponding quiver diagrams have $n + 3$ nodes. For instance the next case, the dP_1 singularity has a square quiver. Furthermore given a vev to a bifundamental between two quivers merge the two nodes into one and reproduces the dP_0 case. This can be interpreted geometrically as the fact that dP_n has $n + 1$ non-trivial 2-cycles and the higgsing corresponds to an independent collapse of one of these cycles.

Other more general classes of singularities have been studied (called toric singularities) that provide a large number of local string models with chiral matter content.

Some of the general properties of these local models are the following:

- The gauge coupling at the singularity is given at tree level by the vev of the dilaton field S so there is unification without necessarily having a simple GUT group. The gauge group is usually a product of simple groups. This avoids the standard problem of D-brane models for which having simple GUT groups leads to vanishing top Yukawas (which has been the main argument to consider F-theory models). Top Yukawas are easily generated if the groups are not simple.

- The maximum number of families (arrows) happens to be three, which is the one observed in nature. However starting with a complicated enough quiver and by higgsing more (but not many) families may be obtained.

- There is always one zero eigenvalue in the spectrum. In the case of dP_0 the massive states have eigenvalues $(M, M, 0)$ where the mass M is determined by the vev of the low-energy Higgs field. Having two families degenerated and one massless is not realistic. However all other dP_ns with three families have eigenvalues $(M, m, 0)$ with $m \ll M$ which have the proper hierarchy observed in nature. Quantum effects are expected to lift the zero-eigenvalue although this is not straightforward.[10]

- Apart from the overall $U(1)$ that can be simply decoupled, there are two anomalous $U(1)$s that obtain their mass by the Stuckelberg mechanism.

The rest of the physics has to be extracted in a model by model case. A large increase on the number of these models comes from the possibility to add D7 branes to each configuration. These allow for many more choices of integers n_i (for instance in dP_0 we can have $(n_1, n_2, n_3) = (1, 2, 3)$ giving the SM gauge group from D3 branes times a hidden or flavour symmetry coming from the D7 branes which cancel the anomalies). The number of D7 branes is restricted by tadpole cancellations which usually are equivalent with anomaly cancellation. This enhance substantially the number of realistic models at singularities.

Quasi realistic models including the SM gauge and matter fields as well as proper Yukawas, CKM and PMNS mixings, proton stability, and unification have been constructed using D3/D7s at singularities. In particular dP_3 models have proved to be manageable enough and at the same time rich enough to address the flavour questions.[32] The issue of gauge coupling unification in these models is no much if there is unification (up to small thresholds as usual) which is automatically achieved but knowing that the gauge couplings are unified how do they evolve to low energies. Since usually the matter content of the models is not just that of the MSSM then the success of the MSSM for gauge unification would be most probably an accident in this class of models (for good or for bad as in most string constructions). Achieving the right values of the couplings at low energies is not easy. There is one simple local model that stands out in this regard. This is the one based on $(n_1, n_2, n_3) = (3, 2, 2,)$ giving rise to a left-right symmetric model $SU(3) \times SU(2)_L \times SU(2)_R \times U(1)_{B-L}$ with three families and three extra pairs of Higgses (after D7-D7 states get vevs):

$$3[(3, \bar{2}, 1) + (\bar{3}, 1, 2) + (1, 2, 1) + (1, 1, \bar{2}) + (1, 2, \bar{2})] + \text{singlets} \qquad (8)$$

The extra matter content together with the fact that the hypercharge normalisation is not standard combine in a way that there is unification with a similar level of precision as the MSSM. The unification scale though is intermediate 10^{11} GeV approximately.[29,30] In order to get realistic Yukawas this model can be embedded into higher dP_ns.

This model built from both D3s and D7s and the trinification model built from only D3s are examples of very simple quasi-realistic chiral models. But being purely local they cannot be called string vacua, since in their construction compactification was not considered. In order to become honest-to-God string compactifications the corresponding singularity has to be embedded in a compact Calabi-Yau manifold (to preserve supersymmetry and chirality). Actually starting from IIB string theory and compactifying on a CY manifold leads to $\mathcal{N} = 2$ supersymmetry and then non-chiral models. The missing ingredient is *orientifolding*. This is essentially a $\mathbb{Z}_2$ twist of the CY compactification exploiting the fact that the worldsheet theory is orientable and has a $\mathbb{Z}_2$ symmetry. It is well known that combining this twist with a CY compactification leads to chiral $\mathcal{N} = 1$ theory in 4D.

A concrete way to embed local singularity models in fully-fledged CY compactifications was outlined in Ref. 34. The idea is to look for CY compactifications with at least three dP_n surfaces. Two of them map to each other under the $\mathbb{Z}_2$ orientifold twist, where the SM would live and a third one to provide the non-perturbative correction to the superpotential. A fourth 4-cycle (not a dP_n) will be the one dominating the volume. This is the minimum set-up. In the figure below we illustrate it.

In the past 2 years concrete realisations of these models were achieved.[35] The details are too technical for this review but it is worth emphasising the main points:

(1) A classification of Calabi-Yau manifolds constructed as hypersurfaces from toric varieties is available from the work of Kreuzer and Skarke.[33] From this large class of models a classification of those with a relatively small number of Kähler moduli (4 and 5) to fulfill the requirements. This gives a few thousand models of which a couple of hundred have dP_n surfaces mapped into each other under a $\mathbb{Z}_2$ (illustrated in the figure)

(2) A configuration of D3 and D7 branes as well as orientifold planes (fixed under orientifold action) is introduced satisfying a highly non-trivial set of consistency conditions: tadpole cancellations (local and global) for all

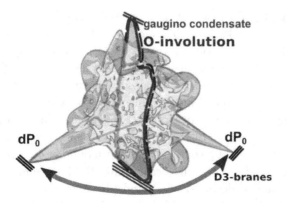

Fig. 2.   An explicit global embedding of local D-brane models. The SM is located at a dP_0 mapped to an identical singularity by a $\mathbb{Z}_2$ twist. Two ther 4-cycles are needed to stabilise Kähler moduli and obtain a global realisation of LVS with chiral matter.

D-brane charges, cancellation of the so-called Freed-Witten anomalies that appear in the presence of fluxes, K-theory charges, etc.

(3) The set-up allows for the SM to be hosted at the singularities on the dP_n's mapped into each other. Realising globally the local examples mentioned before.

(4) Furthermore, the conditions for moduli stabilisation are realised with non-perturbative superpotential generated at the third dP_n cycle (either by Euclidean D3 branes or gaugino condensation.

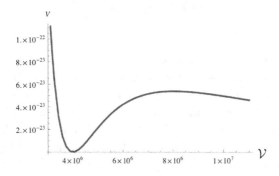

Fig. 3.   de Sitter minimum for a compact fluxed CY compactification with chiral matter at a dP_0 singularity.

(5) The modulus corresponding to the SM cycle is stabilised by D-terms in a way that it is naturally stabilised at zero and therefore in the singu-

larity regime. This part of the potential takes the form $V(T_{SM}, \phi) = V_D + V_{soft}$ with $V_D \sim (\xi_{FI} + q|\phi|^2)^2$ and $V_{soft} \sim m^2|\phi|^2$. It is clear that as long as the soft masses for matter fields ϕ are non-tachyonic this fixes the minimum (at leading order) at $\phi = 0$ and $\xi_{FI} = 0$ since both terms in the potential are positive definite. This is only leading order since these terms are of order $1/\mathcal{V}^2$ so small vevs are generated that compete with next order in $1/\mathcal{V}$ expansion which is the $1/\mathcal{V}^3$ that precisely gives rise to the LVS. These, together with the F-term of the matter fields ϕ end up contributing a term of order $\delta V = cW_0^2/\mathcal{V}^\alpha$ with $1 \leq \alpha < 3$ and $c > 0$. This term uplifts the LVS minimum to higher values including de Sitter space and the superpotential can be tuned to give an almost Minkowski vacuum. Notice that this tuning of W_0 is easy to achieve knowing that the large number of complex structure moduli allows for order 10^{500} solutions for values of W_0 in the small range $0.1 < W_0 < 10^3$.

(6) Having such explicit CY compactifications allows also to explicitly solve the complex structure and dilaton equations $D_U W = D_s W = 0$. However technically this is a huge number of solutions and it has not been done for fluxes in all the 3-cycles. Nevertheless some CYs admit enough discrete symmetries that allow to fix most of the complex structure moduli and only using fluxes in a handful of them. This has been done recently in Refs. 38 and 39. Then we have for the first time a global CY compactification[39] with quasi-realistic visible sector including the SM, with all geometric moduli stabilised leading to de Sitter space using a fully supersymmetric EFT. Supersymmetry broken with computable soft terms.

As a non-trivial check of 'phenomenological consistency' the local LR symmetric model sketched before for a dP_0 singularity has an interesting phenomenological property: having the LR spectrum plus three pairs of Higgses, together with the normalisation of the hypercharge gives precise unification at an intermediate scale 10^{11} GeV with an accuracy similar to that one for the MSSM. However from the global realisation, both the unification scale and the value of the coupling at unification should be outputs of the dynamics of the global model after moduli stabilisation. It is in fact remarkable that for the global realisation of this model the volume is found to be of order $\mathcal{V} \sim 10^{12}$ and the gauge coupling (dilaton vev) precisely of the value that fits with the unification. Two completely independent calculations give rise the same physical quantities (see the second reference in[35] for details). Rather than emphasising the qualities of this model (that has other phenomenological problems regarding flavour and the CMP) this illustrates the challenge for any other attempt to achieve unification including moduli stabilisation.

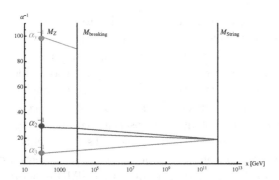

Fig. 4. Unification in Left-Right symmetric model with low energy RG running matching the unification scale and gauge coupling with those values obtained independently from moduli stabilisation.

5. Open Questions

Clearly local models have passed a threshold in the sense that there are now explicit local chiral models that have been properly embedded in a global compactification including moduli stabilisation and realistic values of physical scales and computable soft SUSY breaking terms. This is good progress and a sign of a healthy field that has evolved over the years with continuous progress. Still obtaining a fully realistic model is an open question.

On more concrete open questions: Embed the most realistic local models in a global compactification, such as those based on dP_3 or even higher dP_n including a successful inflation scenario. Also in cosmology, address in detail post-inflationary cosmological issues such as reheating, dark matter and baryogenesis and implement them in explicit models.

Determining dynamically the symmetry breaking pattern that leads to the SM is a next step after geometric moduli stabilisation, that is moduli stabilisation in the open string sector. On SUSY breaking, study in detail all the different scenarios including phenomenological observables at reach at LHC and potential future colliders. In particular the sequestered scenarios.[23] An important challenge is to make consistent the large string and gravitino scale hinted by BICEP with TeV soft terms.

Regarding more calculation developments, next order corrections to the

EFT are needed. In particular quantum corrections to the Kähler potential for the moduli and matter fields. This is crucial to obtain reliable soft terms, including contributions to non-universality. Finally it would be interesting to identify CFT duals of the AdS compactifications to provide them a proper non-perturbative definition.

A key point to keep in mind is to try identify potential observables that can put classes of models to test. The dark radiation issue raised recently is a good example. It has led to put strong constraints on hidden sector models that would give too large an effect but also to identify a cosmic axion background (CAB) with potential observable effects manifested as X-ray excess in galaxy clusters.[40] Similar ideas have been put forward in the past regarding the potential of observing cosmic strings motivated by the way to end brane inflation, etc. These observable effects are very rare and difficult to identify but worth pursuing. Also, potential discoveries in the recent future by LHC, Planck and other CMB experiments, on dark matter, axions searches, etc. should give us guidelines as to how to restrict or identify realistic string models. We may get lucky one of these days.

Acknowledgments

I would like to thank my collaborators over the past few years on the original work reported in this overview: Luis Aparicio, Ralph Blumenhagen, Cliff Burgess, Michele Cicoli, Joe Conlon, Shanta de Alwis, Anamaria Font, Rajesh Gupta, Ehsan Hatefi, Denis Klevers, Sven Krippendorf, Anshuman Maharana, Christoph Mayrhofer, Sebastian Moster, Stefan Theisen, Roberto Valandro. Special thanks to Sven Krippendorf for comments on the manuscript.

References

1. L. E. Ibanez and A. M. Uranga, *String theory and particle physics: An introduction to string phenomenology*, Cambridge, UK: Univ. Pr. (2012).
2. E. Witten, *Reflections on the fate of space-time*, Phys. Today **49N4** (1996) 24.
3. A. Font, F. Quevedo and S. Theisen, *A Comment on Continuous Spin Representations of the Poincare Group and Perturbative String Theory*, arXiv:1302.4771 [hep-th].
4. S. Weinberg, *The Quantum theory of fields. Vol. 1: Foundations*, Cambridge, UK: Univ. Pr. (1995) 609 p
5. G. D. Coughlan, W. Fischler, E. W. Kolb, S. Raby and G. G. Ross, *Cosmological Problems for the Polonyi Potential*, Phys. Lett. B **131** (1983) 59;

T. Banks, D. B. Kaplan and A. E. Nelson, *Cosmological implications of dynamical supersymmetry breaking*, Phys. Rev. D **49** (1994) 779 [hep-ph/9308292];
B. de Carlos, J. A. Casas, F. Quevedo and E. Roulet, *Model independent properties and cosmological implications of the dilaton and moduli sectors of 4-d strings*, Phys. Lett. B **318** (1993) 447 [hep-ph/9308325].

6. S. B. Giddings, S. Kachru and J. Polchinski, *Hierarchies from fluxes in string compactifications*, Phys. Rev. D **66** (2002) 106006 [hep-th/0105097].

7. M. R. Douglas and S. Kachru, *Flux compactification*, Rev. Mod. Phys. **79** (2007) 733 [hep-th/0610102].

8. V. Balasubramanian, P. Berglund, J. P. Conlon and F. Quevedo, *Systematics of moduli stabilisation in Calabi-Yau flux compactifications*, JHEP **0503** (2005) 007 [hep-th/0502058];
J. P. Conlon, F. Quevedo and K. Suruliz, *Large-volume flux compactifications: Moduli spectrum and D3/D7 soft supersymmetry breaking*, JHEP **0508** (2005) 007 [hep-th/0505076].

9. S. Kachru, R. Kallosh, A. D. Linde and S. P. Trivedi, *De Sitter vacua in string theory*, Phys. Rev. D **68** (2003) 046005 [hep-th/0301240].

10. C. P. Burgess, R. Kallosh and F. Quevedo, *De Sitter string vacua from supersymmetric D terms*, JHEP **0310** (2003) 056 [hep-th/0309187].

11. S. L. Parameswaran and A. Westphal, *de Sitter string vacua from perturbative Kahler corrections and consistent D-terms*, JHEP **0610** (2006) 079 [hep-th/0602253];
M. Cicoli, A. Maharana, F. Quevedo and C. P. Burgess, *De Sitter String Vacua from Dilaton-dependent Non-perturbative Effects*, JHEP **1206** (2012) 011 [arXiv:1203.1750 [hep-th]];
J. Blbck, D. Roest and I. Zavala, *De Sitter Vacua from Non-perturbative Flux Compactifications*, arXiv:1312.5328 [hep-th].

12. S. de Alwis, R. Gupta, E. Hatefi and F. Quevedo, *Stability, Tunneling and Flux Changing de Sitter Transitions in the Large Volume String Scenario*, JHEP **1311** (2013) 179 [arXiv:1308.1222 [hep-th], arXiv:1308.1222].

13. C. P. Burgess, A. Maharana and F. Quevedo, *Uber-naturalness: unexpectedly light scalars from supersymmetric extra dimensions*, JHEP **1105** (2011) 010 [arXiv:1005.1199 [hep-th]].

14. C. P. Burgess, M. Cicoli and F. Quevedo, *String Inflation After Planck 2013*, JCAP **1311** (2013) 003 [arXiv:1306.3512, arXiv:1306.3512 [hep-th]].

15. S. Dimopoulos, S. Kachru, J. McGreevy and J. G. Wacker, "N-flation," JCAP **0808** (2008) 003 [hep-th/0507205]; M. Cicoli, K. Dutta and A. Maharana, "N-flation with Hierarchically Light Axions in String Compactifications," arXiv:1401.2579 [hep-th]; E. Silverstein and A. Westphal, Phys. Rev. D **78** (2008) 106003 [arXiv:0803.3085 [hep-th]]; L. McAllister, E. Silverstein and A. Westphal, Phys. Rev. D **82** (2010) 046003 [arXiv:0808.0706 [hep-th]]; A. Avgoustidis, D. Cremades and F. Quevedo, "Wilson line inflation," Gen. Rel. Grav. **39** (2007) 1203 [hep-th/0606031]; A. Avgoustidis and I. Zavala, "Warped Wilson Line DBI Inflation," JCAP **0901** (2009) 045 [arXiv:0810.5001 [hep-th]].

16. M. Cicoli, J. P. Conlon and F. Quevedo, *Dark Radiation in LARGE Volume Models*, Phys. Rev. D **87** (2013) 043520 [arXiv:1208.3562 [hep-ph]];
    T. Higaki and F. Takahashi, *Dark Radiation and Dark Matter in Large Volume Compactifications*, JHEP **1211** (2012) 125 [arXiv:1208.3563 [hep-ph]].
17. M. Cicoli, *Axion-like Particles from String Compactifications*, arXiv:1309.6988 [hep-th].
18. K. Choi, A. Falkowski, H. P. Nilles and M. Olechowski, *Soft supersymmetry breaking in KKLT flux compactification*, Nucl. Phys. B **718** (2005) 113 [hep-th/0503216].
19. J. P. Conlon and F. Quevedo, *Gaugino and Scalar Masses in the Landscape*, JHEP **0606** (2006) 029 [hep-th/0605141];
    J. P. Conlon, S. S. Abdussalam, F. Quevedo and K. Suruliz, *Soft SUSY Breaking Terms for Chiral Matter in IIB String Compactifications*, JHEP **0701** (2007) 032 [hep-th/0610129];
    J. P. Conlon, C. H. Kom, K. Suruliz, B. C. Allanach and F. Quevedo, *Sparticle Spectra and LHC Signatures for Large Volume String Compactifications*, JHEP **0708** (2007) 061 [arXiv:0704.3403 [hep-ph]].
20. A. Hebecker, A. K. Knochel and T. Weigand, *A Shift Symmetry in the Higgs Sector: Experimental Hints and Stringy Realizations*, JHEP **1206** (2012) 093 [arXiv:1204.2551 [hep-th]]. L. E. Ibanez, F. Marchesano, D. Regalado and I. Valenzuela, *The Intermediate Scale MSSM, the Higgs Mass and F-theory Unification*, JHEP **1207** (2012) 195 [arXiv:1206.2655 [hep-ph]]. L. E. Ibanez and I. Valenzuela, *The Higgs Mass as a Signature of Heavy SUSY*, JHEP **1305** (2013) 064 [arXiv:1301.5167 [hep-ph]].
21. L. J. Hall and Y. Nomura, *Grand Unification and Intermediate Scale Supersymmetry*, arXiv:1312.6695 [hep-ph].
22. R. Blumenhagen, J. P. Conlon, S. Krippendorf, S. Moster and F. Quevedo, *SUSY Breaking in Local String/F-Theory Models*, JHEP **0909** (2009) 007 [arXiv:0906.3297 [hep-th]].
23. L. Aparicio *et al.*, to appear.
24. L. B. Anderson, J. Gray, A. Lukas and B. Ovrut, *Stabilizing the Complex Structure in Heterotic Calabi-Yau Vacua*, JHEP **1102** (2011) 088 [arXiv:1010.0255 [hep-th]]; *Stabilizing All Geometric Moduli in Heterotic Calabi-Yau Vacua*, Phys. Rev. D **83** (2011) 106011 [arXiv:1102.0011 [hep-th]]; M. Cicoli, S. de Alwis and A. Westphal, *Heterotic Moduli Stabilisation*, JHEP **1310** (2013) 199 [arXiv:1304.1809 [hep-th]].
25. B. S. Acharya, G. Kane and P. Kumar, *Compactified String Theories – Generic Predictions for Particle Physics*, Int. J. Mod. Phys. A **27** (2012) 1230012 [arXiv:1204.2795 [hep-ph]].
26. R. Blumenhagen, S. Moster and E. Plauschinn, *Moduli Stabilisation versus Chirality for MSSM like Type IIB Orientifolds*, JHEP **0801** (2008) 058 [arXiv:0711.3389 [hep-th]].
27. J. P. Conlon, *Mirror Mediation*, JHEP **0803** (2008) 025 [arXiv:0710.0873 [hep-th]].
28. A. Maharana and E. Palti, *Models of Particle Physics from Type IIB String Theory and F-theory: A Review*, Int. J. Mod. Phys. A **28** (2013) 1330005 [arXiv:1212.0555 [hep-th]].

29. G. Aldazabal, L. E. Ibanez and F. Quevedo, *A D⁻ brane alternative to the MSSM*, JHEP **0002** (2000) 015 [hep-ph/0001083].

30. G. Aldazabal, L. E. Ibanez, F. Quevedo and A. M. Uranga, *D-branes at singularities: A Bottom up approach to the string embedding of the standard model*, JHEP **0008** (2000) 002 [hep-th/0005067].

31. H. Verlinde and M. Wijnholt, *Building the standard model on a D3-brane*, JHEP **0701** (2007) 106 [hep-th/0508089]; M. Buican, D. Malyshev, D. R. Morrison, H. Verlinde and M. Wijnholt, *D-branes at Singularities, Compactification, and Hypercharge*, JHEP **0701** (2007) 107 [hep-th/0610007]; M. Wijnholt, *Geometry of Particle Physics*, Adv. Theor. Math. Phys. **13** (2009) [hep-th/0703047].

32. S. Krippendorf, M. J. Dolan, A. Maharana and F. Quevedo, *D-branes at Toric Singularities: Model Building, Yukawa Couplings and Flavour Physics*, JHEP **1006** (2010) 092 [arXiv:1002.1790 [hep-th]]; M. J. Dolan, S. Krippendorf and F. Quevedo, *Towards a Systematic Construction of Realistic D-brane Models on a del Pezzo Singularity*, JHEP **1110** (2011) 024 [arXiv:1106.6039 [hep-th]].

33. M. Kreuzer and H. Skarke, *Complete classification of reflexive polyhedra in four-dimensions*, Adv. Theor. Math. Phys. **4** (2002) 1209 [hep-th/0002240]; *PALP: A Package for analyzing lattice polytopes with applications to toric geometry*, Comput. Phys. Commun. **157** (2004) 87 [math/0204356 [math-sc]].

34. J. P. Conlon, A. Maharana and F. Quevedo, *Towards Realistic String Vacua*, JHEP **0905** (2009) 109 [arXiv:0810.5660 [hep-th]].

35. M. Cicoli, S. Krippendorf, C. Mayrhofer, F. Quevedo and R. Valandro, *D-Branes at del Pezzo Singularities: Global Embedding and Moduli Stabilisation*, JHEP **1209** (2012) 019 [arXiv:1206.5237 [hep-th]]; *D3/D7 Branes at Singularities: Constraints from Global Embedding and Moduli Stabilisation*, JHEP **1307** (2013) 150 [arXiv:1304.0022 [hep-th]].

36. V. Balasubramanian, P. Berglund, V. Braun and I. Garcia-Etxebarria, *Global embeddings for branes at toric singularities*, JHEP **1210** (2012) 132 [arXiv:1201.5379 [hep-th]].

37. M. Cicoli, C. Mayrhofer and R. Valandro, *Moduli Stabilisation for Chiral Global Models*, JHEP **1202** (2012) 062 [arXiv:1110.3333 [hep-th]].

38. J. Louis, M. Rummel, R. Valandro and A. Westphal, *Building an explicit de Sitter*, JHEP **1210** (2012) 163 [arXiv:1208.3208 [hep-th]]; D. Martinez-Pedrera, D. Mehta, M. Rummel and A. Westphal, *Finding all flux vacua in an explicit example*, JHEP **1306** (2013) 110 [arXiv:1212.4530 [hep-th]].

39. M. Cicoli, D. Klevers, S. Krippendorf, C. Mayrhofer, F. Quevedo and R. Valandro, *Explicit de Sitter Flux Vacua for Global String Models with Chiral Matter*, arXiv:1312.0014 [hep-th].

40. J. P. Conlon and M. C. D. Marsh, *Searching for a 0.1-1 keV Cosmic Axion Background*, Phys. Rev. Lett. **111** (2013) 151301 [arXiv:1305.3603 [astro-ph.CO]]; S. Angus, J. P. Conlon, M. C. D. Marsh, A. Powell and L. T. Witkowski, *Soft X-ray Excess in the Coma Cluster from a Cosmic Axion Background*, arXiv:1312.3947 [astro-ph.HE].

9

F-theory: From Geometry to Phenomenology

Sakura Schäfer-Nameki

Department of Mathematics, King's College London
The Strand, London, WC2R 2LS, UK
sakura.schafer.nameki@gmail.com

We give a lightning overview of recent progress in F-theory model building. After a review of the basic paradigm of geometric engineering of grand unified theories by elliptically fibered Calabi-Yau fourfolds, we explain the connection to local models, construction of fluxes, phenomenological implications as well as the recent advances in building global models with $U(1)$ symmetries.

1. Introduction

String phenomenology has prospered in the last decade, and much progress has been made in developing a connection to phenomenology for the corners of the string theory landscape. A strong emphasis has been put on developing a set of rules and mechanisms that allow the embedding of the Standard model and its supersymmetric extensions into string compactifications.

There are two distinct questions that one can pose when validating string phenomenology: first of all, is it possible, at all, to embed the (Minimal supersymmetric) Standard Model ((MS)SM) into string theory. Secondly, does the class of string compactifications that allow for such an embedding have phenomenological implications, i.e. testable predictions. These questions do not necessarily have to be answered in succession. In fact in F-theory the initial approach has largely been to try to answer the latter by determining the class of effective theories that can arise from F-theory compactifications.

The very starting point for F-theory realizations of the MSSM or grand unified theories (GUTs) with $N = 1$ supersymmetry is a local model, which is analyzed systematically for its phenomenological validity, and as a second step, global constraints are incorporated. A general review of this type of bottom-up model building can be found in[1] in this volume. F-theory is particularly amenable to this approach as the gauge degrees of freedom, which model the four-dimensional effective theory, are localized on 7-branes that wrap a complex surface S and stretch long $\mathbb{R}^{1,3}$, and thus can be to some extend decoupled in a local model. F-theory local models have a beautiful description in terms of Higgs bundle spectral covers, and studying the vacuum structure of such gauge configurations is already highly constraining the possible four-dimensional theories.

In this review we will give a brief overview of the main tools in constructing F-theory models, starting in Section 2 by first explaining the global structure of elliptically fibered Calabi-Yau fourfolds, the relation to 7-branes and the structure of singularities that give rise to gauge fields, matter and Yukawa couplings. In the second part we will turn more concretely to models with $SU(5)$ unification groups and characterize the local limit and its description in terms of Higgs bundle spectral covers in Section 3.3. Embedding into the spectral cover models is already highly constrained, and has non-trivial phenomenological implications.

Often, symmetries are required to make GUT models phenomenologically sound, e.g. by forbidding the perturbative generation of certain phenomenologically disfavorable couplings such as dimension 5 proton decay. Discrete symmetries usually require a high specialization of the surface S, thus defying the point of characterizing generic features of a larger class of consistent models. Thus large parts of the F-theory literature focus on additional $U(1)$ gauge symmetries. Likewise, breaking of the GUT group and lifting the Higgs triplets and non-SM gauge bosons is achieved by switching on background flux in the hypercharge direction $U(1)_Y$. One of the main conclusions in local model building, based on anomaly cancellation, is that combining hypercharge GUT breaking with $U(1)$ symmetries for proton decay protection, are only consistent with certain types of $U(1)$ symmetries.

The global validity of such spectral cover models relies on understanding their lift to global compactification geometries, which in the case of F-theory are elliptically fibered Calabi-Yau manifolds. Construction of global models has been recently the main focus, and will be reviewed in detail. Likewise, construction of background G_4-fluxes is an essential input into the model building, for instance to induce chirality. A systematic exploration of the possible global models with $U(1)$ symmetries will enable to give an exact answer to the question, which four-dimensional effective theories can be embedded into a global construction, potentially giving interesting phenomenological constraints. Construction of global models with additional abelian gauge factors is at present a very active field of research, which is reviewed in Section 4. Clearly this review has as a main focus the spectral problem, i.e. the F-theoretic construction of realistic supersymmetric spectra for GUT models. Many important string phenomenological questions remain only briefly touched upon, and we will give an overview of these, as well as some of the most pressing open questions in the concluding Section 5.

It may be useful for the reader to provide a brief comparison with other string theories amenable to realistic model building. Perturbative IIA and IIB brane constructions have been immensely successful, both from a bottom-up and top down point of view, with one key advantage of these setups is that the effective action of the four-dimensional theories is much better under control than in F-theory. Compared to IIB orientifold constructions, on the other hand, the main advantage of F-theory resides in the existence of exceptional gauge groups, which in GUT models can be essential for the construction of top Yukawa couplings. In IIB these would have to be obtained by instanton effects, whereas in F-theory the exceptional singularities, that realize such couplings, are on the same footing as the classical gauge groups. Heterotic string theory is another classic setting for string phenomenology, where exceptional gauge groups based on $E_8 \times E_8$ and subgroups thereof, are paramount and thus provide a natural setting for realizing GUT models. The key difference to F-theory is that heterotic constructions do not allow a decoupling of gauge theory degrees of freedom from gravity ones, as these have a common origin in the closed heterotic string. This in particular implies that characterizations of heterotic models are much more dependent on a case by case analysis of complete compactifications (including the geometry and vector bundles), than F-theory models. In the latter, gauge degrees of freedom are localized on 7-branes and much generic features can be infered by whole classes of F-theory compactifications, rather than specific example geometries. This point has led to much progress in developing techniques for scanning efficiently through heterotic sub-landscapes, a development that has not had as much attention in F-theory, as other methods are available to infer properties of classes of compactifications (by means of studying and constraining the structure of the elliptic fiber of the compactification).

Much of the progress in F-theory GUTs is based on the development of mathematical tools, and F-theory model building has been progressing hand in hand with exciting new developments in properties of elliptically fibered

Calabi-Yaus and their birational geometry. Therefore some part of this review may seem too mathematical for a review of phenomenological implications. However, what the progress in the last few years has shown is that taking F-theory GUTs seriously, and requiring local and global consistency – which most of the time are based on the intricate mathematical structures underlying F-theory compactifications – usually results in interesting phenomenological implications. One of the prime examples is the study of $U(1)$ symmetries, which are constraint both locally, through anomalies, and also globally, through the general mathematical constraints on elliptic fibrations that realize such abelian symmetries (i.e. elliptic fibrations with extra sections, as we shall review in Section 4). Precisely this interplay between the beautiful mathematics on the one hand, and phenomenological constraints of the corresponding F-theory compactifications is the main strength of this approach to string phenomenology, which has the potential to constrain the viable F-theory models through the intricate structures of elliptic curves and fibrations.

2. F-theory geometric engineering

2.1. F-theory

F-theory[2–4] is a placeholder name for Type IIB superstring theory vacua which are not necessarily perturbative. The axio-dilaton, which is the complex combination of the dilaton ϕ, which sets the string coupling, and the axion C_0 defined by

$$\tau = C_0 + ie^{-\phi}, \tag{1}$$

in F-theory is not necessarily constant, or does $g_s = e^{-\phi}$ have to be small. Under the S-duality group $SL_2\mathbb{Z}$ of Type IIB, the axio-dilaton transforms as

$$\tau \to \frac{a\tau + b}{c\tau + d}, \tag{2}$$

where $ab - cd = 1$ and defines a two-dimensional representation of $SL_2\mathbb{Z}$. The ingenious idea in[2] was to interpret τ geometrically, namely, as the complex structure modulus of a two-dimensional torus, or more precisely, an elliptic curve. Compactifications of F-theory have the elliptic curve incorporated. The standard Type IIB paradigm of considering a spacetime compactification manifold B, gets augmented to an elliptic fibration $Y : \mathbb{E}_\tau \to B$, which assigns to each each point on the base B an elliptic curve, whose complex structure is given by the axio-dilaton. For the purpose of building four-dimensional F-theory vacua, which preserve $N = 1$ supersymmetry, we consider elliptically fibered Calabi-Yau fourfolds, with a complex three-dimensional base B_3. If the fibration is non-trivial (which it will be in the case of interest to us), the base will not be Calabi-Yau, but can e.g. be (weak-) Fano.

We do not know a fundamental description of F-theory. However, there are various indirect ways to study F-theory compactifications, either by taking a weak-coupling limit to Type IIB, the so-called Sen limit,[5,6] or by dualities to other string theories. For the present purpose, two dualities are particularly important:

- F-theory on elliptic K3-fibered Calabi-Yau fourfold is dual to the heterotic string compactified on an elliptic Calabi-Yau three-fold:.[3,4,7] For Calabi-Yau fourfolds with an elliptic $K3$-fibrations, many checks could be enabled using this duality, and much of the local model building is based on matching the spectral data of heterotic vector bundles to the Higgs bundle spectral data for local 7-brane models in F-theory.[7,8]

- M-theory/F-theory duality:[9–12]
  This approach is particularly useful in order to determine the effective action of F-theory. It is based on T-duality between IIB and IIA, as well as lifting IIA to M-theory. The duality chain for a simple T^2 torus compactification with A and B cycles, is as follows:

$$M/S_A^1 \times S_B^1 \quad \overset{R_A \to 0}{\longrightarrow} \quad IIA/S_B^1 \quad \overset{R_B \to 0}{\longrightarrow} \quad IIB$$

$$R_A, R_B \to 0\,, \qquad g_s = R_A/R_B = \text{fixed}$$

More generally, F-theory is obtained from M-theory on an elliptic curve $\mathbb{E}_\tau$ in the following limit

$$\text{Elliptic curve} \quad \mathbb{E}_\tau \sim S_A^1 \times S_B^1 : \quad \begin{cases} \text{Im}(\tau) = g_s = \text{fixed} \\ \text{Vol}(\mathbb{E}_\tau) \to 0 \end{cases}$$

This duality has in particular been very useful in understanding the effective action,[10] as well as the geometry of the F-theory compactification.[13]

2.2. *Gauge theory from geometry*

Gauge degrees of freedom in F-theory are localized on 7-branes, which are generalizations of the standard D7-branes of Type IIB, and have a purely

S. Schäfer-Nameki

geometric characterization. Recall that D7-branes in IIB source F_9, which is dual to dC_0, where C_0 is the RR 0-form axion in 10 dimensions. For a D7-brane located at z_0 in the direction perpendicular to the 7-brane, $z = x^8 + ix^9$, the corresponding source is $d \star F_9 = \delta(z - z_0)$, which gives rise a non-trivial monodromy of the dual dC_0

$$\oint_{S^1} dC_0 = 1 \,. \tag{3}$$

In a local expansion, this can be solved for C_0 and thus τ, by

$$\tau(z) = \tau(z_0) + \frac{1}{2\pi i} \log(z - z_0) + \cdots \tag{4}$$

Along the branch-cut, the axio-dilaton undergoes a monodromy $\tau \to \tau + 1$. More generally, a 7-brane, that couples to the $SL_2\mathbb{Z}$ invariant combination τ, transforming as (p, q) under the $SL_2\mathbb{Z}$, will result in a general monodromy for τ (2). Geometrically we can therefore characterize such (p, q) 7-branes as those loci in the base above which the elliptic fibration has a singular fiber, i.e. the complex structure τ diverges.

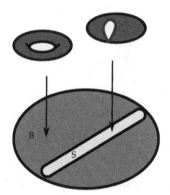

Fig. 1.   Schematic depiction of the elliptic fibers: above a generic point in the base B of the elliptic fibration the fiber is a smooth torus (elliptic curve). Above a surface $S \subset B$, i.e. a codimension one subspace of the base, which is a component of the discriminant, the fiber becomes singular, realizing gauge degrees of freedom on $S \times \mathbb{R}^{1,3}$. The gauge group is determined by the singularity type.

In mathematical terms, the singularities that can occur in elliptic curves, or more generally in elliptic fibrations, have a long history. The main tool for describing elliptic fibrations $\mathbb{E}_\tau \to B_3$ is its realization (which exists for all elliptic curves with a section – more about this in Section 4) in terms of the Weierstrass equation

$$y^2 = x^3 + fxw^4 + gw^6 \,, \tag{5}$$

where $[w, x, y]$ are homogeneous coordinates in the weighted projective space $\mathbb{P}^{123}$, and we usually work in the patch $w = 1$, where y, x are now affine coordinates. Furthermore, f and g are functions (more precisely, sections of line bundles $K_{B_3}^{-4}$ and $K_{B_3}^{-6}$, respectively) on the base B_3. The generic fiber will be an elliptic curve, but as the point in the base varies, the elliptic curve can become singular. To characterize these loci, note that (5) describes a two-sheeted cover of the complex plane, with branch cuts connecting the roots of the cubic equation and infinity. When two such branch-points collapse, the curve becomes singular, which can be detected by the vanishing of the discriminant

$$\Delta = 4f^3 + 27g^2 \,. \tag{6}$$

In terms of the elliptic fibration, we can think of $\Delta = 0$ as an equation in B_3, that characterizes a complex codimension one subspace, i.e. a surface. This is exactly the surface that is wrapped by the 7-branes.

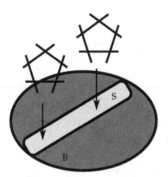

Fig. 2. Fibers in the resolved geometry corresponding to an $SU(5)$ singularity. Above each point in the GUT surface S, which is a component of the discriminant, the resolved fiber is a ring of five $\mathbb{P}^1$s (shown as black lines), which intersect in the affine A_5 Dynkin diagram.

In general this has several components, so let us denote by z a local coordinate on B_3, such that $z = 0$ corresponds to a surface S_{GUT} in B_3, which is an irreducible component of the discriminant. Then both f and g have a local expansion

$$f = \sum_i f_i z^i, \qquad g = \sum_i g_i z^i \,. \tag{7}$$

If the elliptic fibration (5) is singular above $z = 0$, then one considers the expansion of Δ, starting with $n > 1$

$$\Delta = \delta_0 z^n + \delta_1 z^{n+1} + \delta_2 z^{n+2} + \cdots \,. \tag{8}$$

The singular fibers have been classified for elliptic surfaces by Kodaira and Neron and it is believed that this classification holds true in codimension one in the base also for higher dimensional elliptic fibrations and they characterized by the vanishing orders of (f, g, Δ) with respect to z. For instance $SU(n)$ gauge groups are realized by I_n fibers, which correspond to the vanishing orders $(0, 0, n)$. Resolving the singularities in the fiber results in a chain of $\mathbb{P}^1$s that intersect along an affine Dynkin diagram, in the case of $SU(5)$, the resolved fiber is shown in figure 2.

From the point of view of F-theory compactifications, the singularity type determines the gauge group of the effective theory on the 7-brane that wrap the surface $S_{GUT} \times \mathbb{R}^{1,3}$ characterized by $z = 0$. To see this, it is useful to consider the M-theory compactification on the resolved Calabi-Yau fourfold. Consider for instance an I_n singular fiber, where the exceptional $\mathbb{P}^1$ in the resolved fiber gives rise to a $(1,1)$ form, $\omega_i^{(1,1)}$, which allows decomposition of the C_3 three-form in M-theory as

$$C_3 = \sum_i A_i \wedge \omega_i^{(1,1)}, \tag{9}$$

where A_i are the gauge potentials for $U(1)^{n-1}$, corresponding to the Cartan subalgebra of the gauge group. The remaining gauge bosons arise from wrapped M2-branes, which however become massless only in the singular limit, and thus in the F-theory limit.

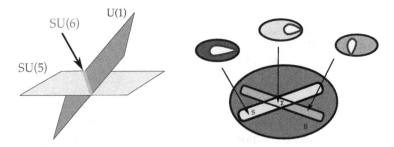

Fig. 3.   Geometric realization of matter: the usual paradigm of matter from intersection of the stack of branes realizing the GUT gauge degrees of freedom with a flavor brane is shown on the left hand side. In F-theory this is realized by further enhancement of the singularity above a curve Σ inside the GUT surface $S \subset B$. These codimension two loci above which the singularity can enhance are encoded in the discriminant. The fiber type above Σ determines the representation of the localized matter.

2.3. Matter and Yukawas from geometry

In perturbative Type IIB, matter arises from the intersections of D7-branes, for instance as shown in figure 3. Consider a D7-brane stack, realizing a gauge theory with gauge group G, and intersect this with a single D7-brane, carrying (flavor) $U(1)$ symmetry. Then the open string degrees of freedom localized along the intersection give rise to matter, where locally one can think of the gauge group as being enhanced to a higher rank group $\tilde{G}$, which is higgsed to the gauge group G. The adjoint of $\tilde{G}$ gives rise to the adjoint of G, as well as (bifundamental) matter in a representation, which depends on the enhanced group $\tilde{G}$, for instance a simple intersection of D7-branes of this type would give rise to matter in the fundamental representation, as shown in figure 3, with $\tilde{G} = SU(n+1)$, $G = SU(n)$[a]

The F-theoretic realization of this corresponds to a further degeneration of the elliptic fibration, i.e. along a curve contained inside S_{GUT} and therefore a codimension two sublocus in the base B_3, the singularity in the fiber gets enhanced to a singularity associated to a higher rank gauge group $\tilde{G}$, and the corresponding matter localized along this curve is obtained by the decomposition of the adjoint of $\tilde{G}$

$$
\begin{aligned}
\tilde{G} &\rightarrow G \times U(1) \\
\mathrm{Adj}(G) &\rightarrow \mathrm{Adj}(G) \oplus \mathrm{Adj}(U(1)) \oplus \mathbf{R}_+ \oplus \mathbf{R}_- \,.
\end{aligned}
\tag{10}
$$

The loci where such singularity enhacement can occur are determined from the discriminant (8). The subleading order term δ_0, indicates exactly where further singularity enhancements are localized along $z = 0$. The fibers that can occur in codimension two $z = \delta_0 = 0$ have a beautiful description that ties in with representation theory of the gauge groups and we refer the reader to[13] for further details.

Like in the case of intersecting D7-branes, the singularity enhancement can be thought of as the collision of two singularities above a codimension two locus, the matter curve, in B_3: each singular fiber corresponds to 7-branes, e.g. an I_n singular fiber above $z = 0$ colliding with a single 7-brane corresponding to an I_1 fiber given by an equation $b = 0$, as shown in figure 3. The equation for the matter curve is then $z = b = 0$ inside the Calabi-Yau fourfold.

Yukawa couplings are generated in codimenion three in B, along which matter curves intersect and thereby lead to a higher vanishing order of the

---

[a]More precisely in perturbative IIB, the gauge groups would be $U(n)$, however the overall $U(1)$ gets a mass of order of the string coupling in F-theory.[14]

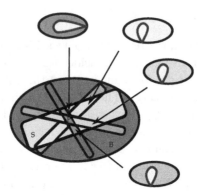

Fig. 4. Yukawa couplings: Yukawa coulings arise from codimension three loci, i.e. points, in the base B where matter curves intersect. Above these Yukawa points, singularity enhances further, and again the type of singular fiber determines what coupling is generated. Note that three divisors in a Calabi-Yau fourfold generically meet in a point.

discriminant. This has a local gauge theoretic interpretation in terms of higgsing a higher rank gauge group

$$G_p \quad \rightarrow \quad G \times U(1) \times U(1), \quad\quad (11)$$

which generates couplings of matter fields consistent with the $U(1)$ charges. Geometrically, the singular fiber above the Yukawa point in B determines the local enhancement type, i.e. G_p. One of the main advantages compared to perturbative Type IIB brane constructions is the existence of exceptional gauge groups in F-theory, and in particular exceptional higher codimension fibers. As we shall see below, this is of quite some importance in $SU(5)$ GUT models building, where the top Yukawa coupling arises from E_6.

3. $SU(5)$ GUTs in F-theory

Following this general overview of geometric engineering in F-theory, we now turn to the concrete phenomenological question of generating an $SU(5)$ GUT model within this framework. The basic phenomenological input, like matter and Yukawa couplings as well as constraints on GUT breaking and proton decay have been summarized in Appendix A.1.

3.1. The $SU(5)$ Tate model

In F-theory we can engineer an $SU(5)$ GUT model from an elliptic fibration in Weierstrass form with vanishing orders along a surface $z = 0$, given by

$f = O(z^0)$, $g = O(z^0)$ and $\Delta = O(z^5)$. A slightly more elegant way to construct such models is to apply Tate's algorithm[15,16] and write the most general such equation in terms of a hypersurface in the weighted projective space $\mathbb{P}^{123}[w, x, y]$

$$y^2 = x^3 + b_1 xy + b_2 zx^2 + b_3 z^2 y + b_4 z^3 x + b_6 z^5, \qquad (12)$$

where we set $w = 1$[b], and the explicit z-dependence guarantees that the singular fiber in codimension one is of I_5 type, realizing $SU(5)$. The b_i replace f, g and are sections of suitable line bundles on B_3, with an expansion $b_i = b_{i,0} + b_{i,1}z + \cdots$. The Tate form has the additional advantage that in a scaling limit it encodes the spectral data of the Higgs bundle that describes the local gauge theory on the 7-brane,[17] and so provides a direct link between local and global models. For the $SU(5)$ Tate model (12) the discriminant has an expansion in z

$$\Delta = z^5 \delta_5 + z^6 \delta_6 + O(z^7), \qquad (13)$$

where δ_i have multiple components. As we have explained, these higher codimension loci correspond to further enhancements of the singularity and allow the generation of matter and Yukawa couplings. In terms of the coefficients b_i the codimension two and three enhanced singular loci are

Codim	Gauge Group	Fiber Type	Equation in B
codim 1	$SU(5)$	I_5	$z = 0$
codim 2	$SU(6)$	I_6	$z = P = 0$
	$SO(10)$	I_1^*	$z = b_1 = 0$
codim 3	$SO(12)$	I_2^*	$z = b_1 = b_3 = 0$
	E_6	$\widetilde{IV^*}$	$z = b_1 = b_2 = 0$

$$(14)$$

Here $P \equiv b_1^2 b_6 - b_1 b_3 b_4 + b_2 b_3^2$ denotes the $\bar{\mathbf{5}}$ matter locus. To determine the actual singularity type and corresponding local enhaced symmetry, one has to resolve the geometry and determine the intersection structure of the $\mathbb{P}^1$ in the fiber in codimesion 2 and 3[17,18c]. Above a generic point in the surface $z = 0$, the fiber in the resolved geometry is an I_5 fiber corresponding to a ring of five $\mathbb{P}^1$s as shown in Figure 2.

Along codimension 2 loci (14), or matter curves, the discriminant vanishes to higher order. The precise nature of the singularity enhancement is determined again by following the resolved fiber components to the codimension 2 locus, above which some of the $\mathbb{P}^1$s will become reducible. The

---

[b]This corresponds to going to the patch where w does not vanish. This is sufficient, as the singularity is exactly located in this coordinate patch on the $\mathbb{P}^{123}$.

[c]There are various topologically inequivalent resolutions, which were all determined in.[19]

irreducible components of the fiber in codimension 2 can be associated to weights of representations of $SU(5)$, determined by a higher rank group $\tilde{G}$ as in (10). For $SU(5)$ the matter is obtained from local fiber enhacements to I_6 and I_1^*, realizing $SU(6)$ and $SO(10)$, with matter in the $\bar{\mathbf{5}}$ and $\mathbf{10}$ representations, respectively. The fibers are depicted in figure 5.

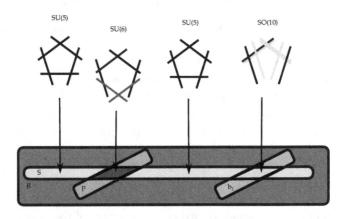

Fig. 5. Elliptic fibration for $SU(5)$ models, with resolved fibers in codimension one and two. The codimension one (in B) fiber above $z = 0$, the GUT surface S, is I_5, i.e. five $\mathbb{P}^1$s intersecting in the affine Dynkin diagram of $SU(5)$. This gives rise to the $SU(5)$ gauge degrees of freedom. Along matter curves in S, i.e. codimension two loci, $z = P = 0$ or $z = b_1 = 0$, the fibers split (with new fiber components shown in blue/green), either into an I_6 and I_1^* fiber, realizing locally $SU(6)$ and $SU(10)$, respectively.

Likewise along the codimension 3 loci (14), the fibers above the matter curves split further. The fiber type above the $SO(12)$ point is precisely I_2^*, which generates the down-type Yukawa coupling. Much interesting mathematics has arisen from the fiber above the E_6 Yukawa point, which is not a standard IV^* fiber, but monodromy reduced,[13,18,20] as shown in figure 6. For $SU(5)$ GUTs this is especially vital as the top Yukawa coupling $\mathbf{10} \times \mathbf{10} \times \mathbf{5}_H$ obtained by decomposing the adjoint of E_6 to $SU(5)$

$$E_6 \quad \rightarrow \quad SU(5) \times U(1)_1 \times U(1)_2$$

$$\mathbf{78} \quad \rightarrow \quad (\mathbf{24}_{0,0} \oplus \mathbf{1}_{0,0} \oplus \mathbf{1}_{0,0}) \oplus (\mathbf{1}_{-5,-3} \oplus \mathbf{1}_{5,3})$$

$$\oplus \left(\mathbf{5}_{-3,3} \oplus \bar{\mathbf{5}}_{3,-3}\right) \oplus \left(\mathbf{10}_{-1,-3} \oplus \overline{\mathbf{10}}_{1,3}\right) \oplus \left(\mathbf{10}_{4,0} \oplus \overline{\mathbf{10}}_{-4,0}\right) .$$

$$\tag{15}$$

The absence of a full IV^* fiber, which implies the absence of a full local E_6 gauge symmetry, does however not affect the generation of a Yukawa coupling, as the fiber components above the codimension two matter curves split consistently with the Yukawas.[17] For instance, a fiber component

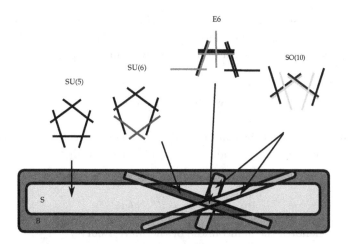

Fig. 6. Fibers in codimension three: as in figure 5, the codimension one and two fibers are shown in terms of intersecting $\mathbb{P}^1$ (colored/black lines). Above the codimension three locus, i.e. Yukawa points in B (purple), with the equation $z = b_1 = b_2 = 0$, the fibers split further and generate a (monodromy-reduced) E_6 fiber.

associated to the weight of a **10** representation can split into components corresponding to weights of the **10** and **5** representations.

3.2. *Chirality from Fluxes*

The structure we discussed so far generates bifundamental matter. Chirality is induced by switching on supersymmetric background G_4 fluxes, which are harmonic (2,2) forms on the Calabi-Yau fourfold. The three-form potential C_3 for the G-flux has to be have one leg on the fiber[d]. From the point of view of the 7-brane theory, the G-flux induces gauge flux by decomposing it in the $(1,1)$-forms ω_i

$$G_4 = dC_3 = F_i \wedge \omega_i\,. \tag{16}$$

The four-form flux $G_4 \in H^{2,2}(Y_4)$, with one leg in fiber has to satisfy the supersymmetry and quantization conditions

$$G \wedge J = 0\,, \qquad G + \frac{1}{2}c_2(Y_4) \in H^4(Y_4,\mathbb{Z})\,, \tag{17}$$

where $c_2(Y_4)$ is the second Chern class of the Calabi-Yau fourfold, which can be computed from the resolved fourfold. The four-form flux can be

---

[d] C_3 that has a component when integrated over a curve in the fiber or is containe in the base will break Lorentz invariance in four dimensions.

integrated over so-called matter surfaces $S_{\mathbf{R}}$, which are obtained by considering a $\mathbb{P}^1$ fiber of the resolved four-fold over a matter curve, corresponding to a representation $\mathbf{R}$ of the gauge group. This computes the chiral index[7,11,17,21–23]

$$n_{\mathbf{R}} - n_{\bar{\mathbf{R}}} = \int_{S_{\mathbf{R}}} G_4 \, . \tag{18}$$

The chiralities can also be computed by computing the cohomologies valued in the line bundle that C_3 induces on a matter curve $\Sigma_{\mathbf{R}}$[21,24]

$$\begin{aligned} n_{\mathbf{R}} &= h^0(\Sigma_{\mathbf{R}}, K^{1/2} \otimes \mathcal{L}_{\mathbf{R}}) \\ n_{\bar{\mathbf{R}}} &= h^1(\Sigma_{\mathbf{R}}, K^{1/2} \otimes \mathcal{L}_{\mathbf{R}}) \, . \end{aligned} \tag{19}$$

Here $K^{1/2}$ denotes the spin bundle of the matter curve.

One class of fluxes is obtained from dualizaing the $(2,2)$ forms to surfaces, which are orthogonal to vertical and horizontal surfaces, i.e. they should be orthogonal to all surfaces that are either the restriction of the elliptic fibration to a curve, or sit entirely inside a section of the fibrations. In addition we will assume that the flux does not break the $SU(5)$ GUT group. Fluxes of this type have been constructed in Refs. 17, 23, 25–27.

More generally one can consider fluxes that correspond to algebraic cycles in Y_4, which are homologically non-trivial already in the singular geometry.[22] More recently these were discussed in the context of the Deligne cohomology and Chow groups in Ref. 28, a direction that certainly will see much more development in the near future.

3.3. Local Models and Spectral Covers

F-theory model building had its advent not directly in terms of constructions of global models, but initially the focus was largely on the local models, i.e. effective theories describing the 8-dimensional gauge theory that lives on the 7-brane compactified on the GUT surface S times $\mathbb{R}^{1,3}$.[8,21,24,29,30] The local model is characterized in terms of a Higgs bundle, i.e. vacuum expectation value of an adjoint scalar field, and a gauge field. The global analog of these are complex structure parameters and G-fluxes, discussed in the previous sections. The spectral cover construction for this Higgs bundle is inspired by heterotic/F-theory duality, however, it has a very precise meaning directly in terms of a local limit of a general F-theory compactification, that is not necessarily dual to a heterotic model.[8,17]

The effective theory on $S_{GUT} \times \mathbb{R}^{1,3}$ is a half-twisted supersymmetric Yang-Mills theory, with maximal rank gauge group E_8, whose gauge degrees of freedom are higgsed to the GUT group $SU(5)$ and its commutant, $SU(5)_\perp$,

$$
\begin{aligned}
E_8 \quad &\rightarrow \quad SU(5)_\perp \times SU(5)_{\mathrm{GUT}} \\
\mathbf{248} \quad &\rightarrow \quad (\mathbf{24}, \mathbf{1}) + (\mathbf{1}, \mathbf{24}) \\
&\quad + (\overline{\mathbf{10}}, \mathbf{5}) + (\mathbf{5}, \overline{\mathbf{10}}) + (\mathbf{10}, \overline{\mathbf{5}}) + (\mathbf{5}, \mathbf{10})
\end{aligned}
\tag{20}
$$

by an $SU(5)_\perp$ adjoint valued scalar field $\langle \Phi_{SU(5)_\perp} \rangle \sim \mathrm{diag}$ $(\lambda_1, \lambda_2, \lambda_3, \lambda_4, \lambda_5)$ with $\sum \lambda_i = 0$. The eigenvalues λ_i are solutions to the characteristic polynomial

$$
\mathcal{C} \equiv \det(s - \langle \Phi_{SU(5)_\perp} \rangle) = b_1 + b_2 s + b_3 s^2 + b_4 s^3 + b_6 s^5 = 0.
\tag{21}
$$

This spectral surface should be thought of as embedded in the bundle $\mathbb{P}^1(K_S \oplus \mathcal{O}_S) \rightarrow S$, and defines a five-fold covering of the GUT surface S. The coefficients b_i are by definition symmetric polynomials of the eigenvalues, for instance

$$
b_1 = b_6 \lambda_1 \lambda_2 \lambda_3 \lambda_4 \lambda_5, \qquad b_2 = b_0 \sum_{i<j<k<l} \lambda_i \lambda_j \lambda_k \lambda_l
$$
$$
b_3 = b_0 \sum_{i<j<k} \lambda_i \lambda_j \lambda_k, \qquad b_4 = b_0 \sum_{i<j} \lambda_i \lambda_j, \qquad b_5 = \sum_i \lambda_i = 0.
\tag{22}
$$

Along special loci in S, the gauge group is enhanced as follows:

$$
\begin{aligned}
SO(10) &: 0 = b_1 \sim \prod_i \lambda_i \\
SU(6) &: 0 = P \sim \prod(\lambda_i + \lambda_j) \\
SO(12) &: 0 = b_1 = b_3 : (\lambda_i + \lambda_j) + (\lambda_k + \lambda_l) + (\lambda_m) = 0 \\
E_6 &: 0 = b_5 = b_4 : (\lambda_i) + (\lambda_j) + (-\lambda_k - \lambda_l) = 0 \\
E_8 &: 0 = b_2 = b_3 = b_4 = b_5.
\end{aligned}
\tag{23}
$$

These loci in turn realize matter and Yukawa couplings, much like the higher codimension loci in the elliptic fibration. This analogy is of course not a coincidence. One obvious question is, to link these local models to globally consistent F-theory compactifications, which can be addressed in the Tate form (12) of the singularity. We can define a divisor $y^2 = x^3$ in the Tate form, the Tate divisor, and in the local limit, i.e. when zooming into the singularity at $x = y = z = 0$ by defining $s = y/x$ and $z/s = $ fixed while taking $s, z \rightarrow 0$, precisely reproduces the Higgs bundle spectral cover $\mathcal{C}$.[17,31,32]

In addition to the spectral cover, the local model is characterized through a gauge configuration given by a non-abelian vector bundle V on S with $c_1(V) = 0$. Denoting the projection map from the spectral cover by

$$p_{Higgs} : \quad C \to S, \tag{24}$$

such a vector bundle can be constructed as push-forward of a line bundle on the spectral cover $\mathcal{N}_{Higgs} \to C$ by

$$V = p_{Higgs,*}\mathcal{N}_{Higgs}, \qquad \mathcal{N}_{Higgs} = \mathcal{O}\left(\gamma + \frac{r}{2}\right), \tag{25}$$

where γ is a divisor in C and r is the ramification divisor of the covering map p_{Higgs}. This combination ensures that $c_1(V) = 0$ as $p_{Higgs,*}\gamma = 0$. The chiralities of matter induced by such spectral cover fluxes are computed by intersection of γ with the matter curves.[8,30]

3.4. GUT breaking

String theory provides alternative mechanisms to break the unification group gauge group to the Standard Model gauge group, which go beyond a GUT Higgs mechanism. For instance, discrete Wilson lines are a well-studied mechanisms in heterotic model building. In F-theory, a flat line bundle in the hypercharge direction, unfortunately, always leads to non-GUT exotics in the zero-mode spectrum of S_{GUT}, and therefore is phenomenologically excluded.[33,34]

A more successful way to break the $SU(5)$ to the Standard Model gauge group is to consider non-trivial background flux in the hypercharge direction.[29,33,35] The requirements in a local models are that the zero modes on S in the presence of the hypercharge flux $\mathcal{L}_Y$ produces only vectors in the adjoint of the Standard Model gauge group, i.e. from the following table, we see that $h^2(S, K_S) = 1$ and all others vanish.

$SU(3) \times SU(2) \times U(1)_Y$	Cohomologies
$(\mathbf{8},\mathbf{1})_0 \oplus (\mathbf{1},\mathbf{3})_0 \oplus (\mathbf{1})_0$	$H^2(S, K_S) \oplus H^0(S, K_S) \oplus H^1(S, K_S)$
$(\mathbf{3},\mathbf{2})_{-5/6}$	$H^0(S, \mathcal{L}_Y^{-1}) \oplus H^1(S, \mathcal{L}_Y) \oplus H^1(S, \mathcal{L}_Y^{-1})$
$(\bar{\mathbf{3}},\mathbf{2})_{+5/6}$	$H^0(S, \mathcal{L}_Y) \oplus H^1(S, \mathcal{L}_Y^{-1}) \oplus H^1(S, \mathcal{L}_Y)$

$$\tag{26}$$

On the other hand the hypercharge flux should restrict trivially to all $\mathbf{10}$ and $\bar{\mathbf{5}}$ matter curves, however it should have non-trivial restricion on the Higgs matter curves in order to lift the zero modes of the Higgs triplets

$$\mathcal{L}_Y|_{\Sigma_{H_{u/d}}} \neq 0. \tag{27}$$

The main phenomenological requirement on such hypercharge flux is however, that it retains masslessness of the hypercharge gauge boson. To see this, recall that in the F-theory effective action, there is a Chern-Simons like coupling with the RR 4-form C_4

$$\int_{Y \times \mathbb{R}^{1,3}} C_4 \wedge G_4 \wedge G_4 . \tag{28}$$

Expanding G_4 with respect to the $(1,1)$ form corresponding to the hypercharge direction into gauge connection and background value $G = (F_Y + c_1(\mathcal{L}_Y)) \wedge \omega_Y$ and $C_4 = C_2^i \wedge \omega_i$, $\omega_i \in H^2(B_3, \mathbb{Z})$, and inserting into the Chern-Simons coupling generates a mass term for the hypercharge F_Y

$$\left(\text{Tr}(T_Y^2) \int_S c_1(\mathcal{L}_Y) \wedge i^* \omega_i \right) \left(\int_{\mathbb{R}^{1,3}} C_2^i \wedge F_Y \right) , \tag{29}$$

where T_Y is the generator of $U(1)_Y$. However this unwanted mass term can vanish as long as

$$\int_S c_1(\mathcal{L}_Y) \wedge i^* \omega_i = 0 \qquad \text{for all} \quad \omega_i \in H^2(B_3, \mathbb{Z}) . \tag{30}$$

This topological condition on the hypercharge flux is equivalent to the existence of three-chain Ω_3 in B_3 ,whose boundary is the dual inside S of $c_1(\mathcal{L}_Y)$. Examples of base geometries, that accomodate such classes have appeared in Refs. 25, 36–38. The characterization of hypercharge flux in global models based on elliptic fibrations is an active area of research, and recent developments have appeared in Ref. 39. Nevertheless, the triviality of the hypercharge flux as a class in B_3 causes it to integrate restrict trivially over matter surfaces in the compact Calabi-Yau. This in particular causes an issue if we would like to combine this with certain $U(1)_{PQ}$ charges, under which the Higgs multiplets are charged non-oppositely. Engineering global models with hypercharge flux and PQ symmetries, in addition remains a challenge.

4. $U(1)$ Symmetries

$U(1)$ symmetries are indispensable when constructing realistic F-theory GUT models, mainly to ensure protection from rapid proton decay by dimension 4 and 5 operators. An F-theory model will have an additional $U(1)$ symmetry only after suitably tuning the complex structure of the elliptic fibration, which in the local model corresponds to tuning the coefficients of the Higgs bundle spectral cover.

One of the striking results of local F-theory GUTs has been the insight that the existence of $U(1)$ symmetries in addition to hypercharge flux GUT breaking induces very stringent phenomenological constraints on the models imposed by anomaly cancellation. Understanding the types of globally consistent $U(1)$ models, and their geometric constructions in terms of explitic elliptic fibrations, has been one of the main directions of research in the last few years, with the most promising phenomenological implications.

4.1. *Local models, $U(1)$s and Anomalies*

$U(1)$ symmetries in local models are engineered by reducing the monodromy of the eigenvalues of the Higgs bundle, which leads to a factorization of the spectral cover. The permutation group S_5 acts on the five sheets of the $SU(5)_\perp$ spectral cover, labeled by the eigenvalues of the Higgs bundle λ_i. Generically, all sheets are permuted into each other and there is no overall $U(1)$ symmetry. However, if $\Gamma \subset S_5$ is a subgroup that does not act transitively on the sheets of the spectral cover, then the λ_i form a reducible representation of Γ, and decomposing them into irreducible orbits results in the factorization of the spectral cover

$$\mathcal{C} = \prod_{i=1}^{N} \mathcal{C}^{(i)} . \tag{31}$$

The factorization leads to a reduced structure group of the Higgs bundle, which now corresponds to an $S(U(n_1) \times \cdots U(n_N))$ group. This leaves unbroken an $SU(5) \times U(1)^{N-1}$ gauge group. From a phenomenological point of view it is crucial to see what type of $U(1)$ symmetries can be retained, when higgsing the E_8 down to $SU(5)_{GUT}$, and a complete survey of such models can be found in.[40]

$U(1)$ symmetries are instrumental for controlling dimension 5 proton decay operators. We define a Peccei-Quinn $U(1)_{PQ}$ symmetry to have charges of up- and down-type Higgs doublets that are not opposite to each other, i.e.

$$q_{PQ}(H_u) + q_{PQ}(H_d) \neq 0 . \tag{32}$$

This perturbatively forbidgs the μ-term, as well as (under the assumptions that the Yukawa couplings are consistent with the $U(1)$ symmetry), the dimension 5 proton decay operator

$$W_\mu \sim \mu H_u H_d \quad \text{and} \quad W_{\text{dim5}} \sim \frac{1}{\Lambda} Q^3 L . \tag{33}$$

A spectral cover model that realizes this is obtained by a $3+2$ factorization of the five-sheeted cover $\mathcal{C}$.[41] Another model of interest, which realizes a $U(1)_{B-L}$ corresponds to a $4+1$ factorization of the cover.[30]

The most striking result, which shows the existence of generically applicable, stringent phenomenological constraints already in the framework of local model building, is that there is a tension between the GUT breaking using hypercharge flux, and the existence of certain $U(1)$ symmetries, which imply constraints on the spectrum.[40,42–45] These constraints were observed initially from spectral cover models and were then derived from anomalies between the additional $U(1)$ symmetries and the SM gauge group.

To formulate the anomaly constraints, let Σ be a matter curve with matter representation $\mathbf{R}$ carrying hypercharge $Y_{\mathbf{R}}$. Then the chiral index for matter with flux determined by a line bundle $\mathcal{L}_\Sigma$ and hypercharge flux $\mathcal{L}_Y$ is computed by

$$n_{\mathbf{R}} - n_{\bar{\mathbf{R}}} = \int_\Sigma c_1(\mathcal{L}_\Sigma \otimes \mathcal{L}_Y^{Y_{\mathbf{R}}}) = \int_\Sigma \left(c_1(\mathcal{L}_\Sigma) + M_{\mathbf{R}} c_1(\mathcal{L}_Y^{Y_{\mathbf{R}}}) \right) . \tag{34}$$

For the $\mathbf{10}$ and $\bar{\mathbf{5}}$ matter we can write the multiplicities in terms of the chiral index for multiplets in irreducible $SU(3) \times SU(2) \times U(1)_Y$ representations

$$n_{(\mathbf{1},\mathbf{1})_{+1}} - n_{(\mathbf{1},\mathbf{1})_{-1}} = M_{\mathbf{10}} + N_{\mathbf{10}}$$

$$n_{(\mathbf{3},\mathbf{2})_{+1/6}} - n_{(\bar{\mathbf{3}},\mathbf{2})_{-1/6}} = M_{\mathbf{10}}$$

$$n_{(\bar{\mathbf{3}},\mathbf{1})_{-2/3}} - n_{\mathbf{3},\mathbf{1})_{+2/3}} = M_{\mathbf{10}} - N_{\mathbf{10}} \tag{35}$$

$$n_{(\bar{\mathbf{3}},\mathbf{1})_{+1/3}} - n_{(\mathbf{3},\mathbf{1})_{-1/3}} = M_{\mathbf{5}}$$

$$n_{(\mathbf{1},\mathbf{2})_{-1/2}} - n_{(\mathbf{1},\mathbf{2})_{+1/2}} = M_{\mathbf{5}} - N_{\mathbf{5}} .$$

The integers M correspond to the number of chiral multiplets that are in full GUT group representations, whereas N measures the hypercharge flux, that threads the corresponding matter curve. Considering a theory with gauge group $SU(5) \times U(1)^n$, we first note the pure G_{MSSM} anomaly cancelation conditions, which in this parametrization read

$$\sum_i M_{\mathbf{10}^i} = \sum_j M_{\bar{\mathbf{5}}^j} , \qquad \sum_i N_{\mathbf{10}^i} = \sum_j N_{\bar{\mathbf{5}}^j} = 0 . \tag{36}$$

The $G_{MSSM}^2 \times U(1)$ mixed anomaly cancellation imply a constraint on the hypercharge restrictions N_i and the $U(1)$ charges q of the respective multiplets[43]

$$\sum_i q(\mathbf{10}^i) N_{\mathbf{10}^i} = \sum_j q(\bar{\mathbf{5}}^i) N_{\bar{\mathbf{5}}^j} . \tag{37}$$

This constraint is automatically satisfied in spectral cover models. Finally, considering the $U(1)_Y \times U(1) \times U(1)'$ anomalies yields[45]

$$3 \sum_{10^i} q(10^i) q'(10^i) N_{10^i} = \sum_j q(\bar{5}^j) q'(\bar{5}^j) N_{\bar{5}^j} \,. \tag{38}$$

The relatively innocent looking (37) has some very stringent constraints on the type of $U(1)$ symmetries that are viable if we require a minimal GUT spectrum:

If we require a minimal $SU(5)$ GUT model then (37) implies

$$q(H_u) + q(H_d) = 0 \,, \tag{39}$$

and the resulting $U(1)$ symmetry is essentially $U(1)_{B-L}$. In particular, this is not a PQ symmetry and does not forbid the dimension 5 proton decay operator.

If one insists on a $U(1)$ symmetry that fobids the couplings W_μ and $W_{\mathrm{dim}\,5}$, then the anomaly constraint (37) implies

$$q(H_u) + q(H_d) \neq 0 \qquad \Rightarrow \qquad \text{some} \quad N_{10}, N_{\bar{5}} \neq 0 \,. \tag{40}$$

So requiring a $U(1)_{PQ}$ implies the existence of non-GUT multiplets in the spectrum – in addition to the standard $H_{u,d}$ doublets. Various approaches to deal with this have been put forward.[42,44] One way to make a virtue out of this unfortunate exotic spectrum is to use the exotics as messengers in a high-scale gauge mediation setup, by giving them a mass by coupling to a charged singlet field, which has interesting, non-minimal type GMSB characteristics.[44] Unification in this context becomes quite subtle, with additional contributions from hypercharge flux, Higgs fields, exotics and high scale threshold corresctions.[33,44,46]

The constraint (38) is not automatic in spectral cover models and furthermore constrains them, allowing only a small subclass of 3+2 and 2+2+1 models. Globally constructed consistent F-theory models with $U(1)$ symmetries and hypercharge flux breaking should elucidate what mechanism cancels this anomaly[e], which brings us to the final topic of $U(1)$ symmetries in global F-theory compactifications.

4.2. *Global realizations: Elliptic fibrations with extra sections*

In view of the severe constraints that the co-existence of $U(1)$ symmetries and hypercharge flux imply in local models, it is one of the most pressing

---

[e]In the perturbative IIB orientifold limit, it was understood in[47] that there is a geometric Stueckelberg type mechanism, which makes the $U(1)$ massive, see also.[48]

questions in F-theory model building to understand the class of global F-theory models, which have $U(1)$ symmetries and can support hypercharge flux. Regarding the altter, some progress has recently been made in Ref. 39. Global models with abelian gauge factors were initially constructed with some inspiration from local factored spectral cover models in Refs. 23, 49, which realize the global lift of the $4+1$ split spectral cover model and fluxes.

However a more systematic approach is necessary if we want to determine the complete set of F-theory vacua with $SU(5) \times U(1)^n$ gauge groups. The geometric characterization of non-abelian gauge symmetries in F-theory can be achieved by singularities of elliptic fibrations, as explained in Section 2. Abelian gauge group factors correspond to additional (rational) sections[f].[3,4] Recall, that an elliptic curve written in Weierstrass form has a section, which when embedded into $\mathbb{P}^{123}[w, x, y]$

$$y^2 = x^3 + fxw^4 + gw^4 \,, \tag{41}$$

is given by $w = 0$. This defines a point $[0 : 1 : 1]$ in the elliptic curve above each point in the base B_3 of the fibration. The rational points[g] of an elliptic curve form a group, the Mordell-Weil group, with the operation defined by the group law on the elliptic curve.

The zero-section, which is present in any Weierstrass model, defines a copy of the base in the fiber. Each additional rational section gives rise to a $U(1)$ gauge symmetry as follows. A section σ_i defines a divisor in the Calabi-Yau fourfold, S_i, and a dual $\omega_i^{(1,1)}$ form in the fiber. Reducing the M-theory three-form on these

$$C_3 = \sum_i \omega_i^{(1,1)} \wedge A^i \,, \tag{42}$$

results in abelian gauge fields A^i. The generator of the $U(1)$ is then determined, in the resolved Calabi-Yau fourfold, by subtracting suitable exceptional classes, so to make the $U(1)$s orthogonal to the non-abelian part of the gauge group – the Shioda map.

A systematic construction of models with $U(1)$ symmetries was initiated by Ref. 51, which showed that an elliptic curve with two rational points have a natural embedding into $\mathbb{P}^{112}$. Following this several works appeared constructing models with multiple $U(1)$ symmetries.[50,52–56] The elliptic curves realizing one and two $U(1)$s can be embedded as follows:

---

[f]For a nice discussion of holomophic versus rational sections see Ref. 50.
[g]Here, we think of the elliptic curve defined over some field, and the points are rational solutions over this field to the elliptic curve equation. Concretely, here this is the function field of the base manifold B_3.

- One extra section, i.e. $U(1)$:
  Embedding into $\mathbb{P}^{112}[w, x, y]$:

$$y^2 + b_0 x^2 y = c_0 w^4 + c_1 w^3 x + c_2 w^2 x^2 + c_3 w x^3 \qquad (43)$$

Sections: $y = w = 0$ and $w = y + b_0 x^2 = 0$.

- Two extra sections, i.e. $U(1)^2$:
  Embedding into $dP_2[w, x, y; l_1, l_2]$:

$$s_1 l_1^2 l_2^2 w^3 + s_2 l_1^2 l_2 w^2 x + s_3 l_1^2 w x^2 + s_5 l_1 l_2^2 w^2 y$$
$$+ s_6 l_1 l_2 w x y + s_7 l_1 x^2 y + s_8 l_2^2 w y^2 + s_9 l_2 x y^2 = 0 \qquad (44)$$

Sections: $l_1 = 0$; $l_2 = 0$; $x = s_9$, $y = -s_7$

Again, as in the Weierstrass or Tate forms in $\mathbb{P}^{123}$, the coefficient sections c_i, b_i, s_i can have vanishing orders along a divisor $z = 0$ in the base, that correspond to a singular elliptic fibration. Resolving these singularities, one can determine the intersection of the $U(1)$ generators with the matter curves.

Several example models with $SU(5) \times U(1)^i$ for $i = 1, 2, 3$ have been constructed based on toric methods.[50,52–57] Their phenomenology has been given some consideration for $i = 1, 2$ in Ref. 58. However, this remains an incomplete analysis as long as the full set of possible global models is not determined first.

To achieve this goal of a determining a comprehensive list of $SU(5) \times U(1)^i$ spectra that have a global realization with such a gauge group, it is vital to find a method that generalizes the Tate type forms, i.e. canonical realizations of a given type of singular Kodaira fiber with a distribution of extra sections. This analysis is carried out in Refs. 59 and 60 and the resulting types of models have been shown to allow for more general models than were obtained in the toric constructions. The possible $SU(5) \times U(1)$ models with one or two **10** curves are shown in Table 1. The toric models are subcases of this set of fiber types. There are further models with three **10** curves which can be found in Ref. 59. Similarly, there is a classification of $SU(5) \times U(1) \times U(1)$ models realized in Ref. 60. Clearly, repeating a survey-type analysis that was done for local models in Refs. 42, 40, 44 for all global models with $U(1)$ is an outstanding problem of considerable interest. For the toric constructions with one and two $U(1)$s this was done in Ref. 58, however these do not include models with multiple **10** matter curves, which can add considerable model building freedom.

Table 1. Codimension two loci, fiber types, and matter and $U(1)$ charges for $SU(5) \times U(1)$ models with one or two $\mathbf{10}$ matter loci, based on Tate's algorithm in $\mathbb{P}^{112}$ in Ref. 59. Superscripts in the fiber types indicate the separation of the two sections, i.e. (01) has the sections on one component of the I_5 fiber, whereas (0|1) on neighboring, etc. The codimension 2 loci are shown schematically, see Ref. 59 for details. Subscript nc indicates that this is a model that that does not have simply realization in terms of vanishing orders of coefficients in (43), but the coefficients satisfy non-trivial relations, much like in the non-standard Tate forms in Ref. 16.

Fiber in Codim 1	Codim 2 locus	Rep and $U(1)$ charge	Codim 2 fiber
$I_5^{(01)}$	$b_{1,0}$	$10_0 + \overline{10}_0$	$I_1^{*(01)}$
	$c_{3,0}$	$5_{-1} + \overline{5}_1$	$I_6^{(0|1)}$
	$c_{3,0} + b_{0,5}b_{1,0}$	$5_1 + \overline{5}_{-1}$	$I_6^{(0|1)}$
	$b_{1,0}^2 c_{0,5} - b_{1,0}b_{2,2}c_{1,3} + b_{2,2}^2 c_{2,1}$	$5_0 + \overline{5}_0$	$I_6^{(0|1)}$
$I_5^{(0|1)}$	$b_{1,0}$	$10_2 + \overline{10}_{-2}$	$I_1^{*(0|1)}$
	$b_{0,0}$	$5_6 + \overline{5}_{-6}$	$I_6^{(01)}$
	$b_{0,0}c_{2,1} - b_{1,0}c_{3,1}$	$5_{-4} + \overline{5}_4$	$I_6^{(0||1)}$
	$b_{1,0}^2 c_{0,4} - b_{1,0}b_{2,2}c_{1,2} - c_{1,2}^2$	$5_1 + \overline{5}_{-1}$	$I_6^{(0|1)}$
$I_{5,nc}^{(0||1)}$	P_1^{10}	$10_1 + \overline{10}_{-1}$	$I_1^{*(0||1)}$
	P_2^{10}	$10_{-4} + \overline{10}_4$	$I_1^{*(01)}$
	$P_1^{\bar{5}}$	$5_{-7} + \overline{5}_7$	$I_6^{(01)}$
	$P_2^{\bar{5}}$	$5_{-2} + \overline{5}_2$	$I_6^{(0|1)}$
	$P_3^{\bar{5}}$	$5_3 + \overline{5}_{-3}$	$I_6^{(0|||1)}$
$I_{5,nc}^{(0|1)}$	P_1^{10}	$10_2 + \overline{10}_{-2}$	$I_1^{*(01)}$
	P_2^{10}	$10_{-3} + \overline{10}_3$	$I_1^{*(0|1)}$
	$P_1^{\bar{5}}$	$5_6 + \overline{5}_{-6}$	$I_6^{(01)}$
	$P_2^{\bar{5}}$	$5_{-4} + \overline{5}_4$	$I_6^{(0|1)}$
	$P_3^{\bar{5}}$	$5_1 + \overline{5}_{-1}$	$I_6^{(0|1)}$

5. Outlook

F-theory model building has come a very long and successful way, developing gauge theoretic and geometric tools that allow engineering of realistic spectra for supersymmetric GUTs. One of the driving forces in F-theory model building has been the focus on generic properties, i.e. properties that not so much depend on the specific compactification, i.e. a specific base manifold, but will be applicable in (almost) any attempt to build models within the framework of F-theory. The rich mathematical structure of elliptic curves is an immensely powerful tool in achieving this – for instance we have seen stringent constraints on what type of $U(1)$ symmetries can be realized geometrically, in a completely general fashion. Another powerful way to approach constraints is using the ideas of bottom-up model building[1] applicable in this context by first analysing the local model that controls the effective theory on 7-branes.

In this review, we have focused thematically on the progress in constructing realistic spectra and the uncovering of generic features in F-theory compactifications to supersymmetric GUTs. Consequently (and due to the finite amount of space in this volume) we have omitted various aspects that are instrumental in building realistic models. Much of the basics of F-theory have been reviewed in Ref. 61 and geared towards GUT models building in Refs. 62, 63. The omissions made here include the following aspects of model building:

- Supersymmetry breaking, moduli stabilization and non-perturbative effects:
  supersymmetry breaking is is intimately tied to non-perturbative effects, such as D3-instantons, which have been studied in Refs. 64–66 and applied in supersymmetry breaking scenarios in Refs. 67, 68. Understanding their global lift remains an open question.
- Flavor:
  in the early days of F-theory model building, the favorable flavor structure, which compared to Type IIB arises from the tree-level generation of the top Yukawa coupling, with a leading order rank one Yukawa matrix, was one of the main motivations for F-theory model building. Much excitement occurred following[69] regarding the subleading terms in the CKM mixing matrix, however much careful computations have followed and generated a comprehensive picture of flavor in F-theory.[70–72] Most recently, T-branes or gluing branes,[73–75] were utilized to generate realistic up-type quark masses.[76]
- Cosmology:
  cosmology in F-theory GUTs was first discussed in local models.[77] More recently, following the BICEP2 measurement of B-modes, the validity of inflationary scenarios in F-theory with 7-branes and axion inflation, have been studied in Refs. 78, 79.

One constant in the developments of this field since the initial papers[21,24,29] has been the interplay between particle physics and geometry. Modeling nature in F-theory goes hand in hand with studying and exploring beautiful mathematical structures, at times beyond what is known in the mathematics literature.

Many open issues remain, before a complete model, including spectrum, moduli stabilization and supersymmetry breaking, as well as cosmological issues, can be attained. In addition to the already mentioned open questions in computing G-fluxes and alternative GUT breaking mechanisms, as well as global aspects of hypercharge flux breaking, the following questions remain to be explored:

- Effective action of F-theory:
  Using M/F-theory duality much progress has been made in the direction of determining the effective action of F-theory.[10,11,80,81] Determining however Kaehler potentials for matter fields would be very important for instance for studying supersymmetry breaking.
  From an alternative point of view, it would be interesting to find the F-theory effective action from first principles, for example using the $SL_2\mathbb{Z}$ duality and making this manifest, similar to T- or U-folds.
- Supersymmetry breaking and moduli stabilization:
  Which mechanisms work in F-theory, and is there a LARGE volume type mechanism for F-theory[h]? Some preliminary results to this effect have been obtained in Ref. 82. What mediation mechanisms are viable in F-theory, beyond local gauge mediation models?
- Landscaping:
  What class of compactifications of F-theory yield four-dimensional $N = 1$ supersymmetric vacua? Some interesting alternative directions to the standard paradigm of elliptically fibered Calabi-Yau compactifications were proposed recently. One interesting direction is to consider genus one fibrations,[83,84] which do not have a section. Another direction of generalization are gluing branes or T-branes, which correspond to local models with non-diagonalizable Higgs bundle, where the spectral data alone does not fully characterize the gauge configuration.[73-75] One open question is to fully understand the global lift of such compactifications[85,86] as well as their full phenomenological implications, such as in flavor physics. Finally, recently an alternative to Calabi-Yau compactifications was proposed by studying F-theory on $Spin(7)$ manifolds, which breaks supersymmetry entirely.[87]
- Geometry and singularity resolution:
  Albeit not necessarily of direct phenomenological relevance, the geometric structures studied in F-theory model building have developed into an exciting direction of research in itself. In particular the resolution of singularities in higher codimension, which are instrumental in realizing matter and Yukawa couplings have uncovered so-far unknown results in

---

[h]See Ref. 1 in this volume for a general review of this mechanism.

the resolution of singular elliptic fibrations,[13,17–19,23,88] which are beautiful structures, that are well worth exploring for their own mathematical sake in the future.

Acknowledgments

Thanks to Joe Marsano and Natalia Saulina for the most exilarating, energetic and fun collaboration and friendship over many years exploring F-theory! Thanks are due also to Matt Dolan, Sheldon Katz, Moritz Kuentzler, Craig Lawrie, Dave Morrison, Hirotaka Hayashi, Jonathan Heckman, Jamie Sully and Cumrun Vafa for collaborations in this exciting field. I would like to thank Andreas Braun for discussions and comments on this review. This work is supported in part by the STFC grant ST/J002798/1 and the European COST action "The String Theory Universe".

A.1. Phenomenological input

The standard minimal phenomenological requirements that are imposed on F-theory GUTs correspond to an $N = 1$ supersymmetric four-dimensional GUT with gauge group $SU(5)$ and the following additional data:

- Gauge group: minimal unification group $SU(5)$
- Matter content of the MSSM packaged into representations of $SU(5)$, i.e. three generations of

$$
10_M = \begin{pmatrix} Q \sim (\mathbf{3}, \mathbf{2})_{+1/6} \\ U^c \sim (\bar{\mathbf{3}}, \mathbf{1})_{-2/3} \\ E^c \sim (\mathbf{1}, \mathbf{1})_{+1} \end{pmatrix}, \qquad \bar{\mathbf{5}}_M = \begin{pmatrix} D^c \sim (\bar{\mathbf{3}}, \mathbf{1})_{+1/3} \\ L \sim (\mathbf{1}, \mathbf{2})_{-1/2} \end{pmatrix}
$$
(A.1)

- Higgs sector:

$$
5_H = \begin{pmatrix} H_u \sim (\mathbf{1}, \mathbf{2})_{+1/2} \\ H_u^{(3)} \sim (\mathbf{3}, \mathbf{1})_{-1/3} \end{pmatrix}, \qquad \bar{\mathbf{5}}_H = \begin{pmatrix} H_d \sim (\mathbf{1}, \mathbf{2})_{-1/2} \\ H_d^{(3)} \sim (\bar{\mathbf{3}}, \mathbf{1})_{+1/3} \end{pmatrix}
$$
(A.2)

- Superpotential couplings

$$
W \sim (\lambda_t)_{ij}\, 5_H \times 10_M^i \times 10_M^j + (\lambda_b)_{ij}\, \bar{\mathbf{5}}_H \times \bar{\mathbf{5}}_M^i \times 10_M^j
$$
(A.3)

- A mechanism to break the GUT group to the MSSM gauge group, lifting the unwanted XY-bosons:

$$SU(5) \to SU(3) \times SU(2) \times U(1)_Y$$

$$\mathbf{24} \to (\mathbf{8,1})_0 \oplus (\mathbf{1,3})_0 \oplus (\mathbf{1,1})_0 \oplus (\mathbf{3,\bar{2}})_{-5} \oplus (\mathbf{\bar{3},2})_{+5} \qquad \text{(A.4)}$$

Gauge Fields   Exotics

- Doublet-triplet splitting: mechanism to lift the Higgs triplets

$$SU(5) \to SU(3) \times SU(2) \times U(1)_Y$$

$$\mathbf{5}_H \to (\mathbf{1,2})_{+1/2} \oplus (\mathbf{3,1})_{-1/3}$$

H_u Exotics

- R-parity to protect from dimension 4 proton decay operators.
- $U(1)$ symmetry to disallow dimension 5 proton decay operators

$$W \sim \frac{Q^3 L}{\Lambda}, \qquad \Lambda > 10^{27} \text{ GeV}. \qquad \text{(A.5)}$$

References

1. F. Quevedo, *Local String Models and Moduli Stabilisation*, 1404.5151.
2. C. Vafa, *Evidence for F-Theory*, *Nucl. Phys.* **B469** (1996) 403–418, [hep-th/9602022].
3. D. R. Morrison and C. Vafa, *Compactifications of F-Theory on Calabi–Yau Threefolds – I*, *Nucl. Phys.* **B473** (1996) 74–92, [hep-th/9602114].
4. D. R. Morrison and C. Vafa, *Compactifications of F-Theory on Calabi–Yau Threefolds – II*, *Nucl. Phys.* **B476** (1996) 437–469, [hep-th/9603161].
5. A. Sen, *F-theory and Orientifolds*, *Nucl. Phys.* **B475** (1996) 562–578, [hep-th/9605150].
6. A. Sen, *Orientifold limit of f-theory vacua*, *Phys. Rev.* **D55** (1997) 7345–7349, [hep-th/9702165].
7. H. Hayashi, R. Tatar, Y. Toda, T. Watari, and M. Yamazaki, *New Aspects of Heterotic–F Theory Duality*, *Nucl. Phys.* **B806** (2009) 224–299, [0805.1057].
8. R. Donagi and M. Wijnholt, *Higgs Bundles and UV Completion in F-Theory*, 0904.1218.
9. K. Intriligator, D. R. Morrison, and N. Seiberg, *Five-dimensional supersymmetric gauge theories and degenerations of Calabi–Yau spaces*, *Nuclear Phys.* B **497** (1997) 56–100, [arXiv:hep-th/9702198].
10. T. W. Grimm, *The N=1 effective action of F-theory compactifications*, *Nucl.Phys.* **B845** (2011) 48–92, [1008.4133].
11. T. W. Grimm and H. Hayashi, *F-theory fluxes, Chirality and Chern-Simons theories*, *JHEP* **1203** (2012) 027, [1111.1232]. 53 pages, 5 figures/ v2: typos corrected, minor improvements.
12. K. Intriligator, H. Jockers, P. Mayr, D. R. Morrison, and M. R. Plesser, *Conifold Transitions in M-theory on Calabi-Yau Fourfolds with Background Fluxes*, 1203.6662. 94 pages.

13. H. Hayashi, C. Lawrie, D. R. Morrison, and S. Schafer-Nameki, *Box Graphs and Singular Fibers*, 1402.2653.
14. T. W. Grimm, M. Kerstan, E. Palti, and T. Weigand, *Massive Abelian Gauge Symmetries and Fluxes in F-theory*, JHEP **1112** (2011) 004, [1107.3842].
15. M. Bershadsky, K. A. Intriligator, S. Kachru, D. R. Morrison, V. Sadov, et. al., *Geometric singularities and enhanced gauge symmetries*, Nucl.Phys. **B481** (1996) 215–252, [hep-th/9605200].
16. S. Katz, D. R. Morrison, S. Schafer-Nameki, and J. Sully, *Tate's algorithm and F-theory*, JHEP **1108** (2011) 094, [1106.3854].
17. J. Marsano and S. Schafer-Nameki, *Yukawas, G-flux, and Spectral Covers from Resolved Calabi-Yau's*, JHEP **1111** (2011) 098, [1108.1794].
18. M. Esole and S.-T. Yau, *Small resolutions of SU(5)-models in F-theory*, 1107.0733.
19. H. Hayashi, C. Lawrie, and S. Schafer-Nameki, *Phases, Flops and F-theory: SU(5) Gauge Theories*, JHEP **1310** (2013) 046, [1304.1678].
20. A. P. Braun and T. Watari, *On Singular Fibres in F-Theory*, JHEP **1307** (2013) 031, [1301.5814].
21. R. Donagi and M. Wijnholt, *Model Building with F-Theory*, 0802.2969.
22. A. P. Braun, A. Collinucci, and R. Valandro, *G-flux in F-theory and algebraic cycles*, Nucl.Phys. **B856** (2012) 129–179, [1107.5337]. 55 pages.
23. S. Krause, C. Mayrhofer, and T. Weigand, *G₄ flux, chiral matter and singularity resolution in F-theory compactifications*, Nucl.Phys. **B858** (2012) 1–47, [1109.3454]. 53 pages, 2 figures.
24. C. Beasley, J. J. Heckman, and C. Vafa, *GUTs and Exceptional Branes in F-theory - I*, JHEP **01** (2009) 058, [0802.3391].
25. J. Marsano, N. Saulina, and S. Schafer-Nameki, *F-theory Compactifications for Supersymmetric GUTs*, JHEP **08** (2009) 030, [0904.3932].
26. J. Marsano, N. Saulina, and S. Schafer-Nameki, *On G-flux, M5 instantons, and U(1)s in F-theory*, 1107.1718.
27. J. Marsano, N. Saulina, and S. Schafer-Nameki, *A Note on G-Fluxes for F-theory Model Building*, JHEP **11** (2010) 088, [1006.0483].
28. M. Bies, C. Mayrhofer, C. Pehle, and T. Weigand, *Chow groups, Deligne cohomology and massless matter in F-theory*, 1402.5144.
29. C. Beasley, J. J. Heckman, and C. Vafa, *GUTs and Exceptional Branes in F-theory - II: Experimental Predictions*, JHEP **01** (2009) 059, [0806.0102].
30. J. Marsano, N. Saulina, and S. Schafer-Nameki, *Monodromies, Fluxes, and Compact Three-Generation F-theory GUTs*, JHEP **08** (2009) 046, [0906.4672].
31. M. Kuntzler and S. Schafer-Nameki, *G-flux and Spectral Divisors*, JHEP **1211** (2012) 025, [1205.5688].
32. R. Donagi, S. Katz, and M. Wijnholt, *Weak Coupling, Degeneration and Log Calabi-Yau Spaces*, 1212.0553.
33. R. Donagi and M. Wijnholt, *Breaking GUT Groups in F-Theory*, 0808.2223.
34. J. Marsano, H. Clemens, T. Pantev, S. Raby, and H.-H. Tseng, *A Global SU(5) F-theory model with Wilson line breaking*, JHEP **1301** (2013) 150, [1206.6132].

35. M. Buican, D. Malyshev, D. R. Morrison, H. Verlinde, and M. Wijnholt, *D-branes at singularities, compactification, and hypercharge*, JHEP **01** (2007) 107, [hep-th/0610007].

36. R. Blumenhagen, T. W. Grimm, B. Jurke, and T. Weigand, *F-theory uplifts and GUTs*, JHEP **0909** (2009) 053, [0906.0013].

37. R. Blumenhagen, T. W. Grimm, B. Jurke, and T. Weigand, *Global F-theory GUTs*, Nucl.Phys. **B829** (2010) 325–369, [0908.1784].

38. T. W. Grimm, S. Krause, and T. Weigand, *F-Theory GUT Vacua on Compact Calabi-Yau Fourfolds*, JHEP **1007** (2010) 037, [0912.3524].

39. A. P. Braun, A. Collinucci, and R. Valandro, *Hypercharge flux in F-theory and the stable Sen limit*, 1402.4096.

40. M. J. Dolan, J. Marsano, N. Saulina, and S. Schafer-Nameki, *F-theory GUTs with U(1) Symmetries: Generalities and Survey*, 1102.0290.

41. J. Marsano, N. Saulina, and S. Schafer-Nameki, *Compact F-theory GUTs with $U(1)_{PQ}$*, JHEP **04** (2010) 095, [0912.0272].

42. E. Dudas and E. Palti, *On hypercharge flux and exotics in F-theory GUTs*, JHEP **09** (2010) 013, [1007.1297].

43. J. Marsano, *Hypercharge Flux, Exotics, and Anomaly Cancellation in F-theory GUTs*, Phys.Rev.Lett. **106** (2011) 081601, [1011.2212].

44. M. J. Dolan, J. Marsano, and S. Schafer-Nameki, *Unification and Phenomenology of F-Theory GUTs with U(1)PQ*, JHEP **1112** (2011) 032, [1109.4958].

45. E. Palti, *A Note on Hypercharge Flux, Anomalies, and U(1)s in F-theory GUTs*, Phys.Rev. **D87** (2013), no. 8 085036, [1209.4421].

46. R. Blumenhagen, *Gauge Coupling Unification in F-Theory Grand Unified Theories*, Phys. Rev. Lett. **102** (2009) 071601, [0812.0248].

47. C. Mayrhofer, E. Palti, and T. Weigand, *Hypercharge Flux in IIB and F-theory: Anomalies and Gauge Coupling Unification*, JHEP **1309** (2013) 082, [1303.3589].

48. A. P. Braun, A. Collinucci, and R. Valandro, *The fate of U(1)'s at strong coupling in F-theory*, 1402.4054.

49. T. W. Grimm and T. Weigand, *On Abelian Gauge Symmetries and Proton Decay in Global F-theory GUTs*, Phys.Rev. **D82** (2010) 086009, [1006.0226].

50. M. Cvetic, A. Grassi, D. Klevers, and H. Piragua, *Chiral Four-Dimensional F-Theory Compactifications With SU(5) and Multiple U(1)-Factors*, 1306.3987.

51. D. R. Morrison and D. S. Park, *F-Theory and the Mordell-Weil Group of Elliptically-Fibered Calabi-Yau Threefolds*, JHEP **1210** (2012) 128, [1208.2695].

52. J. Borchmann, C. Mayrhofer, E. Palti, and T. Weigand, *Elliptic fibrations for SU(5) x U(1) x U(1) F-theory vacua*, Phys.Rev. **D88** (2013) 046005, [1303.5054].

53. J. Borchmann, C. Mayrhofer, E. Palti, and T. Weigand, *SU(5) Tops with Multiple U(1)s in F-theory*, 1307.2902.

54. M. Cvetic, D. Klevers, and H. Piragua, *F-Theory Compactifications with Multiple U(1)-Factors: Constructing Elliptic Fibrations with Rational Sections*, JHEP **1306** (2013) 067, [1303.6970].

55. M. Cvetic, D. Klevers, and H. Piragua, *F-Theory Compactifications with Multiple U(1)-Factors: Addendum*, JHEP **1312** (2013) 056, [1307.6425].

56. M. Cvetic, D. Klevers, H. Piragua, and P. Song, *Elliptic Fibrations with Rank Three Mordell-Weil Group: F-theory with U(1) x U(1) x U(1) Gauge Symmetry*, 1310.0463.

57. V. Braun, T. W. Grimm, and J. Keitel, *New Global F-theory GUTs with U(1) symmetries*, JHEP **1309** (2013) 154, [1302.1854].

58. S. Krippendorf, D. K. M. Pena, P.-K. Oehlmann, and F. Ruehle, *Rational F-Theory GUTs without exotics*, 1401.5084.

59. M. Kuentzler and S. Schafer-Nameki, *Tate Trees for elliptic fibrations with rank one Mordell-Weil group*.

60. C. Lawrie and S. Schafer-Nameki, *Tate Trees for elliptic fibrations with rank two Mordell-Weil group*.

61. F. Denef, *Les Houches Lectures on Constructing String Vacua*, 0803.1194.

62. T. Weigand, *Lectures on F-theory compactifications and model building*, Class.Quant.Grav. **27** (2010) 214004, [1009.3497].

63. A. Maharana and E. Palti, *Models of Particle Physics from Type IIB String Theory and F-theory: A Review*, Int.J.Mod.Phys. **A28** (2013) 1330005, [1212.0555].

64. R. Blumenhagen, M. Cvetic, S. Kachru, and T. Weigand, *D-Brane Instantons in Type II Orientifolds*, Ann.Rev.Nucl.Part.Sci. **59** (2009) 269–296, [0902.3251].

65. J. J. Heckman, J. Marsano, N. Saulina, S. Schafer-Nameki, and C. Vafa, *Instantons and SUSY breaking in F-theory*, 0808.1286.

66. J. Marsano, N. Saulina, and S. Schafer-Nameki, *An Instanton Toolbox for F-Theory Model Building*, JHEP **01** (2010) 128, [0808.2450].

67. J. Marsano, N. Saulina, and S. Schafer-Nameki, *Gauge Mediation in F-Theory GUT Models*, Phys. Rev. **D80** (2009) 046006, [0808.1571].

68. J. J. Heckman and C. Vafa, *F-theory, GUTs, and the Weak Scale*, JHEP **0909** (2009) 079, [0809.1098].

69. J. J. Heckman and C. Vafa, *Flavor Hierarchy From F-theory*, Nucl.Phys. **B837** (2010) 137–151, [0811.2417].

70. A. Font and L. E. Ibanez, *Matter wave functions and Yukawa couplings in F-theory Grand Unification*, JHEP **09** (2009) 036, [0907.4895].

71. H. Hayashi, T. Kawano, Y. Tsuchiya, and T. Watari, *Flavor Structure in F-theory Compactifications*, 0910.2762.

72. F. Marchesano and L. Martucci, *Non-perturbative effects on seven-brane Yukawa couplings*, Phys.Rev.Lett. **104** (2010) 231601, [0910.5496].

73. S. Cecotti, C. Cordova, J. J. Heckman, and C. Vafa, *T-Branes and Monodromy*, 1010.5780.

74. R. Donagi and M. Wijnholt, *Gluing Branes, I*, JHEP **1305** (2013) 068, [1104.2610].

75. R. Donagi and M. Wijnholt, *Gluing Branes II: Flavour Physics and String Duality*, JHEP **1305** (2013) 092, [1112.4854].

76. A. Font, F. Marchesano, D. Regalado, and G. Zoccarato, *Up-type quark masses in SU(5) F-theory models*, JHEP **1311** (2013) 125, [1307.8089].

77. J. J. Heckman, A. Tavanfar, and C. Vafa, *Cosmology of F-theory GUTs*, *JHEP* **1004** (2010) 054, [0812.3155].
78. T. W. Grimm, *Axion Inflation in F-theory*, 1404.4268.
79. E. Palti and T. Weigand, *Towards large r from [p,q]-inflation*, 1403.7507.
80. T. W. Grimm, D. Klevers, and M. Poretschkin, *Fluxes and Warping for Gauge Couplings in F-theory*, *JHEP* **1301** (2013) 023, [1202.0285].
81. M. Cvetic, T. W. Grimm, and D. Klevers, *Anomaly Cancellation And Abelian Gauge Symmetries In F-theory*, *JHEP* **1302** (2013) 101, [1210.6034].
82. T. W. Grimm, R. Savelli, and M. Weissenbacher, *On alpha' corrections in N=1 F-theory compactifications*, *Phys.Lett.* **B725** (2013) 431–436, [1303.3317].
83. V. Braun and D. R. Morrison, *F-theory on Genus-One Fibrations*, 1401.7844.
84. D. R. Morrison and W. Taylor, *Sections, multisections, and U(1) fields in F-theory*, 1404.1527.
85. J. Marsano, N. Saulina, and S. Schafer-Nameki, *Global Gluing and G-flux*, *JHEP* **1308** (2013) 001, [1211.1097].
86. L. B. Anderson, J. J. Heckman, and S. Katz, *T-Branes and Geometry*, 1310.1931.
87. F. Bonetti, T. W. Grimm, and T. G. Pugh, *Non-Supersymmetric F-Theory Compactifications on Spin(7) Manifolds*, *JHEP* **1401** (2014) 112, [1307.5858].
88. C. Lawrie and S. Schafer-Nameki, *The Tate Form on Steroids: Resolution and Higher Codimension Fibers*, *JHEP* **1304** (2013) 061, [1212.2949].

10

Compactified String Theories — Generic Predictions for Particle Physics

Piyush Kumar

Sloane Physics Laboratory, Department of Physics
Yale University, New Haven, CT 06511, USA

In recent years it has been realized that in string/M theories compactified to four dimensions which satisfy cosmological constraints, it is possible to make some generic predictions for particle physics: a non-thermal cosmological history before primordial nucleosynthesis, a scale of supersymmetry breaking which is "high" as in gravity mediation, and scalar superpartners too heavy to be produced at the LHC (although gluino production is predicted in many cases). When the matter and gauge spectrum below the compactification scale is that of the MSSM, a robust prediction of about 125 GeV for the Higgs boson mass, as well as predictions for future precision measurements, can be made. As a prototypical example, M theory compactified on a manifold of G_2 holonomy leads to a good candidate for our "string vacuum", with the TeV scale emerging from the Planck scale, a de Sitter vacuum, robust electroweak symmetry breaking, and solutions of the weak and strong CP problems. In this article we review how these and other results are derived, from the key theoretical ideas to the final phenomenological predictions.

1. Introduction

We are living in an exciting era of high-energy physics. The Large Hadron Collider (LHC) has had a monumental achievement – the discovery of the Higgs Boson. The question about the nature of electroweak symmetry breaking (EWSB) has thus been answered. However, it is well known that the Higgs mass in the Standard Model is subject to quadratic divergences from quantum corrections. In addition to this prominent aesthetic defect, the Standard Model is incapable of explaining the origins of Dark Matter, the asymmetry between the abundance of matter and antimatter, and the origin of neutrino masses, thereby strongly motivating the presence of beyond-the-Standard-Model (BSM) physics. Experiments in a variety of fields in high-energy physics, ranging from the energy frontier to the precision and cosmic frontiers, are involved in serious explorations of physics beyond the Standard Model (SM), already ruling out large regions of parameter space of many models. The hope is that at least some of the above experiments would signal a clear evidence for BSM physics.

There is a second reason why we are entering an exciting era. Developments over the past few years have led us to an understanding that, under some very general, simple and broad assumptions, string/M theory provides a *framework* that (in practice) is capable of addressing key, fundamental questions about particle physics and cosmology. Moreover, the framework addresses them in a unified way. At first sight, this may appear to be a surprising statement since we presumably still have a lot to learn about string theory. Furthermore, there are an enormous number of solutions to string/M theory which describe effectively four-dimensional Universes – the string landscape. The enormity of the landscape has led to the popular, but incorrect, view that string theory has no predictive power and virtually *any* low-energy theory could be a part of the landscape.

In this article we provide a brief review some of the results which demonstrate that, on the contrary, *generic predictions* do arise from string/M theory which can be directly tested at current and future particle physics experiments as well as with astrophysical and cosmological observations. This article is a shorter and updated version of the review [1] which appeared a few years ago.

2. Basic Idea

The basic assumption we make is the following: *Our Universe is described by a solution of string/M theory.* Once one believes in the above, then the task is to figure out ways to test predictions of various solutions of string/M theory against current and future data. To this end, we elicit from the theory the simplest, generic consequences which could describe our Universe and are relevant for high energy physics and cosmological observables. In order to make our task tractable and because low-energy supersymmetry and grand unification are extremely well motivated, we focus on string/M theory solutions with low energy supersymmetry and grand unification(at around[a] 10^{16} GeV). However, in principle one can also carry out the exercise below relaxing the above criteria.

---

[a]We will use 'natural' units in which $c = \hbar = 1$.

Given the above, the physics below the GUT scale can be effectively described by a four dimensional supergravity theory whose field content is at least that of the minimal supersymmetric Standard Model (MSSM) [2]. This four dimensional theory can be thought of as any other quantum field theory, but with an essential difference. The theory also contains what are moduli and axion fields, which parametrize the size and shape of extra dimensions, as well as their couplings to matter and to each other. A detailed description of moduli and axions is provided in one of the other chapters in this book. The moduli and axions are essentially the only low energy remnants of the string/M theory origin of the effective theory, and all other string/M theory modes are decoupled from the four dimensional theory. There are typically large numbers of moduli and axion fields and, moreover, the axion decay constants are of order the GUT scale [3]. We would now like to ask: if we consider a *generic* solution of string/M theory with low energy supersymmetry and grand unification, what phenomena does it describe? This is completely analogous to asking within the framework of quantum field theory, for example: what are the generic predictions of chiral gauge theories with hierarchical Yukawa couplings and spontaneous symmetry breaking? Essentially, if we threw a dart at the set of all solutions of string/M theory which reduce to the Standard Model for physics processes below the TeV scale, what would the properties of that solution be?

The key to answering this question lies in the physics of the moduli and axion fields and the effective supergravity theory. In the supergravity theory, the mass of gravitino (the superpartner of the graviton), $m_{3/2}$, sets the scale for the masses of *all scalar fields* unless symmetries or special dynamics prevent this. This is borne out by explicit string/M theory calculations. One of the key results that underlies many of the predictions is a connection between the lightest moduli mass and the gravitino mass. Essentially, the gravitino mass $m_{3/2}$ is related to the lightest modulus mass (the smallest eigenvalue of the extended moduli mass matrix) by an $\mathcal{O}(1)$ factor. Details of the derivation are given in [4]. In fact, both the MSSM scalars and the moduli fields will have masses of order $m_{3/2}$. This is not true of the axion fields a_i due to the shift symmetries $a_i \to a_i + c_i$ that originate from gauge invariance in higher dimensions. We will see that this basic result (with some mild assumptions) has extremely profound and robust implications for a large number of observables for particle physics and cosmology.

3. Assumptions

In this section we set out clearly the broad set of working assumptions under which our arguments and claims are valid. Strictly speaking, none of the assumptions are inevitable consequences of string theory compactifications to four dimensions. However, we will explain why each assumption is well motivated and sufficiently generic such as to hold true for a large class of solutions.

The first working assumption is that the vacuum structure of a compactified string theory is determined by an effective potential V_{eff} which is a function of all the moduli fields. The task is to include all relevant classical and quantum effects in determining V_{eff} and then determine its local and/or global minima. This seems to be a well justified approach in most cases, but is difficult to make precise within a theory of quantum gravity (see [5]). A nice summary of these issues is provided in [6]. The philosophy we adopt is that we are only interested in string/M theory solutions which could describe our world, so we do not need to study the most general set of solutions. With this point of view, the Wilsonian effective action paradigm seems quite natural.

The second assumption we will make is that the solution to the cosmological constant problem is largely decoupled from particle physics considerations. Note that we will still require that the vacuum energy vanishes approximately, but assume that additional mechanisms responsible for giving rise to the exceptionally tiny value of the cosmological constant have virtually no effect on particle physics. This assumption seems to be quite natural and conservative as there is no known, measurable, particle physics process in which the *precise* value of the cosmological constant is important. While we cannot be sure until the solution to the cosmological constant problem is agreed on, it seems unlikely that knowing its solution will help in calculating the Higgs boson mass or the relic density of dark matter, etc, and it seems unlikely that not knowing the solution will prevent us from doing such calculations. As mentioned earlier, we restrict to compactifications with low scale supersymmetry and with standard grand unification at the Kaluza-Klein scale, $M_{KK} \sim M_{GUT} \sim 0.1 M_{st}$, though many of our results can be extended to other cases. Here M_{st} is the string scale.

Regarding cosmological assumptions, we assume that the Hubble parameter in the very early Universe, such as during inflation, is larger than $\mathcal{O}(m_{3/2})$. Since we consider compactifications with low-scale supersymmetry, this implies that the Hubble scale during inflation H_I satisfies $H_I \gtrsim 10^5$ GeV. Having a large H_I seems to be a natural assumption for the following reason. It is known that the slow-roll parameter for simple (single-field slow-roll) models of inflation, $\epsilon \equiv m_{pl}^2 \left(\frac{V_I'}{V_I}\right)^2$, where V_I is the inflaton potential and V_I' is its derivative with respect to the inflaton, can be written in terms of H_I as:

$$\epsilon \approx 10^{10} \left(\frac{H_I}{m_{pl}}\right)^2 \tag{1}$$

using the value of the primordial density perturbations $\frac{\delta\rho}{\rho} \sim 10^{-5}$. The requirement $\epsilon \lesssim 10^{-2}$ for ~ 60 e-foldings of inflation to solve the flatness and horizon problems implies that $H_I \lesssim 10^{-6} \, m_{pl}$, which is the standard fine-tuning in slow-roll inflation models. A smaller value of H_I will make ϵ even smaller, implying an even larger fine-tuning than is necessary for inflation. Hence, it is natural for H_I to be as large as allowed by data, giving rise to H_I (much) larger than $\mathcal{O}(m_{3/2})$. Another assumption we make relevant for cosmology is that not *all* the moduli are stabilized close to an enhanced symmetry point. It is clear that in a generic case the above assumption will be satisfied, only under extremely special circumstances could it be violated, if at all.

We finally outline assumptions about model-building. There has been considerable progress in string phenomenology in this regard, and large classes of string models have been constructed with quasi-realistic gauge groups and matter content. The origin of flavor has also been addressed in several different classes of solutions [7, 8]. We will see that some of our broad predictions do not depend on the precise spectrum for the visible sector, while some of the more detailed predictions do. For the latter, we will make the reasonable assumption that the string/M theory compactification is such that the *visible* sector at low energies consists of the SM gauge group with matter content that of the MSSM[b]. The precise unification of gauge couplings, as well as successful radiative electroweak symmetry breaking in the MSSM provide strong support for such an assumption. Moreover, explicit string solutions with *precisely* the MSSM content have been constructed [9]. The latter assumption can be easily relaxed to include more complicated matter as well as gauge sectors. We assume that the visible sector is weakly coupled, i.e. all the SM fermions as well as the Higgs fields are elementary (as opposed to composite) so that the standard

---

[b]We allow the possibility of "dark sectors" weakly coupled to the visible sector, see discussion in section 8.1.

Higgs mechanism gives rise to electroweak symmetry breaking in the effective low-scale theory. As far as supersymmetry breaking is concerned, because of the moduli problems discussed above, we assume gravity mediation with a "hidden sector" of supersymmetry breaking. This is generic in eleven dimensional M theory compactifications in which the extra dimensions form a G_2-manifold. There, non-Abelian gauge fields are localized along three-dimensional submanifolds of the seven extra dimensions. In seven dimension, two three-manifolds generically don't intersect, there are no matter fields charged under both the Standard Model and hidden sector gauge symmetries. For other compactifications, this is more model-dependent but can be satisfied; in any case, generically it must be satisfied in order to avoid the moduli problem.

Finally, we assume that there are no other R or non-R global symmetries (such as a PQ symmetry) at low energies. This is a natural assumption since global symmetries are generically broken in the presence of gravity by Planck suppressed operators, and within gravity mediation these Planck suppressed operators are relevant. Also, the vanishingly tiny value of the cosmological constant implies that the superpotential does not vanish in the vacuum obtained after moduli stabilization, which explicitly breaks any potential R-symmetry.

4. Moduli Stabilization and Supersymmetry Breaking

Based on the philosophy and assumptions outlined in the previous section, we summarize the results about moduli stabilization and supersymmetry breaking which will be relevant for low energy particle physics. We are interested in compactifications of string/M theory to four dimensions which preserve $\mathcal{N} = 1$ supersymmetry. In the limit in which the string coupling, the string length and the size of the extra dimensions are small, the low energy four-dimensional theory obtained in $\mathcal{N} = 1$ compactifications of *all* corners of string/M theory is $\mathcal{N} = 1$, $D = 4$ supergravity. The internal volumes, although small, must be large enough such that the supergravity approximation is valid. Here "low energy" refers to energies far below the compactification scale, or Kaluza-Klein (KK) scale. As explained earlier, we will consider cases where $M_{KK} \sim M_{GUT}$. Since M_{KK} is only determined *after* moduli stabilization, the above condition has to be checked self-consistently.

$\mathcal{N} = 1$, $D = 4$ supergravity is completely specified at the two-derivative level by three functions:[c]

- The superpotential W, which is a holomorphic function of the chiral superfields. W is not renormalized in perturbation theory but receives non-perturbative corrections in general.
- The gauge kinetic functions f_a for each gauge group G_a, which are also holomorphic functions of the chiral superfields.
- The Kähler potential K, which is a real non-holomorphic function of the chiral superfields and their complex conjugates. Unlike W and F, K receives corrections to all orders in perturbation theory. The finiteness of the string scale gives rise to corrections in powers of $\left(\frac{l_s}{V^{1/6}}\right)$ where l_s is the string length and V is the (stabilized) volume of the extra dimensions [d]. For values of V which correspond to the unification scale M_{GUT}, these corrections are small, as in the ones discussed below.

Different compactifications of string/M theory give rise to different functional forms for K, W and f in general, although we will see later that phenomenologically realistic solutions arising from different corners of string theory share many common features.

The fields in the four dimensional theory include the moduli, axions and charged matter fields (both visible and hidden) as well as their superpartners. However, since the moduli (and some hidden sector matter) fields generically acquire large *vevs* ($\sim M_{st}$) while the visible matter fields must have vanishing *vevs*[e], it is a good approximation to first study the moduli and hidden matter potential, subsequently adding visible sector matter fields as an expansion around the origin of field space. It is important to also make sure that effects which could induce *vevs* for SM-charged matter fields are not present [11].

The gravitino mass $m_{3/2}$ is given in $\mathcal{N} = 1$ supergravity by:

$$m_{3/2} = e^{K/2}\frac{\langle W \rangle}{m_{pl}^2} = \frac{\sqrt{\sum_i \langle F^i F_i \rangle}}{\sqrt{3}\, m_{pl}} \tag{2}$$

where F_i are the F-terms (defined as derivatives of K and W wrt to the scalar fields: $F_i = e^{K/2}(\partial_i W + \partial_i K\, W)$); a non-zero expectation value for any of the F_i implies supersymmetry is broken. m_{pl} is the reduced Planck

---

[c]See [10] for a review. Strictly speaking, the lagrangian depends upon two functions, but it is convenient to use three.

[d]This can be easily generalized to the 11D M theory case.

[e]The Higgs *vev*, although non-zero, is much smaller than the typical moduli *vevs*.

scale $m_{pl} \equiv M_{pl}/\sqrt{8\pi}$, and in the second equality we have used the fact that the cosmological constant is vanishingly small. Hence in such vacua the gravitino mass is the order-parameter of supersymmetry breaking. Now, since we are interested in classes of vacua with low-energy supersymmetry to provide a solution to the hierarchy problem, $m_{3/2}$ must be much smaller than m_{pl}. This implies[f] that $\langle W \rangle$ or $\bar{F} \equiv \sqrt{\sum_i \langle F^i F_i \rangle}$ has to be much suppressed relative to m_{pl}. In order to discuss these dynamical issues, it will be useful to illustrate the results with an example. We will then see how the results generalize to other compactifications as well.

4.1. *M theory compactifications on G_2 manifolds*

Compactifications of 11D M theory to four dimensions preserve $\mathcal{N} = 1$ supersymmetry if the metric on the seven extra dimensions has holonomy group equal to the exceptional Lie group G_2 [12].

Phenomenologically relevant compactifications with non-abelian gauge symmetry and chiral fermions can only arise from G_2 manifolds endowed with special kinds of singularities. In particular, non-abelian gauge fields are localized along three-dimensional submanifolds inside the internal space [13] while chiral fermions are supported at points in the extra dimensions where there are conical singularities of particular kinds [14]. The gauge fields and chiral fermions of course also propagate in the four large space-time dimensions. Although many examples of smooth G_2 manifolds have been constructed [15], an explicit construction of *compact* G_2 manifolds with all the singularities required to give rise to phenomenologically relevant solutions has proven so far to be too challenging technically, though many such manifolds are strongly believed to exist. "String dualities" imply the existence of many examples: for instance, the duality between M theory and heterotic string and Type IIA compactifications. We will thus assume that singular G_2 manifolds supporting non-abelian gauge theories and chiral fermions exist and use the fact that enough is known about the K's, W's and f's which arise from G_2-manifolds in order to proceed. Many properties of the four-dimensional $\mathcal{N} = 1$ theory relevant for particle physics can be derived from a Kaluza-Klein reduction of 11D supergravity to four dimensions, which is the low energy limit of M theory.

At low energies, M theory is described by eleven dimensional supergravity theory which contains a metric, a 3-form gauge field (C) and a gravitino. We will not be interested in solutions in which there is a non-trivial flux

---

[f]A large and negative K corresponds to a Kaluza-Klein scale much less than M_{GUT}.

for the field strength of C along the extra dimensions: although fluxes can stabilize moduli [16], they do not generate a hierarchy between the Planck scale and the gravitino mass. We will see in Section 4.2 that unlike M theory, fluxes do play an important role in Type IIB string theory.

In M theory compactifications on a G_2-manifold, all the moduli fields are geometric — they arise as massless fluctuations of the metric of the extra dimensions [17]. Since these moduli s_j are *real* scalar fields[g], in order to reside in the *complex* chiral supermultiplets required by supersymmetry, additional real scalar fields must also be present. These additional fields are the axions a_j which arise as the harmonic fluctuations of C along the G_2-manifold. The moduli and axions pair up to form *complex* scalar fields which are the lowest components of chiral superfields Φ_i in the effective 4d supergravity theory:

$$\Phi_j = a_j + is_j + \text{fermion terms}. \tag{3}$$

4.1.1. *Moduli and scales in M theory on a G_2-manifold*

For future reference this subsection summarizes the relations between the G_2-moduli, the volume of the extra dimensions and the gauge couplings. More precise relations are given in [18]. The volume V_7 of a G_2-manifold is a homogeneous function of the moduli of degree 7/3. For instance, if the volume of a G_2 manifold is dominated by a single modulus field, then $V_7 \sim s^{7/3}$. Roughly speaking, *it is useful to think of the moduli vevs as parametrizing the volumes of a set of independent three-dimensional submanifolds* of the G_2-manifold, in units of the 11d Planck length. So, if V_7 is dominated by a single term, we think of the volume of the G_2-manifold as being dominated by a single three-cycle Q with volume $Vol(Q) \sim s$. Non-Abelian gauge fields are localized along three-dimensional submanifolds, hence the effective gauge coupling g_{YM}^2 is related to the volume of the three-manifold as $\frac{4\pi}{g_{YM}^2} = 1/\alpha = Vol(Q)$. Consider then a G_2-manifold such that the three-manifold which supports the Standard Model (unified) gauge group dominates V_7. We then have that

$$V_7 \sim \frac{1}{\alpha_{GUT}^{7/3}} \sim (25)^{7/3} \tag{4}$$

where we use the fact that these volumes are understood to be given at the GUT scale and that $\alpha_{GUT} \sim 1/25$ with the MSSM field content. We then

---

[g]The subscripts $i, j, k, ..$ are used to enumerate the moduli fields and their superpartners.

further infer that

$$m_{pl}^2 \sim V_7 M_{11}^2 = \frac{M_{11}^2}{\alpha_{GUT}^{7/3}} \tag{5}$$

and, because the volume of Q is given by $1/\alpha_{GUT}$ that

$$M_{KK} \sim M_{GUT} \sim M_{11}\alpha_{GUT}^{1/3}. \tag{6}$$

Thus, a value of $\alpha_{GUT} \sim 1/25$ gives a set of relations consistent with Newtons constant, $M_{GUT} \sim 2 \times 10^{16}$GeV and $M_{11} > M_{GUT}$. The latter fact is required for validity of the low energy effective field theory approximation.

More generally if we assume that the G_2-manifold is more or less isotropic then we expect the vevs of all moduli to be of the same order, and hence the above scalings with α_{GUT} will still hold true. We now return to discuss the potential for the moduli in more detail.

4.1.2. Hierarchies are generic in M theory

All the moduli Φ_j of M theory are invariant under shift symmetries [17]:

$$\Phi_j \to \Phi_j + c_j \tag{7}$$

with c_j being an arbitrary constant. The origin of the shift symmetries can be understood as follows. The real parts of the moduli Φ_j, denoted by a_j in (3), arise from the Kaluza-Klein (KK) reduction of the three-form (antisymmetric tensor field with three indices) in eleven dimensions to four dimensions. The underlying gauge symmetry of this three-form in higher dimensions reduces to shift symmetries for the individual axions in four dimensions: $a_j \to a_j + c_j$. With $\mathcal{N} = 1$ supersymmetry, the a_j combine with the modes arising from the KK reduction of the metric in eleven dimensions to form chiral superfields whose scalar components are Φ_j, and which are invariant under (7).

The above symmetries imply that the effective superpotential in four dimensions, W, which must be a holomorphic function of the Φ_j, does not contain 'perturbative' terms (i.e. terms polynomial in the Φ's). Hence *the perturbative superpotential for the moduli vanishes exactly.* This is a key point which distinguishes M theory on a G_2-manifold from other compactifications such as Type IIB and heterotic string theories on a Calabi-Yau manifold. For instance, Calabi-Yau manifolds generically have complex structure moduli; since these moduli are already complex fields, the corresponding supermultiplets do not have a shift symmetry and, consequently, the superpotential can contain perturbative contributions dependent on these fields.

However, since axionic shift symmetries are generically broken by non-perturbative effects the superpotential will not be zero in general. For instance, if there is an asymptotically free gauge interaction present then the corresponding strong gauge dynamics at low energies will *necessarily* generate a non-perturbative superpotential proportional to (the cube of) the dynamically generated strong coupling scale (Λ): $W \sim \Lambda^3 \sim e^{\frac{-b}{\alpha_Q}} m_{pl}^3$ (in this case $1/b$ is the one loop β-function coefficient of the hidden sector gauge theory and α_Q is its fine-structure constant). More generally, 'pure' membrane instantons can generate terms in the superpotential [19]. In fact, *every term in the superpotential can be associated with a 3-cycle and will be proportional to* $e^{ibN^j \Phi_j}$. Here, the N^j are the $b_3(X)$ integers specifying the homology of the 3-cycle and b is a number characterising the given instanton contribution. Obviously, different instanton contributions will have different values for b and N_j.

For solutions of M theory for which the 4d supergravity approximation is valid, the KK scale is below the 11d Planck scale M_{11} and all of these non-perturbative contributions, which are of order $e^{-b\frac{M_{11}^3}{M_{KK}^3}}$, are *exponentially small*. Thus, on general grounds, one expects M theory compactifications without flux to generate a very small expectation value for W, which in turn implies an exponential *hierarchy* between $m_{3/2}$ and the Planck scale m_{pl}. Thus, we see that M theory on a G_2-manifold without flux is an ideal framework for addressing the hierarchy problem. The key questions then become: a) can the moduli potential generated by strong hidden sector gauge dynamics also stabilize the moduli? b) does this potential spontaneously break supersymmetry? These questions were answered affirmatively in [20, 21], thereby providing a proof of the 'lore' that hidden sector strong dynamics could i) generate the hierarchy between the Planck and weak scale, ii) stabilize the moduli fields and iii) spontaneously break supersymmetry.

It is further important to note that, in a region of field space where the supergravity approximation is valid, volumes are larger than one in 11d units, so the contributions to the potential from strong gauge dynamics is exponentially larger than the purely membrane instanton effects; this is because in the former case b is proportional to $\frac{1}{d}$ where d is a one-loop beta-function coefficient – typically an integer larger than unity – whereas such a 'suppression' of b is not present for membrane instantons. Hence, if strong gauge dynamics is present at low energies it will dominate the moduli potential.

We will now review the results of [20–22] in more detail. The simplest possibility is to consider a G_2-manifold with a single hidden sector interaction which becomes strongly coupled at some scale much smaller than M_{KK}. For instance, this could be given by $SU(N)$ super Yang-Mills theory with no light charged matter. In terms of the supergravity quantities, the inverse gauge coupling α_h^{-1} can be identified with the imaginary part of the gauge kinetic function for the hidden sector gauge theory (f_h), while the strong coupling scale Λ can be identified with the non-perturbative superpotential alluded to above [17]:

$$W \sim \Lambda^3 = e^{\frac{2\pi i}{N} f_h} m_{pl}^3. \tag{8}$$

In M theory, the Kähler potential $K = -3 \log(V_7)$ [23]. Substituting K, W, f_h into the formula for the supergravity potential and minimising indeed shows *formally* that all the moduli can be stabilized in this case. However, the vacuum is located in a region where the supergravity approximation is not valid since the the 3-cycle volume is negative.

Following this, we considered two hidden sector gauge theories – both super Yang-Mills without light charged matter. There are thus two dominant terms in the superpotential, characterised by two integers P and Q, the one-loop β-function coefficients.

$$W = A_1 e^{\frac{2\pi i}{P} f_{h1}} + A_2 e^{\frac{2\pi i}{Q} f_{h2}} \tag{9}$$

where we introduced two *constant* normalizations A_1 and A_2 and set $m_{pl} = 1$. The normalizations are constant in M theory due to the axionic shift symmetries. Further simplification arises by assuming that f_{h1} and f_{h2} are proportional to one another, though the more general case was analyzed in [24]. We will describe the results of the simplified cases here for ease of exposition, thus we set $f_{h1} = f_{h2} = f = \sum_{i=1}^{N} N^i \Phi_i$, since the proportionality constant can be absorbed into re-defining Q. Here i runs over all the N moduli in general and N^i are positive integers.

With two strong hidden sector interactions the supergravity potential has many stable vacua in which all the moduli are stabilized. The fact that there are many vacua should not come as a surprise since a sufficiently generic potential for N fields will possess of order 2^N critical points. Moreover many of these vacua are in regions where the supergravity approximation is applicable and in these vacua the hidden sector coupling $\alpha_h \ll 1$. One can (semi-analytically) study the potential close to the minima in an expansion in α_h. Let us consider the supersymmetric (anti-de Sitter) vacuum[h].

Here, one finds

$$\frac{1}{\alpha_h} = \frac{1}{2\pi} \frac{PQ}{Q-P} \log\left(\frac{A_1 Q}{A_2 P}\right) \tag{10}$$

and the (dimensionless) moduli vevs are fixed to be of order

$$s_i \sim \frac{1}{N_i} \frac{1}{\alpha_h}; \quad i = 1, 2, .., N. \tag{11}$$

The physical meaning of these moduli vevs is that these are the volumes (in units of the inverse eleven dimensional Planck scale) of various three-manifolds inside the seven-dimensional G_2 manifold. They are related to moduli fields $\hat{s}_i$ in four dimensions with dimension one, via $\hat{s}_i = m_{pl} s_i$. We can now calculate the gravitino mass as a function of P, Q and α_h up to numerical constants:

$$\frac{m_{3/2}}{m_{pl}} = e^{K/2} \frac{W}{m_{pl}^3} \sim A_2 \frac{e^{\frac{2\pi i}{c} f_h}}{V_7^{3/2}} = A_2 \frac{|Q-P|}{Q} \alpha_h^{7/2} e^{-\frac{2\pi}{Q\alpha_h}} \tag{12}$$

so that a value of $\alpha_h \sim 1/25$ with $A_2 = 1, Q = 8, P = 7$ for example, gives $m_{3/2} \sim 100$ TeV. Note that $\alpha_h \sim 1/25$ can arise from values of P and Q ≤ 10 and normalization constants such that the logarithm in the formula for α_h is ≥ 1.

Another requirement for a realistic compactification is a de Sitter (dS) vacuum i.e. positive cosmological constant. It is well known that de Sitter vacua do not arise in the classical limit of string/M theory [25]. A review of de Sitter space in string/M theory is given in [26]. One can interpret this result as "the classical potential for moduli fields does not possess de Sitter vacua". In the examples studied above, though the potential is generated through quantum effects, it is exponentially close to the classical limit in the sense that all terms in the potential are exponentially small. Hence, it is not surprising that all of the vacua found had negative cosmological constant. Therefore, we expect that in a would be de Sitter vacuum that the vacuum energy is dominated by a field which is not a modulus of the G_2-manifold. In [21, 22], it was shown that including matter fields charged under the hidden sector gauge symmetries leads straightforwardly to a vacuum with a positive cosmological constant in which the dominant contribution to the vacuum energy arises from the F-term of a hidden matter field. This turns

---

[h]The formulae for the non-supersymmetric vacua are very similar

out to be quite relevant for many phenomenological features, as will be seen below.

Finally, it is important to mention the stabilization of axions, which are the imaginary parts of the complex moduli fields. The moduli stabilization mechanism stabilizes *all* the moduli but gives a mass of $\mathcal{O}(m_{3/2})$ to only one combination of axions. The masses of the other axions are generated by higher order instanton effects which make them exponentially suppressed relative to $m_{3/2}$ [27]. This is crucial for a solution to the strong CP-problem discussed in Section 8.2.

4.2. *Type IIB and other compactifications*

Here we consider moduli stabilization and supersymmetry breaking in other branches of string theory. We will discuss Type IIB compactifications here as progress towards phenomenologically viable moduli stabilization was first made for these compactifications [28–30], stimulating a lot of activity [31–34]. The Type IIB compactifications are also better understood from a technical point of view and compactification manifolds with the required properties to stabilize moduli can be explicitly constructed. We will also see that the moduli stabilization mechanism described above in the M theory case can be essentially carried over to the Type IIB case with minimal differences. Since it is possible to construct explicit compactifications in Type IIB satisfying the criteria for moduli stabilization, this proves the robustness of the physical ideas which are crucial in stabilizing all moduli in M theory and Type IIB compactifications. At the end, we will briefly comment on Type IIA and Heterotic compactifications.

In Type IIB compactifications to four dimensions, non-abelian gauge theories can arise on the worldvolumes of D-branes — such as D7-branes wrapping a four-dimensional manifold inside the six-dimensional internal manifold. Chiral fermions arise from open strings at the intersection of two D7-branes [35]. There are three different kinds of moduli in these compactifications — complex structure, dilaton and Kähler moduli. Unlike M theory compactifications where all moduli were invariant under a shift symmetry, in this case only the Kähler moduli are invariant. Therefore, the perturbative superpotential can depend upon the complex structure and dilaton. It is possible to stabilize these moduli supersymmetrically at a high scale ($\sim M_{KK}$) by an appropriate choice of fluxes [6, 30]. However, the Kähler moduli are not stabilized by this mechanism. Non-perturbative effects can stabilize the Kähler moduli just as in the M theory case. Therefore, these moduli are generically much less massive than the complex structure moduli. It is convenient to first integrate out the heavier moduli, which gives a constant contribution to the superpotential — W_0. This has to be combined with the non-perturbative contributions to stabilize the Kähler moduli. As explained earlier, in order to solve the Hierarchy problem the value of the superpotential in the vacuum must be much smaller than m_{pl}^3 (if one does not want the extra dimensions to be extremely large). Hence, W_0 must be very small (or zero). This can be arranged by a proper choice of fluxes in Type IIB, but involves some tuning [36, 37]. Note that in the M theory case $W_0 = 0$ precisely, so the entire superpotential is non-perturbative naturally.

Vacua also exist in Type IIB theory with $W_0 = \mathcal{O}(1)$. In these vacua, if one includes the leading perturbative corrections to the Kähler potential, the moduli are stabilized at large values in which the volume of the Calabi-Yau manifold is exponentially large. This is the so-called LARGE Volume Scenario (LVS) and was developed in [32, 38]. LVS vacua exist partly because of a balancing between the perturbative and non-perturbative contributions to the potential which give rise to an exponentially large volume for the extra dimensions or, equivalently, an intermediate string scale. One obtains a hierarchy between $m_{3/2}$ and m_{pl} precisely because of the exponentially large volume – which corresponds to a large and negative expectation value for the Kähler potential. A variety of different possible phenomenological scenarios are possible in the LVS scheme resulting in different mass hierarchies between supersymmetric particles. The existence of LVS vacua is also closely tied to the fact that, in the classical limit, the low energy effective supergravity theory describing Type IIB compactifications exhibits what is called "no-scale structure". This implies, among other things, that the vacua of the classical potential have zero vacuum energy and is the reason why the perturbative corrections have such a significant effect when W_0 is not tuned to be small. With a lower string scale, LVS vacua do not generically give rise to grand unification at around 10^{16}GeV. For this and related reasons, some of the generic predictions we make may not always apply to LVS solutions.

A generic Type IIB compactification has many Kähler moduli in general, but most of the moduli stabilization mechanisms in many explicit examples work only for a few Kähler moduli. There is however one robust mechanism, valid for small W_0, which stabilizes *all* Kähler moduli in a compactification with many Kähler moduli with minimal ingredients, and

is inspired by results obtained in M theory.[i] It was shown in [39] that if
the non-perturbative superpotential depends on a linear combination of all
Kähler moduli, it is possible to stabilize all Kähler moduli as long as the
four-manifold supporting the instanton or gaugino condensate satisfies a
certain mathematical property, namely it is a "rigid ample divisor". Thus,
the qualitative result of stabilizing all moduli with low-scale supersymmetry can be obtained in this class of compactifications as well. Explicit
Calabi-Yau manifolds satisfying the above criteria were constructed in [39].
These compactifications share the interesting feature with the M theory
case that all but one axion are stabilized with exponentially suppressed
masses relative to $m_{3/2}$, which is crucial for solving the strong-CP problem. Finally, in order to generate a vacuum with positive cosmological
constant, vacuum energy contributions from non-moduli sources must be
included, as in M theory. In Type IIB compactifications, in addition to
possible F-term contribution arising from a hidden matter sector as in the
M theory [40], there could be contributions arising from D-terms [41] or
from explicit supersymmetry-breaking effects as well [30]. However, many
consequences for phenomenology do not depend on the details as long as
certain simple conditions are satsified, as we will explain in the following
subsection.

Finally, let us briefly comment on moduli stabilization in Type IIA and
Heterotic compactifications. In these cases, fluxes can stabilize some moduli
[42–44] but generically fail to generate the hierarchy. However, a better
understanding of these compactifications may eventually lead to progress
in demonstrating the existence of vacua with low energy supersymmetry in
particular classes, see for example [45, 46].

5. Moduli Spectra – Consequences

In this section, we provide some insight into the spectrum of the moduli
in general, and the lightest modulus mass in particular, relative to the
gravitino mass. We first explain a general result about the lightest modulus
mass which works in all cases satisfying the supergravity approximation
and is independent of the details of moduli stabilization. This result was
derived in [4], building on the work of [47]. The basic argument is as

---

[i]Another possible approach to stabilize more than one Kähler moduli in an LVS-like
scenario is to use a diagonal del-Pezzo divisor to stabilize the overall volume. The
remaining Kähler moduli are stabilized by a combination of D-term constraints and
string loop corrections [34].

follows. One considers the mass-matrix $\mathcal{M}^2$ for all the scalar fields in the true dS vacuum with broken supersymmetry, which is positive definite by assumption. Then, one can use the theorem that its smallest eigenvalue $\mathcal{M}^2_{min}$ is smaller than $\xi^\dagger \mathcal{M}^2 \xi$ for any unit vector ξ. Then, choosing a direction in scalar field space which corresponds to that of the sGoldstino (the superpartner of the Goldstino), one can show that:

$$\mathcal{M}^2_{min} = m^2_{3/2} \left(2 + \frac{|r|}{m^2_{pl}} \right) \tag{13}$$

where r is the "holomorphic sectional curvature" in the space of scalar fields [47], evaluated in the sGoldstino directions. Now, if the only scales in the problem are set by m_{pl}, then $\frac{r}{m^2_{pl}} = \mathcal{O}(1)$. For example, this is the case in M theory compactifications where all the moduli arise from the metric and the only scale is set by the 11D planck scale M_{11}, which determines both m_{pl} and $m_{3/2}$ in terms of dimensionless constants after moduli stabilization. In this case it can be shown that r is of order m^2_{pl}. This gives the result:

$$\mathcal{M}_{min} = \mathcal{O}(1)\, m_{3/2} \tag{14}$$

which we set out to prove. In other string compactifications, however, there are different kinds of moduli, such as the dilaton, complex structure and Kähler moduli. We focus on the Type IIB case for concreteness. In this case, the Kähler moduli are similar to the moduli in M theory, but it is possible in general that additional scales $\Lambda \ll m_{pl}$ may be present for the sGoldstino, due to the existence of other kinds of moduli. In this case, $|r|$ is enhanced by the ratio $\frac{m^2_{pl}}{\Lambda^2}$, so that $\frac{|r|}{m^2_{pl}} = \mathcal{O}(1)\left(\frac{m^2_{pl}}{\Lambda^2}\right)$ [4]. In these cases, the general supergravity result, although correct, does not provide a useful bound.

However, in realistic cases where all the moduli are stabilized by the mechanism explained in sections 4.1.2 and 4.2, quite remarkably the result (14) holds even in the presence of additional scales. A more detailed derivation is carried out in [72]. Thus, (14) seems to be a rather robust result in realistic compactifications with stabilized moduli.

5.1. *Cosmological consequences*

The above result that the lightest modulus mass is close to $m_{3/2}$ has a profound impact on pre-BBN cosmology. Current cosmological data can only directly constrain the early Universe when it is colder than about a MeV,

which is the onset of Big-Bang Nucleosynthesis (BBN). The most popular
assumption for cosmological history before BBN is a "thermal" history, i.e.
in which the early Universe starts out with a radiation dominated phase
due to reheating after inflation. In this case, the early Universe consists of
a plasma of relativistic particles at a very high temperature. The results
obtained above, however, question this assumption strongly under very general conditions. For a Hubble parameter during inflation which is bigger
than about 20 TeV [j], the moduli are displaced from their late-time minima during the early Universe, they start oscillating about their late-time
minima when the Hubble parameter becomes comparable to their masses.
Since they redshift like matter, they quickly dominate the energy density
of the Universe giving rise to a matter-dominated phase. As mentioned
earlier, the moduli interact gravitationally with all matter and hence have
very long lifetimes. Requiring that the moduli decay reheats the Universe
to temperatures above a few MeV thus puts a lower bound on their mass
to be about 20 TeV. This gives rise to what is known as a "non-thermal"
cosmological history. The related gravitino problem is also solved in the
following manner. The decay of the lightest modulus produces a lot of entropy, so the initial thermal abundance of the gravitinos is diluted away.
Furthermore, the lightest modulus is lighter than $2\,m_{3/2}$ in most examples [22, 39], so that its branching ratio to gravitinos is also kinematically
forbidden/suppressed [48], and a large abundance of gravitinos is not regenerated.

A non-thermal history of the Universe before BBN has very important
implications for many cosmological observables, and also for the origin and
abundance of Dark Matter (DM). Before moving on to issues related to DM
discussed in the next section in detail, we comment on possible cosmological
observables which follow from the existence of a non-thermal cosmological
history. One such observable could be the detection of gravitational waves
produced during inflation, as pointed out in [49]. Another observable is
related to the growth of substructures in the early Universe. As shown
in [50], the existence of a matter-dominated phase in the pre-BBN Universe
leads to a significantly different pattern in the growth of structure. More
studies are required to extract possible observable consequences.

---

[j]This is part of our set of assumptions, which seems quite natural. See Section 3.

5.2. Range of $m_{3/2}$

We saw that there is a lower bound on the gravitino mass of around 20 TeV from cosmological constraints. A gravitino mass around this value can also be obtained from realistic compactifications, as pointed out in [21, 22]. Is there also an upper bound on $m_{3/2}$ consistent with experimental constraints? The answer is yes for the following reason. As will be seen later, axions arising in these string compactifications naturally have a relic abundance comparable to the observed DM abundance. Moreover, the axion relic abundance is proportional to a positive power of $m_{3/2}$ and naturally gives rise to an $\mathcal{O}(1)$ fraction of DM with minimal tuning only when $m_{3/2} \lesssim 100$ TeV [27], so one expects an upper bound on $m_{3/2}$ of around 100 TeV. We discuss these issues related to dark matter in more detail the following section. *A point worth noting is that the upper limit on $m_{3/2}$ arises from phenomenological input rather than theoretical constraints.* On the other hand, for the lower bound, in addition to the phenomenological requirement above, there is also a theoretical argument [22].

6. Particle Phenomenology

We now discuss important aspects of the broad phenomenology arising in the setup considered. We will elaborate on general features of the superpartner spectra discussed briefly in previous sections, followed by a discussion of aspects of electroweak symmetry breaking, and the supersymmetric flavor and CP-problems. As discussed in Section 2, within the general setup considered, supersymmetry breaking in the hidden sector must be mediated to the visible sector by gravitational interactions, since otherwise there is a serious moduli problem. Thus, the gravitino mass sets the scale of all superpartners so one generically expects all the supersymmetry breaking mass parameters — the scalar masses, the trilinear parameters, and the gaugino masses to be $\mathcal{O}(m_{3/2})$. We will now consider generic features of these parameters and the resulting consequences below.

6.1. Scalar mass-squared and trilinear parameters

The expression for the soft scalar masses in $\mathcal{N} = 1$ supergravity coupled to matter is given by [51]:

$$m_{\bar{\alpha}\beta}^2 = m_{3/2}^2 \tilde{K}_{\bar{\alpha}\beta} - \Gamma_{\bar{\alpha}\beta} \tag{15}$$

where $\tilde{K}_{\bar{\alpha}\beta}$ is the Kähler metric for the matter fields α and β, and $\Gamma_{\bar{\alpha}\beta} \sim F^{\bar{i}}F^{j}\partial_{\bar{i}}\partial_{j}\tilde{K}_{\bar{\alpha}\beta}$ with F^i as the F-term for the modulus or hidden matter field labelled by i. Here m_{pl} has been set to unity. The precise expression for $\Gamma_{\bar{\alpha}\beta}$ can be found in [51]. In this basis the kinetic terms for the visible matter scalars are not canonical since $\tilde{K}_{\bar{\alpha}\beta}$ is non-trivial. To go to the canonically normalized basis, one does a unitary transformation $\mathcal{U}$ to make the Kähler metric diagonal, $(\mathcal{U}^{\dagger}\tilde{K}\mathcal{U})_{\bar{\alpha}\beta} = \hat{K}_{\alpha}\delta_{\bar{\alpha}\beta}$, and then does an appropriate rescaling for each field labelled by α to scale away the $\hat{K}_{\alpha}$. In the canonical basis, the mass-squared for the visible matter scalars is denoted by $\hat{m}_{\bar{\alpha}\beta}^{2}$ and is given by:

$$\hat{m}_{\bar{\alpha}\beta}^{2} = m_{3/2}^{2}\delta_{\bar{\alpha}\beta} - \left(\frac{1}{\sqrt{\tilde{K}}}\mathcal{U}^{\dagger}\Gamma\frac{1}{\sqrt{\tilde{K}}}\mathcal{U}\right)_{\bar{\alpha}\beta}. \tag{16}$$

Since $\Gamma_{\bar{\alpha}\beta}$ depends on the derivatives of $\tilde{K}_{\bar{\alpha}\beta}$, the second term in (16) is in general *not* proportional to $\delta_{\bar{\alpha}\beta}$. Both $\tilde{K}_{\bar{\alpha}\beta}$ and $\Gamma_{\bar{\alpha}\beta}$, however, depend on the values of the stabilized moduli, and are generically $\mathcal{O}(1)$ in string or 11D units. Hence, this gives rise to

$$\hat{m}_{\bar{\alpha}\beta}^{2} = \mathcal{O}(1)\, m_{3/2}^{2}. \tag{17}$$

A similar analysis for the trilinears gives $\hat{A}_{\alpha\beta\gamma} = \mathcal{O}(1)\, m_{3/2}$ in the canonically normalized basis.

We have seen that scalar masses and trilinears are generically $\mathcal{O}(1)\, m_{3/2}$ in a general supergravity theory. Although phenomenological models have been considered in which the mass-squared parameters above are parametrically separated from $m_{3/2}$ and/or each other, it is very hard to realize these within phenomenologically realistic string/M theory compactifications. In M theory compactifications with no background closed string fluxes turned on, sequestering does not seem to be possible [22]. The situation in Type IIB string compactifications is more subtle. With partial moduli stabilization, it was argued that sequestering may be possible in the presence of strong warping [52], or due to the visible sector being localized in the extra dimensions [53]. However,after taking into account the stabilization of all moduli, there arise couplings between the moduli and the visible matter sector in the superpotential which do not allow for phenomenologically viable sequestering[k] [54]. Thus, we conclude that both scalar masses and trilinear parameters are generically of $\mathcal{O}(m_{3/2})$ in viable examples. In the M theory and Type IIB moduli stabilization mechanisms considered, one can

---

[k] "sort-of-sequestering", defined in [54], may still be possible but that does not help.

go much further, as one can compute $\tilde{K}_{\bar{\alpha}\beta}, \Gamma_{\bar{\alpha}\beta}$ in terms of the microscopic parameters. Moreover, these functions satisfy homogeneity properties at leading order in supergravity, and it can be shown that [22]:

$$\Gamma_{\bar{\alpha}\beta} \propto \tilde{K}_{\bar{\alpha}\beta} + \text{higher order corrections.} \tag{18}$$

If these higher order corrections are small, as will be assumed in the following, then in the un-normalized basis one has:

$$m_{\bar{\alpha}\beta}^2 \simeq m_{3/2}^2 \left(1 - \frac{7}{3} \frac{(m_{1/2}^{tree})^2}{m_{3/2}^2} \right) \tilde{K}_{\bar{\alpha}\beta}$$

$$\simeq m_{3/2}^2 \tilde{K}_{\bar{\alpha}\beta}, \tag{19}$$

which gives rise to the following in the canonically normalized basis:

$$\hat{m}_{\bar{\alpha}\beta}^2 \simeq m_{3/2}^2 \delta_{\bar{\alpha}\beta} \tag{20}$$

since $(m_{1/2}^{tree})^2 \ll m_{3/2}^2$ for these compactifications as will be shown below, and in the second line we have written the mass-squared matrix for the sfermion fields in the canonically normalized basis. Thus, scalar masses are very close to $m_{3/2}$ within the framework with stabilized moduli. A similar statement can be made for trilinears [22].

6.2. *Flavor and CP*

The above results have implications for flavor and CP issues in the beyond-the-SM (superpartner) sector. Note that this issue is distinct from the origin and pattern of the quark and lepton Yukawa couplings in the Standard Model. That still remains a mystery, although progress has been made in understanding at least some of the issues involved from a top-down point of view [7, 8].

It is well known that *absent* any underlying structure, gravity mediation models generically lead to too large predictions for flavor and CP-violating observables. However, within the context of an underlying supersymmetry breaking mechanism arising in string theory, additional underlying structures, which help shed more light on these issues, are often present. One possibility is that the underlying string compactification preserves flavor symmetries which could effectively suppress flavor and CP-violation [96]. Another possibility arises from the structure of the underlying hidden sector and moduli dynamics associated with supersymmetry breaking and its mediation to the visible sector. We will focus on the latter as this is more directly connected to the physics of moduli stabilization. Also, for concreteness, we will focus on the M theory case as these issues have been studied in detail in this context. However, we expect the qualitative results to hold for other classes of compactifications as well, such as the class of Type IIB compactifications described in Section 4.2.

As mentioned above, within realistic moduli stabilization mechanisms as described in Section 4, $\tilde{K}_{\bar{\alpha}\beta}$ satisfies homogeneity properties at leading order that are broken by higher order derivative corrections. We assume that these higher order corrections are either small and/or have approximately the same flavor structure as that for the leading order [22], see also the discussion in Section 6.1 above Eq. (19). Then, $\Gamma_{\bar{i}j}$ is proportional to $\tilde{K}_{\bar{i}j}$ to a good approximation. This gives rise to approximately flavor-diagonal and universal soft mass-squared matrices, as shown in [22]. Note, however, that this is only true at the scale where the boundary conditions for the RG evolution are imposed, in this case the unification scale. RG effects and rotation to the super-CKM basis in general introduce a small amount of flavor violation. This can be parametrized by the quantities :

$$(\delta_{XY})_{ij} = \frac{(\hat{m}_{XY}^2)_{ij}}{\sqrt{(\hat{m}_{XY}^2)_{ii}(\hat{m}_{XY}^2)_{jj}}}. \tag{21}$$

Here $X, Y \in \{L, R\}$ and a hat denotes a matrix in the super-CKM basis.

The moduli stabilization mechanism discussed in Sections 4.1 and 4.2 also has an important consequence for the supersymmetric (weak) CP problem. An important feature of the moduli stabilization mechanism is that it gives rise to a *real* superpotential in the vacuum at leading order, i.e. it does not contain any CP violating phases. This has important consequences for CP violation in the flavor diagonal and off-diagonal sector as we will see. The reason is as follows. As explained in Section 5, at leading order the potential stabilizes all the real parts of the complex moduli but only stabilizes a few axions (the imaginary part of the moduli). For example, for the case considered in detail in Section 5, there are two terms in the superpotential at leading order. Then, it can be shown that one axionic combination t is stabilized such that $\cos t = -1$, implying that the terms in the superpotential align with the same phase (apart from a sign) [27, 55]. Since the overall phase of the superpotential can be rotated away and is not observable, this means that the superpotential in the vacuum is real at leading order. As will be seen in Section 8.2, all remaining axions are stabilized by effects by other non-perturbative terms in the superpotential which are exponentially suppressed relative to the leading terms[1], making them exponentially lighter than the gravitino mass and hence solve the strong CP-problem. The same also implies that once these remaining axions are stabilized, there may be terms in the superpotential with different phases, however since these terms are exponentially suppressed relative to the leading terms, they can be neglected to an excellent approximation. *It is worth emphasizing that the solutions to both the weak and strong CP problems have a common origin.*

Using the above, one can show that the soft supersymmetry breaking parameters in the Lagrangian are real at the unification scale to an excellent approximation [55]. This implies that in particular the gaugino masses and reduced trilinears $A^k_{ij} \equiv \tilde{A}^k_{ij}/Y^k_{ij}$ are real as well. Using the homogeneity properties of $\tilde{K}_{\bar{i}j}$, it is possible to show that $\tilde{A}^k_{ij}$ is roughly proportional to Yukawa couplings at the unification scale. Again, RG effects and rotation to the super-CKM basis introduce CP phases in the trilinear parameters in both the flavor-diagonal and off-diagonal sector. The most stringent constraints arise from observables like ϵ_K, $\text{Re}(\epsilon'/\epsilon)$ and electric dipole moments (EDMs) [56]. CP-violation in the flavor off-diagonal sector affects observables like ϵ_K mainly through chirality-conserving interactions, while that in the flavor diagonal sector affects EDMs through chirality-flipping interactions. The real part of ϵ'/ϵ gets dominant contributions from chirality-flipping flavor-violating effects such as $(\delta_{LR})_{12}$ and $(\delta_{RL})_{12}$. Utilizing the properties mentioned above, the contributions to all the above flavor and CP-violating observables were computed in detail in [55, 56], and it was found that all such constraints are satisfied with hierachical Yukawa couplings, with scalar masses and trilinears $\gtrsim 30$ TeV, and with gaugino masses $\lesssim$ TeV. Predictions were also made for various EDM measurements in [55]. In Section 7.2, we discuss possible experiments at the precision frontier which could test and constrain the framework.

Note that there is a qualitative difference between the electron and hadronic EDMs. The former is virtually vanishing in the SM, but the latter does receive a contribution from the θ angle in QCD, which in fact is the origin of the strong CP-problem. Therefore, once EDMs are observed for the electron, neutron, mercury, etc., it will be important to separate the supersymmetric (BSM) contribution from the contributions proportional to θ. We will show in Section 8.2 that there is a natural solution to the

---

[1]This can happen quite naturally since the arguments of these exponential terms are essentially given by the volume of sub-manifolds. So, if these volumes are just $\mathcal{O}(1)$ larger than those in the leading exponential, these terms will be highly suppressed.

strong CP problem within the string/M theory framework, and the value of θ is in principle determined by microscopic constants which also affect astrophysical observables. So, this gives rise to an extremely interesting (albeit indirect) connection between astrophysics and precision observables, which should be explored further.

6.3. Gaugino masses

What about gaugino masses? We will work within the framework of gravity mediated supersymmetry breaking, which is preferred due to BBN constraints as explained in Section 3. One could try to suppress gaugino masses relative to $m_{3/2}$ by a symmetry (R-symmetry in this case). However, since R-symmetry is broken by the rather large gravitino mass ($> \mathcal{O}(\text{TeV})$) in realistic string compactifications, it is not possible to suppress them arbitrarily. However, they can still be somewhat suppressed relative to the gravitino mass by the *dynamics* of moduli stabilization and supersymmetry breaking. This can be understood as follows. In many mechanisms of moduli stabilization, the geometric moduli which appear in the gauge kinetic function, T_i^{vis}, are stabilized "close" to a supersymmetric point. The dominant supersymmetry breaking contributions which give rise to a dS vacuum are provided by other sources. Hence, the gaugino masses, which are proportional to the F-terms for T_i^{vis}, are suppressed relative to the gravitino mass in these situations. This is true for the case of M theory compactifications [20–22]. In particular, using the results for moduli stabilization in Section 4.1.2 leads to a rather simple expression for the gaugino masses at tree level for phenomenologically viable cases:

$$
M_{1/2}^{tree} = \sum_{i=1}^{N} \frac{F^i \, \partial_i \, f_{vis}}{2 \, i \, \text{Im}(f_{vis})}
$$
$$
\simeq -\frac{\alpha_h \, Q}{3 \, \pi} \, m_{3/2} \, (1 + \mathcal{O}(\alpha_h)) \tag{22}
$$

where F_i is the susy breaking F-term for moduli i (the sum is over all N moduli), and f_{vis} is the visible sector gauge kinetic function which is an integer linear combination of all moduli, $f_{vis} = \sum_i^N N^i \, s_i$. α_h and Q are defined in Section 4.1.2, with the former related to the hidden sector gauge coupling $\alpha_h \equiv \frac{g_h^2}{4\pi}$, and Q being an integer related to the rank of the hidden gauge group. Note that the result is completely independent of the number of moduli N as well as the integer coefficients N^i! As explained above, since f_{vis} only depends on the moduli and not on the hidden field which is the dominant source of supersymmetry breaking, gaugino masses do not receive contributions from this dominant source and are hence suppressed relative to the gravitino mass. At one-loop, there are anomaly mediated contributions to the gaugino masses which turn out to be roughly of the same order [20–22], hence they should be included as well.

Suppressed gaugino masses also arise in many classes of Type IIB compactifications [30, 39, 57]. However, within Type IIB compactifications, it could also happen that the F-terms for T_i^{vis} are not suppressed, if, for example, they are stabilized by string-loop effects or perturbative effects in the Kähler potential [58]. In these cases, the gaugino masses are expected to be of $\mathcal{O}(m_{3/2})$. However, as will be explained in Section 8, in this case the LSP abundance severely overcloses the Universe, so these are ruled out unless R-parity is sufficiently violated. Furthermore, we will argue in Section 6.4 below that constraints from neutrino masses disfavor any form of R-parity violation in string/M theory frameworks with $SU(5)$ GUTs.[m] Therefore, only frameworks with suppressed gaugino masses will be considered.

6.4. μ and R-parity conservation

It is non-trivial to obtain a phenomenologically viable μ parameter in string theory. The phenomenologically viable value of μ in the canonically normalized basis of fields is around the TeV scale. It receives contributions both from supersymmetric terms in the superpotential and supersymmetry breaking terms arising from the Kähler potential. The natural options are, therefore, to have an approximate symmetry which suppresses the coefficient of the the holomorphic term $H_u H_d$ term in the superpotential by a large amount relative to the string/Planck scale (such as by exponential effects) — the Kim-Nilles/Casas-Munoz mechanism, or forbid the superpotential contribution altogether by a symmetry and generate a viable μ term by Kähler potential effects — the Giudice-Masiero mechanism. However, the symmetry has to be such that large masses for color triplet fields *are* allowed. Recall that the spectrum is assumed to arise from a GUT, hence the Higgs fields are part of a multiplet which also includes color triplet fields. These color triplet fields must get a large mass $\gtrsim M_{GUT}$ in order to not mediate proton decay at observably fast rates, which is the well-known "doublet-triplet splitting problem". Different string/M theory solutions can contain different solutions to the doublet-triplet splitting owing to the different origin of matter and gauge degrees of freedom as well as differences

---

[m]Many of the arguments can be generalized to other GUTs under certain conditions.

in the underlying structure of these compactifications. We are interested in those solutions which give rise to $\mu \lesssim m_{3/2}$. See [59] for heterotic examples, [60] for perturbative Type IIA and IIB cases, [61] for F-theory and [62] for M theory. In M theory compactifications one can further constrain μ by combining the requirements of moduli stabilization and the solution to the doublet-triplet splitting problem proposed by Witten [63]. This generically gives rise to $\mu \sim 0.1\, m_{3/2}$, but slightly smaller values may be possible as well.

GUT frameworks arising within String/M theory that have a mechanism for solving the doublet-triplet splitting and generating a phenomenologically viable μ (and $B\mu$), also have an important bearing on the issue of R-parity violation. It was shown in [64] that in contrast to the more model-dependent constraints on the proton lifetime, the limits on neutrino masses provide a robust, stringent and complementary constraint on all $SU(5)$ GUT-based R-parity violating models. Furthermore, imposing the neutrino mass bounds on models within this framework disfavors any R-parity violation altogether [64]. This can be roughly described as follows. In R-parity violating $SU(5)$ GUT models with the above properties, it can be shown that bilinear R-parity violation, through the term $\kappa\, LH_u + h.c.$ in the superpotential, is always present. In addition, barring extreme fine-tuning, κ is generated at the same order as μ unless R-parity violation is forbidden by a symmetry. However, limits on neutrino masses impose an upper bound on κ/μ to be around 10^{-3} even in the least stringent cases. Therefore, these arguments suggest that R-parity should be conserved in these classes of $SU(5)$ GUT models. The same qualitative result holds for minimal $SO(10)$ GUT models within the above framework as well. For more details, the reader is referred to [64].

6.5. The "Little" hierarchy

It is well known that electroweak symmetry is broken by RG effects in a natural manner in the MSSM once one imposes soft supersymmetry breaking boundary conditions at around the unification scale. This is known as "radiative electroweak symmetry breaking". This is because the RG equation for $m_{H_u}^2$ has a dependence on the top Yukawa coupling y_t which is larger than all other Yukawa couplings. Hence, it is natural for $m_{H_u}^2$ to be driven to small or negative values, thereby destabilizing the point $H_u = H_d = 0$ and giving rise to a Higgs vev. Thus, the higgs vev (or equivalently m_Z) becomes connected to the soft parameters and μ.

Although radiative EWSB is an extremely appealing feature of the MSSM, it turns out that obtaining the correct value of the Z mass by choosing $O(1)$ values of soft parameters relative to a common scale m_{soft} requires either a) $m_{soft} \sim m_Z$, or b) cancellation between soft parameters (essentially $m_{H_u}^2$ and μ^2 when $\tan\beta$ is not small) of order m_{soft}^2, if m_{soft} is larger than m_Z. The former option turns out to be incompatible with direct constraints on superpartner masses as well as the Higgs mass bounds from LEP, leaving the latter as the only option. This is the infamous "little hierarchy" problem in the MSSM. Note that this is not just true for the MSSM, the bounds on masses of new physics particles from direct production as well as bounds from indirect electroweak precision data imply that the problem is generically present in all other approaches to electroweak symmetry breaking such as warped extra dimensional models, composite higgs models, little higgs models, etc. and to a lesser extent even in weakly coupled models with an extended matter sector such as the NMSSM.

Since the framework considered here assumes the MSSM matter and gauge spectrum, the fact that the scalar superpartners are heavier than around 220 TeV0 TeV would naively seem to suggest a much more severe fine-tuning compared to MSSM models with scalar masses $\lesssim$ TeV. However, this turns out to be incorrect in models where μ is also suppressed, as we explain below. The basic reason is that in gravity mediation with no sequestering of the visible sector fields relative to the hidden sector, which seems to arise naturally within string theory solutions providing a solution to the moduli problem, *both* scalar masses (M_0) and trilinears (A_0) are close to each other, of $\mathcal{O}(m_{3/2})$. Since M_0 and A_0 appear in the RG equation for the Higgs mass-squared parameter $m_{H_u}^2$ with opposite signs, this gives rise to a near cancellation between the two terms, giving rise to a $m_{H_u}^2$ which is naturally suppressed relative to $m_{3/2}^2$ [65]. More concretely, $m_{H_u}^2$ at any given scale $t \equiv \log(Q/Q_0)$, Q_0 being the unification scale, is given by:

$$m_{H_u}^2(t) \simeq f_{M_0}(t)\, M_0^2 - f_{A_0}(t)\, A_0^2 + R(t). \tag{23}$$

The quantities f_{M_0} and f_{A_0} are determined by SM Yukawa couplings and gauge couplings at leading order. R, on the other hand, is determined primarily by the gluino mass parameter M_3 and hence gives a negligible contribution if $M_0, A_0 \gg M_3$, as is the case here. Then one finds that for $M_0 \simeq A_0 \simeq m_{3/2}$, f_{M_0} and f_{A_0} at the electroweak scale are naturally of order 0.1 and also nearly cancel each other [65], implying that:

$$m_{H_u}^2(Q_{EWSB}) \sim 10^{-2}\, m_{3/2}^2 \sim \text{TeV}^2. \tag{24}$$

Thus, in compactifications where μ is "small' ($\mu^2 \sim 10^{-2} m_{3/2}^2$) as for the frameworks discussed in the previous section, the naive fine-tuning is significantly reduced. For more details, please refer to [65]. Note that this mechanism, dubbed the "Intersection-Point" in [65], is quite different from the "Focus-point" region in the constrained MSSM [66], where A_0 at the unification scale is much smaller than the large soft mass parameters. When μ is "large", i.e. of the same order as $m_{3/2}$, the fine-tuning is quite severe as expected.

Even for small μ, since $m_{H_u}^2 \sim \text{TeV}^2$ rather than $m_Z^2 \sim 100\,\text{GeV}^2$, some degree of fine-tuning remains, at least from an electroweak scale point of view, still remains. However, from a top-down point of view, two possibilities exist. It is, of course, a possibility that a fine-tuning is intrinsically present in Nature, but it is also possible that the fine-tuning is "apparent" and is just a manifestation of our less-than-perfect understanding of the underlying theory at the high scale. Finally, it is interesting to note that in a different context it has been argued that essentially no physics would change if the higgs vev, which is equivalent to m_Z, were several times larger than the experimental value [67].

6.6. *The Higgs mass*

Within the Standard Model, the Higgs mass is just a parameter and cannot be predicted. However, within a top-down framework of beyond-the-Standard Model physics, it is computable in terms of microscopic parameters. Within the framework considered, the Higgs mass can be computed, rather remarkably, in terms of only a few parameters, if the matter and gauge spectrum in the visible sector below the compactification scale is the MSSM. The computation was first presented at the International String Phenomenology Conference at the University of Wisconsin, Madison in August 2011; a more detailed prediction appeared in [68]. It is remarkable that the discovery of a Higgs boson by ATLAS and CMS near 125 GeV is naturally consistent with the prediction of the Higgs mass within the framework.

Since supersymmetric models require two Higgs doublets for anomaly cancellation, by the "Higgs mass" we mean the mass of the lightest CP-even neutral scalar in the Higgs sector. A remarkable fact about the Higgs mass even in general supersymmetric theories is that an upper limit on M_h of order $2 M_Z$ exists just from the requirement of validity of perturbation theory up to the high scale of order 10^{16} GeV [69]. This is due to the fact that the Higgs mass at tree-level only depends on SM gauge couplings (which have been measured), and possibly other Yukawa or gauge couplings (which are bounded from above by perturbativity). However, in addition to the gauge and matter spectrum, the precise value of the Higgs mass depends crucially on radiative effects, which in turn depend on all the soft parameters including the μ and $B\mu$ parameters. Nevertheless, we will see that for the framework considered the Higgs mass depends essentially on two parameters — the overall scale $m_{3/2}$, and the parameter $\tan\beta = \frac{v_u}{v_d}$. $\tan\beta$ is correlated with μ and $B\mu$ via electroweak symmetry breaking.

Since in the string/M theory frameworks above, scalar superpartner masses are generically much larger than the electroweak scale and viable models have gaugino masses which are $\lesssim$ TeV, it is useful to integrate out the scalars below their characteristic mass scale ($\gtrsim 30$ TeV), and study the effective theory consisting of the Standard Model particles, charginos and neutralinos. As discussed earlier, higgsinos may or may not be suppressed depending on the value of μ, we consider both cases when computing the Higgs mass. Note that all Higgs scalars except the lightest CP-even Higgs h are quite heavy, with masses close to $m_{3/2}$. Thus, we are in the so-called "decoupling limit" of the two-Higgs doublet model. In this case, the Higgs mixing angle[n] α is related to β as $\alpha = \beta - \frac{\pi}{2}$, and the lightest CP-even Higgs behaves very close to the Higgs in the Standard Model.

The lightest CP-even Higgs mass, M_h, is given by: $M_h = \sqrt{2}\,\lambda\,v$, where λ is the Higgs quartic coupling and $v = 174$ GeV is the Higgs vev. In the MSSM, the quartic coupling is given by $\lambda_0 \equiv \frac{g^2 + g'^2}{8}\cos^2(2\beta)$ at tree level[o], which is small. Once the squarks, sleptons and heavy Higgs scalars are integrated out around their mass-scale, this gives rise to a threshold correction to the quartic coupling $\delta\lambda$ at the scalar superpartner mass scale $Q = \sqrt{m_{\tilde{t}_1} m_{\tilde{t}_2}}$: $\lambda(Q) = \lambda_0(Q) + \delta\lambda(Q)$[p]. This quartic coupling can then be RG evolved to the electroweak scale, where corrections arising from loops of supersymmetric fermions at around the electroweak scale (more precisely the $\bar{M}S$ scale m_t), denoted by $\delta\tilde{\lambda}$, must be added as well. The lightest Higgs mass M_h is thus given by:

$$M_h = \sqrt{2}\,v\,\sqrt{\lambda(m_t) + \delta\tilde{\lambda}(m_t)}. \tag{25}$$

The corrections $\delta\lambda$ and $\delta\tilde{\lambda}$ also depend on $\tan\beta \equiv \frac{v_u}{v_d}$ in general. For

---

[n] for its definition, please refer to Section 8.1 in [70].

[o] β is defined as $\beta \equiv \tan^{-1}(\frac{v_u}{v_d})$ where v_u and v_d are the vevs of the two Higgses in the supersymmetric two-Higgs doublet model.

[p] $m_{\tilde{t}_1}, m_{\tilde{t}_2}$ stand for the masses of the stop squarks.

details on the Higgs mass computation, see [68] and references therein. The computation in [68] used one-loop threshold corrections for $\delta\lambda(Q)$ and $\delta\tilde{\lambda}(m_t)$ and two-loop beta functions to RG evolve λ from $Q = \sqrt{m_{\tilde{t}_1} m_{\tilde{t}_2}}$ to m_t. However, following the results of [71] it is possible to further refine the result. In particular, it is possible to include two-loop threshold corrections for $\delta\lambda(Q)$ and $\delta\tilde{\lambda}(m_t)$ and three-loop beta functions to RG evolve λ from $Q = \sqrt{m_{\tilde{t}_1} m_{\tilde{t}_2}}$ to m_t. This analysis is being currently carried out in [72]. The preliminary results are shown in Figure 1, and we gratefully acknowledge Bob Zheng for providing the Figure.

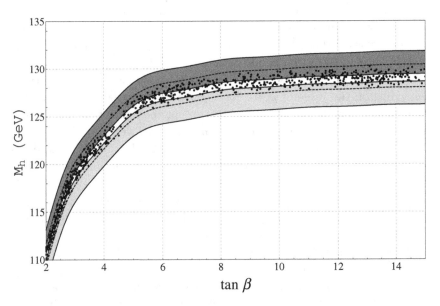

Fig. 1. The prediction for the Higgs mass for realistic string/M theory vacua as described in the text, as a function of $\tan\beta$ for three different values of the gravitino mass $m_{3/2}$, and varying the theoretical and experimental inputs as described below. For precise numbers and more details, see [68]. The central band within the dashed curves for which scatter points are plotted corresponds to $m_{3/2} = 50$ TeV. This band includes the total uncertainty in the Higgs mass arising from the variation of three theoretical inputs at the unification scale, and from those in the top mass m_t and the $SU(3)$ gauge coupling α_s within the allowed uncertainties. The innermost (white) band bounded by solid curves includes the uncertainty in the Higgs mass for $m_{3/2} = 50$ TeV only from theoretical inputs. The upper (dark gray) band bounded by solid curves corresponds to the total uncertainty in the Higgs mass for $m_{3/2} = 100$ TeV while the lower (light gray) band bounded by solid curves corresponds to that for $m_{3/2} = 25$ TeV.

The Higgs mass prediction holds for all compactifications with an MSSM matter and gauge spectrum below the compactification scale and with scalars heavier than around 20 TeV and gaugino masses $\lesssim$ TeV. In addition to the dependence on $\tan\beta$, there is a mild dependence on the overall scale $m_{3/2}$ which can vary from ~ 20 TeV to ~ 100 TeV within the framework. For a given $m_{3/2}$ and $\tan\beta$, there is a small spread in the Higgs mass prediction arising from variation of theoretical inputs like the trilinears and the gluino mass parameter within reasonable ranges at the GUT scale, as well as from experimental uncertainties in the top mass and the strong gauge coupling. A precise experimental value for the Higgs mass will constrain $m_{3/2}, \mu$ and $\tan\beta$ significantly. Note that other authors have earlier proposed that interpreting data in the context of supersymmetry was suggestive of scalars heavier than might have been naively expected [73]. The string theory derivation leads to definite predictions for the heavy scalars (tens of TeV at the high scale) which gives a fairly sharp prediction for the Higgs mass.

The framework also makes precise predictions about Higgs properties. Since we are in the decoupling limit of the Higgs sector in the MSSM, the Higgs behaves very similarly to the SM Higgs. In particular, the Higgs production cross-section in the gluon fusion channel is virtually indistinguishable from that in the SM due to the stops being much heavier than the tops. Furthermore, including the effects of superpartrners, the branching ratios to $b\bar{b}$ (which dominates the total width) and other modes such as $\gamma\gamma$, $Z\gamma$ do not deviate from the SM by more than a few percent. This is completely consistent with the current results of ATLAS and CMS.

7. High Energy Physics Signals

It is natural to ask how the framework we have studied manifests itself at high energy physics experiments. These can be broadly divided into two categories — the energy frontier and the precision frontier. We discuss both of these below.

7.1. *Energy frontier*

The LHC has achieved significant milestones in its performance and has amassed a wealth of high-quality data. It has already ruled out a significant region of parameter space of many beyond-the-Standard-Model frameworks. Here we discuss the broad LHC predictions for the framework under consideration and the constraints on the parameter space arising from current LHC data.

Since scalars are expected to be heavier than about 30 TeV, they cannot be directly produced at the LHC. However, the framework predicts gaugino masses $\lesssim$ TeV, so they should be accessible at the LHC. Therefore, the most promising channel is pair production of gluinos followed by their decay to a realtively high mutliplicity of third generation fermions such as top and/or bottom quarks. The gluinos have a large production cross-section because they carry color and are fermions. However, their cross-section is suppressed relative to the case with comparable squark masses. The gluinos decay via virtual squarks into $q\bar{q}\chi_1^0$ or $q\bar{q}\chi_1^{\pm}$ since the squarks are heavier than the gluinos. Since the rate scales as $m_{\tilde{q}}^{-4}$, the lightest squarks dominate the process. If all the scalar masses are roughly equal at the unification scale (close to $m_{3/2}$), then RG effects drive the third generation squarks to be lighter than the first two. Thus, the gluino decay channels $\tilde{g} \rightarrow t\bar{t}\chi_1^0$, $t\bar{b}\chi_1^{\pm}$, $b\bar{b}\chi_1^0$ dominate over a large region of parameter space[q]. These lead to b-rich and lepton-rich final states with excellent prospects for discovery. Detailed studies of these kind of models have been carried out for the 14 TeV LHC in [74], and for the 7 TeV LHC in [75]. LHC studies of phenomenological models with a similar spectrum have also been performed in [76]. In particular, it has been shown that the 1 lepton, $\geq 4\,b$ channel is particularly sensitive to this class of models even with moderate amounts of data. In some cases, the same-sign (SS) dilepton channel can also be a competitive model for discovery since it encounters fewer backgrounds from SM processes. The current LHC data can be used to put constraints on the allowed parameter space within the framework. For instance, using a combined analysis of different branching fractions as espoused in [77], the latest CMS results put a lower bound of about 900-950 GeV for the gluino mass.

What about signals of the chargino and neutralino sector? If one assumes the simple case of just a single visible sector consisting of the supersymmetric Standard Model, the LSP in the framework is generically wino-like with a small bino component. The higgsino component depends on the value of μ and could be either small or significant. The lightest chargino $\tilde{\chi}_1^{\pm}$ and the lightest neutralino $\tilde{\chi}_1^0$ are quasi-degenerate if the LSP is mostly wino, with $m_{\tilde{\chi}_1^{\pm}} - m_{\tilde{\chi}_1^0} \lesssim 200$ MeV. In this case, the charginos decay to the LSPs emitting soft pions. Thus gluino decays to charginos could lead to

---

[q]Gluino decays to $\chi_1^{\pm}q\bar{q}'$, and $\chi_1^0 q\bar{q}$ are also significant.

the charginos traveling through two or three layers of the tracker and then decaying, giving rise to disappearing high p_T charged tracks. Observing this signal is challenging and requires a dedicated analysis, but should be possible [78]. Electroweak production of charginos and neutralinos has a significant cross-section [74] and should also be observable eventually, and can help provide experimental information about the nature of the LSP. The tree-level production for $\tilde{\chi}_1^{\pm} + \tilde{\chi}_2^0$ vanishes for a pure bino LSP, so the cross-section is sensitive to the bino component of the LSP. Similarly, the rate for production of $\tilde{\chi}_1^{\pm} + \tilde{\chi}_1^0$ is about two times larger for a wino LSP than for a higgsino LSP and can thus help determine the LSP type.

However, as was pointed out in Section 8.1, the simple picture above of a single visible sector is under significant tension from astrophysical observations, the most prominent among them being the FERMI diffuse photon signal coming from the Galactic Center. A natural extension of the simple framework above includes an additional (dark) sector which contains the LSP and is very weakly connected to the visible sector. In this case, the *visible* LSP, when produced in colliders, decays into the hidden sector which ultimately cascade decays into the (dark) LSP and possibly some visible SM particles due to the portal interaction between the two sectors. The collider phenomenology of this framework deserves further exploration and is currently under study [79].

To summarize, the framework gives rise to many falsifiable predictions at the LHC which are being probed currently. The collider phenomenology of the LSP may be especially rich as mentioned above. However, a robust feature of the framework is that gauginos must be eventually observed at the LHC with enhanced branching fractions to the third generation, else the case with suppressed gaugino masses will be ruled out.

7.2. *Precision frontier*

Precision mesurements are sensitive to new particles running inside loops, and hence can indirectly probe BSM physics. Within our framework, the fact that the scalar masses and trilinears are $\gtrsim 30$ TeV while the gauginos are $\lesssim$ TeV helps keep the flavor and/or CP violating effects under control, as explained in Section 6.2. However, some measurements can be sensitive to new physics within the framework in the near future. For example, an improvement in the constraints from $b \to s\,\gamma$ by about an order of magnitude will start probing the framework [56]. Similarly, EDM predictions from the framework naturally turn out to be one-to-two (for the mercury

EDM), two-to-three (for the neutron EDM), and about four (for the electron EDM) orders of magnitude smaller than the current limits [55]. So, an improvement in these limits in the future will be able to test the framework. In addition, even though the decay width of $B_s \to \mu\mu$ is proportional to $\tan\beta^6$, we find that the prediction is still virtually indistinguishable from the SM (Our current understanding of the theory suggests $\tan\beta \lesssim 20$ [62]).

8. Dark Matter

A non-thermal cosmological history requires us to reassess our standard notions of DM vis-a-vis the nature of DM candidates and the parameter space of masses and interactions required to provide the entire DM content of our Universe. The two most attractive candidates for DM are the weakly interacting massive particle (WIMP), and the axion(s). We will find that the generic string/M theory prediction is that both of these serve as excellent candidates and could each provide an $\mathcal{O}(1)$ fraction of DM. Interestingly, the current experimental and observational constraints point to a rather different set of parameters to provide the correct abundance than what is usually considered.

8.1. *WIMPs*

We will first consider the simplest possible setup within our framework, that of a single visible sector which consists of some supersymmetric extension of the Standard Model. Such a model with an exact or sufficiently conserved stabilizing symmetry (such as R-parity) naturally contains a WIMP DM candidate — the lightest superpartner charged under the particular symmetry, the LSP. The LSP in supersymmetric models with gravity mediation is typically a neutralino but other particles, such as staus or sneutrinos, are possible as well. However, as explained above using the result on the lightest modulus mass, one finds that the gravitino is generically heavier than around 20 TeV. This also generically sets the scale for squarks, sleptons and sneutrinos to be around 20 TeV. Gaugino masses, on the other hand, may or may not be suppressed relative to $m_{3/2}$ depending on the nature of moduli stabilization. Examples of both kinds of models exist in the literature. For compactifications with unsuppressed gaugino masses, see [58]. The μ parameter, which determines the higgsino mass, can also be either of the same order or suppresed relative to $m_{3/2}$ depending upon the situation. See Section 6.1 for more discussion.

If gaugino masses are not suppressed then they, along with the squarks and sleptons, are too heavy to provide a viable WIMP DM candidate. In fact, this case can only be viable if the stabilizing symmetry of the LSP (R-parity) is violated so that the would-be LSP decays sufficiently rapidly. However, as argued in Section 6.4, constraints from neutrino masses disfavor any form of R-parity violation in string/M theory frameworks with $SU(5)$ GUTs and possibly other GUT groups as well. Hence, the case with suppressed gaugino masses is, therefore, significantly more interesting. This case can occur naturally in mechanisms of moduli stabilization in which the F-terms of moduli which determine the SM gauge couplings are suppressed relative to the dominant F term, as in [20, 21, 39, 57]. Then the lightest neutralino can be in the sub-TeV range, and can give rise to roughly the correct relic-abundance as follows. As explained above, the lightest modulus X starts oscillating when $H \sim m_X$ and decays when $H = \Gamma_X$, the decay width of the modulus. The decay width of the modulus is given by:

$$\Gamma_X = \frac{D_X \, m_X^3}{m_{pl}^2} \qquad (26)$$

where D_X is a numerical coefficient which depends on the details of moduli stabilization and the compactification. With no prior knowledge, D_X is generically assumed to be $\mathcal{O}(1)$ but in a given compactification, it can be larger, ranging from ~ 10 to $\sim 10^4$ depending upon the micrsocopic details. A large D_X is possible if the modulus vev measures the volume of a submanifold inside the internal manifold which is parametrically larger than unity in string or 11D length units. A large D_X is equivalent to a modulus decay constant which is smaller than the Planck scale, $f_X = \frac{m_{pl}}{\sqrt{D_X}}$, i.e. it can be thought of as replacing the Planck scale in (26) by a smaller scale such as the string scale or the compactification scale. It is a natural possibility, therefore, that the decay constants of the moduli are similar to those of the axions, which are also close to the compactification scale or the GUT scale, as will be seen in Section 8.2[r].

The decay of the modulus reheats the Universe with a reheating temperature T_R which has to be greater than a few MeV to satisfy BBN constraints. Since the canonically normalized X is generally a linear combination of moduli whose values determine the gauge couplings, it has an $\mathcal{O}(1)$ coupling to gauginos. Thus, the branching ratio to the lightest neutralino

---

[r]The precise values for the moduli and axion decay constants can differ by $\mathcal{O}(1)$ due to different mixing between the "flavour" and mass eigenstates for the two cases.

is not small[8], giving rise to the number density of DM particles χ from X decay as:

$$n_\chi^X \sim \frac{\Gamma_X^2 \, m_{pl}^2}{m_X} \sim \frac{D_X^2 \, m_X^5}{m_{pl}^2}. \tag{27}$$

This is to be compared with the critical density for annihilations of χ-particles at the decay time $(H = \Gamma_X)$:

$$n_c \sim \frac{\Gamma_X}{\langle \sigma v \rangle} \sim \frac{D_X \, m_X^3}{m_{pl}^2 \, \langle \sigma v \rangle}. \tag{28}$$

For typical weak scale values of masses and cross-sections, n_χ^X is much larger than n_c, hence the DM particles annihilate after being produced until their number density becomes of order n_c. Thus, the final abundance of χ is given by:

$$n_\chi \sim n_c(T_R) \sim \frac{H(T_R)}{\langle \sigma v \rangle} \Omega_\chi \, h^2 \approx \Omega_\chi \, h_{(thermal)}^2 \left(\frac{T_F}{T_R} \right)$$

where T_F is the thermal freezeout temperature of the LSP. This gives rise to the following parametric dependence of $(\frac{\rho_\chi}{s} \equiv \frac{m_\chi n_\chi}{s})$ on the various quantities [4]:

$$\frac{\rho_\chi}{s} \simeq \frac{0.25 \, \gamma_\chi}{D_X^{1/2} \, m_{3/2}^{1/2} \, m_{pl}^{1/2} \langle \sigma v \rangle}, \tag{29}$$

which has to be normalized to the present value of the quantity, $(\rho_\chi/s)_0 = 3.6 \times 10^{-9}$ GeV to get the LSP relic abundance $\Omega_\chi \, h^2$. Here $\gamma_\chi \equiv \frac{m_\chi}{m_{3/2}}$ is the ratio of the low-scale LSP mass relative to $m_{3/2}$.

For a weak-scale LSP, T_F is $\mathcal{O}(1 - 10)$ GeV while T_R is $\mathcal{O}(1 - 10)$ MeV, giving rise to two-to-three orders of magnitude enhancement of the abundance over that of the thermal one for a given mass and cross-section. This implies that particles which naturally have a *larger* cross-section compared to that in the thermal case are good WIMP DM candidates. Hence, within the supersymmetric standard model, a wino-like or higgsino-like LSP is a great candidate for DM in this context as it naturally provides the above enhancement in cross-section so as to roughly give rise to the correct abundance, see [48, 80]. A wino-like LSP can also be naturally obtained in explicit models with moduli stabilization [22, 81].

Although the above picture is extremely appealing, it seems to be in strong tension with the current data from various astrophysical observa-

---

[8]Note that the modulus, being R-even, will decay to SM particles and superpartners, with all superpartners quickly cascade-decaying to the LSP. This will also add to the branching ratio to the LSP.

tions. More precisely, FERMI-LAT observations of continuum photons coming from the Galactic Center put a rather robust upper bound of $\simeq 10\%$ for the relic abundance fraction of winos with masses of order 100 GeV [82]. Furthermore, this relic abundance can be obtained within a non-thermal cosmological history only for lightest modulus masses that are $10^2 - 10^4$ times heavier than the gravitino mass [82]. However, the results of [1] show that this can happen only for extremely non-generic and fine-tuned cases; generically one expects that the lightest modulus has a mass comparable to $m_{3/2}$. Similar qualitative statements can be made for higgsino DM.

The above suggests that if Dark Matter is a WIMP, then the setup of a single visible sector may be too simplistic. However, another setup for WIMP DM, which is also well motivated from a string theory point of view, may be viable. This setup consists of a visible sector consisting of a supersymmetric Standard Model which is very weakly connected to another sector (call it the 'Dark sector') through some interactions. R-parity, or some other stabilizing symmetry, is supposed to act globally on all the sectors such that the lightest R-odd superpartner belongs to the Dark sector. In this case, it may be possible to evade the various astrophysical bounds with the Dark LSP still comprising an $\mathcal{O}(1)$ fraction of Dark Matter. The details of this setup are currently being explored [79].

8.2. *Axions*

String compactifications to four dimensions generically give rise to a plethora of axions, as they reside in chiral supermultiplets along with the moduli fields. Stabilizing the moduli with a sufficiently large mass so as to evade BBN constraints has interesting implications for axion physics. In order for one of the axions to solve the strong CP-problem, i.e. to serve as the QCD axion, it should predominantly receive its mass from QCD instantons, and not additional stringy or supergravity effects. However, compactifications with moduli stabilized by only superpotential effects give axions masses comparable to $m_{3/2}$;[t] hence none of the axions in these compactifications can solve the strong-CP problem.

On the other hand, the M theory and Type IIB moduli-stabilization mechanisms discussed above have moduli stabilized by a combination of Kähler potential and superpotential effects, as explained in detail in Section 5. In this case, only the moduli are stabilized at leading order while most of the axions are left unstabilized. Integrating out the moduli and taking higher order effects into account, the axions are then stabilized with masses

---

[t] As in KKLT-type models [30].

exponentially suppressed relative to $m_{3/2}$ by higher order non-perturbative effects. This gives rise to a spectrum of axions with masses distributed roughly evenly on a logarithmic scale [27], which was dubbed the "Axiverse" in a more phenomenological approach [83]. In models with $M_{st} \gtrsim M_{GUT}$, one finds that tha axions can span a huge mass range from $m_a \sim H_0 \sim 10^{-33}$ eV to $m_a \sim 1$ eV. One of these light axions could naturally serve as the QCD axion if its mass is less than about 10^{-15} eV, hence solving the strong CP-problem [27].

One of the most important effects of these axions is their contribution to the total energy budget of the Universe. Axions start oscillating when $H \sim m_a$, and the energy in coherent oscillations could provide an $\mathcal{O}(1)$ fraction of the dark matter of the Universe[u]. Within a thermal cosmological history, the WMAP bound on the relic abundance puts an upper bound on the axion decay constant $\hat{f}_a$ to be around 10^{11-12} GeV for an $\mathcal{O}(1)$ misalignment angle. On the other hand, with a non-thermal history, the computation of the axion relic abundance is different, and is schematically given by [84]:

$$\Omega_{a_k} h^2 \simeq 10 \left(\frac{\hat{f}_{a_k}}{2 \times 10^{16}\,\text{GeV}} \right)^2 \left(\frac{T_R}{10\,\text{MeV}} \right) \langle \theta_k^2 \rangle \qquad (30)$$

for an axion a_k which starts to oscillate in the moduli-dominated era. Here $\hat{f}_{a_k}$ is the axion decay constant and T_R is the reheat temperature after the decay of the lightest modulus, for more details see [27]. This can naturally give rise to roughly the correct value for a much larger decay constant $\hat{f}_{a_k} \sim 10^{15}$ GeV and $T_R \gtrsim 5$ MeV arising from the decay of a modulus heavier than around 25 TeV. With around 1 to 10% tuning of the misalignment angle (which may also arise from some hitherto unknown dynamical mechanism), $\hat{f}_a \sim M_{GUT} \approx 10^{16}$ GeV can also be accommodated. Within a GUT-motivated frameowork, this seems a much more natural possibility since the decay constant in string/M theory solutions with unification tends to be around the GUT scale. One can view this both as a solution of the 'cosmological axion decay constant problem' in string theory and also as a demonstration that axion physics is self-consistently much less fine-tuned with the non-thermal cosmological history generically predicted by string/M theory, than with a thermal one.

The Axiverse is subject to cosmological constraints and also has falsifiable predictions as discussed in [27, 83]. For example, axions in the mass

---

[u]Since there are many axions, they will start oscillating at different times. Also, the axions are so light that none of them have decayed yet.

window $10^{-28} \lesssim m_a \lesssim 10^{-18}$ eV could give rise to step-like features in the matter power spectrum at small scales. On the other hand axions in the mass window $10^{-10} \lesssim m_a \lesssim 1$ eV can form bound states with black-holes, thereby significantly affecting their dynamics by graviton emission [83, 85]. An important signal of axions is "isocurvature" fluctuations, arising from the fluctuations of axions during inflation. The size of these fluctuations is linearly proportional to the Hubble parameter H_I during inflation [86]. Therefore, the upper bound on isocurvature fluctuations from PLANCK puts an *upper bound* on H_I. In this context, the potential observation of B-modes attributable to primordial gravitational waves by the BICEP2 collaboration [87] is very important. If the signal holds up, and if it is attributable to primordial gravitational waves, then this provides a *lower bound* on H_I within simple (single-field slow-roll) models of inflation that is inconsistent with the upper bound on H_I from that of isocurvature fluctuations. Taken at face value, this would imply that the axiverse is ruled out unless the misalignment angle is severely fine-tuned [88]. However, it is still too early to make definitive statements. First, the signal has not been completely corroborated yet. Even if the signal turns out to be correct, there are a number of ways in which constraints on axion dark matter may be avoided or alleviated. For example, in more complicated models of inflation, it is possible that the Hubble parameter during inflation, H_I, is relatively small and gravitational waves giving rise to B-modes are produced by some other sources. In fact, it is possible that axions themselves can give rise to B-modes through their coupling to $\vec{E} \cdot \vec{B}$. The pattern of B-modes produced from axions, however, should be distinguishable from that of primordial gravitational waves. Also, non-trivial axion dynamics during inflation, such as that arising from a non-minimal coupling to gravity [91], or if the inflaton directly couples to the sector producing non-perturbative effects relevant for axions [89–91], can evade the bounds.

The above shows that observations in the near-future will have crucial implications for axion dark matter within string frameworks, and it is extremely important to explore these in more detail.

9. The Matter-Antimatter Asymmetry

Finally, we discuss the origin of the matter-antimatter asymmetry of the Universe within the framework and its connection to the LSP abundance. The presence of light moduli imply a period of moduli domination shortly after the end of inflation. This era lasts until the lightest modulus decays providing enough reheating temperature for successful nucleosynthesis. However, the decay also produces a large amount of entropy greatly diluting any pre-existing baryon asymmetry in the Universe. In light of this, two possibilities arise in order to generate the baryon asymmetry within this framework. The first is that a large baryon asymmetry (much larger than the observed amount) is generated in the early Universe and gives rise to the correct asymmetry after entropy dilution. The second possibility is that the decay of the modulus itself generates the asymmetry at temperatures around 10 MeV.

The second possibility requires baryon number violating decays, since the modulus is a gauge singlet with vanishing baryon number and only couples gravitationally to all SM fields. It is not clear at present if this possibility could naturally occur within a string framework. On the other hand, the first possibility is realized quite naturally. It is well known that the Affleck-Dine (AD) mechanism can generate a large (even $\mathcal{O}(1)$) baryon asymmetry in a robust manner [92, 93]. In particular, this can generate via B and L-violating flat-directions in the MSSM denoted by Φ in general. Within our framework, these flat-directions are also displaced from their late-time minima during inflation, just like the light moduli. The subsequent coherent oscillation and decay of these flat-directions could then generate a baryon asymmetry. In the simple MSSM models realizing this possibility, however, there are two issues. First, as mentioned before, there is the danger of producing *too much* baryon asymmetry and second, the origins of the baryon asymmetry and the DM abundance seem to be decoupled from each other. The presence of light moduli in our framework can provide a resolution to both these issues, in the following manner.

The essential point is that the decay of the lightest modulus generates the LSP abundance (see Section 8.1), and at the same time provides the dilution factor for computing the final asymmetric baryon abundance, thereby relating the two. A careful analysis then gives rise to the following ratio for the two abundances [94]:

$$\frac{\Omega_B}{\Omega_\chi} \simeq \mathcal{O}(1) \frac{m_{proton} \, m_{pl} \, T_R^2 \, \langle \sigma v \rangle}{m_{3/2} \, m_\chi} \left(\frac{\Phi_0}{X_0} \right)^2 . \tag{31}$$

Here Φ_0 and X_0 denote the initial displacements of the flat-direction and the lightest modulus during inflation, respectively. Their ratio above arises in the ratio of the corresponding energy densities and determines how much baryon asymmetry is left after the dilution. Furthermore, as shown in [94], flat directions corresponding to the highest dimension operators in the MSSM (which yield the largest Φ_0) naturally give rise to $\frac{\Phi_0}{X_0}$ in the range $10^{-3} - 10^{-2}$. Then, for natural values of other quantities in (31) arising within the framework and consistent with other constraints, such as $m_{3/2}$ in the 20-100 TeV range, giving rise to T_R around few to 100 MeV (this depends on the modulus decay constant), m_χ around 200 GeV, and $\langle \sigma v \rangle$ around few $\times 10^{-6}$ GeV^2, the above ratio can be naturally close to the observed value. Note that Ω_χ in (31) is *not* the full DM abundance since axions also contribute to the DM abundance. Therefore, the ratio $\frac{\Omega_B}{\Omega_\chi}$ has to be somewhat larger than 0.2.

10. Comments and Outlook

In this review, we have outlined the typical or generic predictions of a string/M theory vacuum *given* our Universe is a solution of string/M theory with low energy supersymmetry and grand unification. We have carefully laid out the broad set of working assumptions under which our arguments and claims are valid, in Section 3. In addition to the requirement of stabilizing all moduli in a vacuum which solves the gauge hierarchy problem with supersymmetry, these essentially amount to assuming that the supergravity approximation is valid, the Hubble parameter during inflation (or whatever solves the horizon and flatness problems in the early Universe) is larger than $m_{3/2}$, and that the particles in the visible sector are weakly coupled from the TeV scale until a high scale like the unification scale. Then, many broad predictions can be made for beyond-the-SM physics. It the particle content of the visible sector is precisely that of the minimal supersymmetric standard model, then *precise* predictions, such as for the Higgs mass, can also be made. We now comment on a few relevant issues.

It is worth addressing complaints which critical readers might have about the whole approach. For example, some may complain that the approach considered here has not tackled any of the deep fundamental problems, such as the understanding of the cosmological singularity at very early times, or the solution of the horizon and flatness problems in the early Universe, or the extremely tiny value of the cosmological constant. On a more mundane but technical level, others may complain that although moduli stabilization has been understood at the effective supergravity level, explicit compactifications with the required properties to stabilize all moduli, *and* a realistic matter and gauge spectrum (such as that of the MSSM), *and* a realistic texture of Yukawa couplings for the quarks, leptons and neutrinos, do not exist.

Our philosophy regarding these issues is the following. It is clear that our understanding of these deeper issues is rudimentary at best. However, our main assumption is that our Universe is a solution of string/M theory. If this assumption is correct, then there must be mechanisms present in the theory (albeit unknown to us) which would have solved the first two fundamental issues at very early times and presumably at very high scales. Our focus in this work is on the broad features of beyond-the-SM physics which depend on our understanding of the Universe at much later times, essentially from around the time of BBN to the present time. Hence, these features are largely decoupled from the first two issues. The cosmological constant, on the other hand is a fundamental problem which persists even at late times. So, regarding the cosmological constant our philosophy is as follows. We only require that the cosmological constant approximately vanishes, with the implicit assumption that the (unknown) mechanism which gives rise to the extremely tiny value of the cosmological constant has no bearing on BSM particle physics. This appears to be a rather conservative assumption since there is no known particle physics process whose outcome depends on the *precise* value of the cosmological constant.

For the technical complaints it is worth noting that explicit compactifications which stabilize all moduli by incorporating the underlying physical ideas exist for Type IIB compactifications [39]. For M theory compactifications, although explicit manifolds with such properties do not exist yet, dualities from other corners of string theory suggest that essentially the same mechanism should go through for these compactifications. Similarly, many explicit compactifications realizing a realistic matter and gauge spectrum such as that of the MSSM have been constructed in various corners of string theory [9]. While explicit string compactifications realizing *all* these features in a *single* vacuum may have not yet been constructed, the fact that these features exist (separately) in a large class of vacua lends support to the expectation that there should exist 4D string/M theory vacua in the landscape realizing all these features.

The approach we have espoused is very useful even if predictions do *not* agree with data, since depending upon the nature of experimental data, it could provide insights as to which of the assumptions need to be relaxed. Let us explain this with a few examples. For concreteness, we assume that the gauge hierarchy problem is solved by supersymmetry, since otherwise it is clear that the entire framework is invalid.

One way in which some of the above conclusions could be modified is if the matter and gauge spectrum below the compactification scale is more extended than that of the MSSM. This could give rise to a different prediction for the Higgs mass as well as its properties in general (although in practice the differences may be small depending on the model). For example, this could happen if the Higgs couples to additional particles through Yukawa or gauge interactions.

As another example, if squarks and sleptons are observed at the LHC, this would be in contradiction with some of the basic assumptions of the framework. This would imply one of four possibilities – a) the moduli potential is very non-generic which makes *all* moduli masses much larger than $m_{3/2}$, b) the moduli masses are close to $m_{3/2}$ but the Hubble parameter during inflation is $\lesssim m_{3/2}$, so that the moduli are not displaced from their late-time minima, c) the moduli masses are close to $m_{3/2}$ and the Hubble parameter during inflation is larger than $m_{3/2}$, but there exists a period of thermal inflation at late-times to dilute the entropy production from the decay of moduli [95], or d) the moduli masses are close to $m_{3/2} \gtrsim 30$ TeV, but the squarks and slepton masses are also suppressed relative to $m_{3/2}$ in a phenomenologically consistent way. None of these possibilities seem to be generic with our current understanding (in fact, they seem rather non-generic), and none have been shown to occur in a convincing manner in realistic string compactifications. So, if experiments observe squark and slepton masses, it would be very challenging to realize such a setup within a string framework.

If the predictions of this framework agree with data on the other hand, it would be an extremely important step in connecting string/M theory to the real world and would open up more opportunities for learning about the string vacuum we live in.

Acknowledgments

PK would like to express his heartfelt gratitude to Prof. Gordon Kane for his continuous encouragement, inspiration and infectious enthusiasm about physics. PK would like to particularly thank Prof. Bobby Acharya for various illuminating discussions and collaborations and his insights about many aspects of theoretical high-energy physics. PK would also like to thank Konstantin Bobkov, Ran Lu, Jing Shao, and Bob Zheng for many interesting and helpful discussions. Finally, PK acknowledges Bob Zheng for providing an up-to-date version of the Higgs Mass plot in Figure 1. The work of P.K. is supported by the DoE Grant DE-FG-02-92ER40704.

References

[1] B. S. Acharya, G. Kane and P. Kumar, Int. J. Mod. Phys. A **27**, 1230012 (2012) [arXiv:1204.2795 [hep-ph]].

[2] S. Dimopoulos and H. Georgi, Phys. Lett. B **117**, 287 (1982). S. Weinberg, Phys. Rev. D **26**, 287 (1982).

[3] M. Green, J. Schwarz, and E. Witten, *Superstring Theory, Vol. II*, Cambridge Monographs on Mathematical Physics.

[4] B. S. Acharya, G. Kane and E. Kuflik, arXiv:1006.3272 [hep-ph].

[5] T. Banks, hep-th/0412129.

[6] M. R. Douglas and S. Kachru, Rev. Mod. Phys. **79**, 733 (2007) [hep-th/0610102].

[7] J. J. Heckman and C. Vafa, Nucl. Phys. B **837**, 137 (2010) [arXiv:0811.2417 [hep-th]]. V. Bouchard, J. J. Heckman, J. Seo and C. Vafa, JHEP **1001**, 061 (2010) [arXiv:0904.1419 [hep-ph]].

[8] G. K. Leontaris and G. G. Ross, JHEP **1102**, 108 (2011) [arXiv:1009.6000 [hep-th]]. O. Lebedev, H. P. Nilles, S. Raby, S. Ramos-Sanchez, M. Ratz, P. K. S. Vaudrevange and A. Wingerter, Phys. Rev. D **77**, 046013 (2008) [arXiv:0708.2691 [hep-th]]. B. C. Allanach, S. F. King, G. K. Leontaris and S. Lola, Phys. Rev. D **56**, 2632 (1997) [hep-ph/9610517]. J. P. Conlon, A. Maharana and F. Quevedo, JHEP **0809**, 104 (2008) [arXiv:0807.0789 [hep-th]]. R. Blumenhagen, M. Cvetic, D. Lust, R. Richter, 2 and T. Weigand, Phys. Rev. Lett. **100**, 061602 (2008) [arXiv:0707.1871 [hep-th]]. R. Tatar and T. Watari, Nucl. Phys. B **747**, 212 (2006) [hep-th/0602238].

[9] V. Braun, Y. -H. He, B. A. Ovrut and T. Pantev, JHEP **0605**, 043 (2006) [hep-th/0512177]. G. Honecker and T. Ott, Phys. Rev. D **70**, 126010 (2004) [Erratum-ibid. D **71**, 069902 (2005)] [hep-th/0404055]. O. Lebedev, H. P. Nilles, S. Raby, S. Ramos-Sanchez, M. Ratz, P. K. S. Vaudrevange and A. Wingerter, Phys. Lett. B **645**, 88 (2007) [hep-th/0611095]. V. Braun, P. Candelas, R. Davies and R. Donagi, arXiv:1112.1097 [hep-th].

[10] H. P. Nilles, Phys. Rept. **110**, 1 (1984).

[11] R. Blumenhagen, S. Moster and E. Plauschinn, JHEP **0801**, 058 (2008) [arXiv:0711.3389 [hep-th]].

[12] G. Papadopoulos and P. K. Townsend, Phys. Lett. B **357**, 300 (1995) [hep-th/9506150].

[13] B. S. Acharya, Adv. Theor. Math. Phys. **3**, 227 (1999) [hep-th/9812205].

[14] B. S. Acharya and E. Witten, hep-th/0109152.

[15] D.D. Joyce, Oxford University Press, 2000; A. Kovalev, math.dg/0012189.

[16] B. S. Acharya, hep-th/0212294.

[17] B. S. Acharya and S. Gukov, Phys. Rept. **392**, 121 (2004) [hep-th/0409191].

[18] T. Friedmann and E. Witten, Adv. Theor. Math. Phys. **7**, 577 (2003) [hep-th/0211269].

[19] J. A. Harvey and G. W. Moore, hep-th/9907026.

[20] B. S. Acharya, K. Bobkov, G. Kane, P. Kumar and D. Vaman, Phys. Rev. Lett. **97**, 191601 (2006) [hep-th/0606262].

[21] B. S. Acharya, K. Bobkov, G. L. Kane, P. Kumar and J. Shao, Phys. Rev. D **76**, 126010 (2007) [hep-th/0701034].

[22] B. S. Acharya and K. Bobkov, JHEP **1009**, 001 (2010) [arXiv:0810.3285 [hep-th]].

[23] C. Beasley and E. Witten, JHEP **0207**, 046 (2002) [hep-th/0203061]. A. Lukas and S. Morris, Phys. Rev. D **69**, 066003 (2004) [hep-th/0305078]. B. S. Acharya, F. Denef and R. Valandro, JHEP **0506**, 056 (2005) [hep-th/0502060].

[24] B. S. Acharya and M. Torabian, Phys. Rev. D **83**, 126001 (2011) [arXiv:1101.0108 [hep-th]].

[25] J. M. Maldacena and C. Nunez, Int. J. Mod. Phys. A **16**, 822 (2001) [hep-th/0007018].

[26] M. Rummel and A. Westphal, JHEP **1201**, 020 (2012) [arXiv:1107.2115 [hep-th]].

[27] B. S. Acharya, K. Bobkov and P. Kumar, JHEP **1011**, 105 (2010) [arXiv:1004.5138 [hep-th]].

[28] K. Dasgupta, G. Rajesh and S. Sethi, JHEP **9908**, 023 (1999) [hep-th/9908088].

[29] S. B. Giddings, S. Kachru and J. Polchinski, Phys. Rev. D **66**, 106006 (2002) [hep-th/0105097].

[30] S. Kachru, R. Kallosh, A. D. Linde and S. P. Trivedi, Phys. Rev. D **68**, 046005 (2003) [hep-th/0301240].

[31] F. Denef, M. R. Douglas, B. Florea, A. Grassi and S. Kachru, Adv. Theor. Math. Phys. **9**, 861 (2005) [hep-th/0503124].

[32] V. Balasubramanian, P. Berglund, J. P. Conlon and F. Quevedo, JHEP **0503**, 007 (2005) [arXiv:hep-th/0502058 [hep-th]].

[33] R. Blumenhagen, S. Moster and E. Plauschinn, Phys. Rev. D **78**, 066008 (2008) [arXiv:0806.2667 [hep-th]].

[34] M. Cicoli, C. Mayrhofer and R. Valandro, arXiv:1110.3333 [hep-th].

[35] M. Berkooz, M. R. Douglas and R. G. Leigh, Nucl. Phys. B **480**, 265 (1996) [hep-th/9606139].

[36] M. R. Douglas, hep-th/0405279.

[37] M. Dine, E. Gorbatov and S. D. Thomas, JHEP **0808**, 098 (2008) [hep-th/0407043].

[38] J. P. Conlon, F. Quevedo and K. Suruliz, JHEP **0508**, 007 (2005) [hep-th/0505076].

[39] K. Bobkov, V. Braun, P. Kumar and S. Raby, JHEP **1012**, 056 (2010) [arXiv:1003.1982 [hep-th]].

[40] O. Lebedev, H. P. Nilles and M. Ratz, Phys. Lett. B **636**, 126 (2006) [hep-th/0603047]. E. Dudas, C. Papineau and S. Pokorski, JHEP **0702**, 028 (2007) [hep-th/0610297].

[41] C. P. Burgess, R. Kallosh and F. Quevedo, JHEP **0310**, 056 (2003) [hep-th/0309187]. Z. Lalak, O. J. Eyton-Williams and R. Matyszkiewicz, JHEP **0705**, 085 (2007) [hep-th/0702026 [HEP-TH]]. S. L. Parameswaran and A. Westphal, Fortsch. Phys. **55**, 804 (2007) [hep-th/0701215].

[42] O. DeWolfe, A. Giryavets, S. Kachru and W. Taylor, JHEP **0507**, 066

(2005) [hep-th/0505160].

[43] B. S. Acharya, F. Benini and R. Valandro, JHEP **0702**, 018 (2007) [hep-th/0607223]. G. Milanesi and R. Valandro, JHEP **0712**, 085 (2007) [arXiv:0710.1296 [hep-th]].

[44] S. Gukov, S. Kachru, X. Liu and L. McAllister, Phys. Rev. D **69** (2004) 086008 [hep-th/0310159].

[45] L. B. Anderson, J. Gray, A. Lukas and B. Ovrut, Phys. Rev. D **83**, 106011 (2011) [arXiv:1102.0011 [hep-th]].

[46] B. Dundee, S. Raby and A. Westphal, Phys. Rev. D **82**, 126002 (2010) [arXiv:1002.1081 [hep-th]]. S. L. Parameswaran, S. Ramos-Sanchez and I. Zavala, JHEP **1101**, 071 (2011) [arXiv:1009.3931 [hep-th]].

[47] F. Denef and M. R. Douglas, JHEP **0503**, 061 (2005) [hep-th/0411183]. M. Gomez-Reino and C. A. Scrucca, JHEP **0605**, 015 (2006) [hep-th/0602246].

[48] B. S. Acharya, P. Kumar, K. Bobkov, G. Kane, J. Shao and S. Watson, JHEP **0806**, 064 (2008) [arXiv:0804.0863 [hep-ph]].

[49] R. Durrer and J. Hasenkamp, Phys. Rev. D **84**, 064027 (2011) [arXiv:1105.5283 [gr-qc]].

[50] A. L. Erickcek and K. Sigurdson, Phys. Rev. D **84**, 083503 (2011) [arXiv:1106.0536 [astro-ph.CO]].

[51] A. Brignole, L. E. Ibanez and C. Munoz, In *Kane, G.L. (ed.): Perspectives on supersymmetry II* 244-268 [hep-ph/9707209].

[52] S. Kachru, L. McAllister and R. Sundrum, JHEP **0710**, 013 (2007) [hep-th/0703105 [HEP-TH]].

[53] R. Blumenhagen, J. P. Conlon, S. Krippendorf, S. Moster and F. Quevedo, JHEP **0909**, 007 (2009) [arXiv:0906.3297 [hep-th]].

[54] M. Berg, D. Marsh, L. McAllister and E. Pajer, JHEP **1106**, 134 (2011) [arXiv:1012.1858 [hep-th]].

[55] G. Kane, P. Kumar and J. Shao, Phys. Rev. D **82**, 055005 (2010) [arXiv:0905.2986 [hep-ph]].

[56] K. Kadota, G. Kane, J. Kersten and L. Velasco-Sevilla, arXiv:1107.3105 [hep-ph].

[57] K. Choi, hep-ph/0511162. K. Choi, A. Falkowski, H. P. Nilles and M. Olechowski, Nucl. Phys. B **718**, 113 (2005) [hep-th/0503216]. J. P. Conlon and F. Quevedo, JHEP **0606**, 029 (2006) [hep-th/0605141].

[58] M. Cicoli, J. P. Conlon and F. Quevedo, JHEP **0810**, 105 (2008) [arXiv:0805.1029 [hep-th]].

[59] J. A. Casas and C. Munoz, Phys. Lett. B **306**, 288 (1993) [hep-ph/9302227]. H. M. Lee, S. Raby, M. Ratz, G. G. Ross, R. Schieren, K. Schmidt-Hoberg and P. K. S. Vaudrevange, Nucl. Phys. B **850**, 1 (2011) [arXiv:1102.3595 [hep-ph]].

[60] L. E. Ibanez and R. Richter, 2, JHEP **0903**, 090 (2009) [arXiv:0811.1583 [hep-th]]. M. Cvetic, J. Halverson and R. Richter, 2, JHEP **0912**, 063 (2009) [arXiv:0905.3379 [hep-th]]. D. Green and T. Weigand, arXiv:0906.0595 [hep-th].

[61] J. J. Heckman and C. Vafa, JHEP **0909**, 079 (2009) [arXiv:0809.1098 [hep-

th]]. J. J. Heckman and C. Vafa, arXiv:0809.3452 [hep-ph].

[62] B. S. Acharya, G. Kane, E. Kuflik and R. Lu, JHEP **1105**, 033 (2011) [arXiv:1102.0556 [hep-ph]].

[63] E. Witten, hep-ph/0201018.

[64] B. S. Acharya, G. L. Kane, P. Kumar, R. Lu and B. Zheng, arXiv:1403.4948 [hep-ph].

[65] D. Feldman, G. Kane, E. Kuflik and R. Lu, Phys. Lett. B **704**, 56 (2011) [arXiv:1105.3765 [hep-ph]].

[66] J. L. Feng, K. T. Matchev and T. Moroi, Phys. Rev. Lett. **84**, 2322 (2000) [hep-ph/9908309]. J. L. Feng, K. T. Matchev and T. Moroi, Phys. Rev. D **61**, 075005 (2000) [hep-ph/9909334].

[67] J. F. Donoghue, K. Dutta, A. Ross and M. Tegmark, Phys. Rev. D **81**, 073003 (2010) [arXiv:0903.1024 [hep-ph]].

[68] G. Kane, P. Kumar, R. Lu and B. Zheng, arXiv:1112.1059 [hep-ph].

[69] G. L. Kane, C. F. Kolda and J. D. Wells, Phys. Rev. Lett. **70**, 2686 (1993) [hep-ph/9210242]. J. R. Espinosa and M. Quiros, Phys. Lett. B **302**, 51 (1993) [hep-ph/9212305].

[70] S. P. Martin, In *Kane, G.L. (ed.): Perspectives on supersymmetry II* 1-153 [hep-ph/9709356].

[71] P. Draper, G. Lee and C. E. M. Wagner, Phys. Rev. D **89**, 055023 (2014) [arXiv:1312.5743 [hep-ph]].

[72] G. Kane, R. Lu and B. Zheng, To appear.

[73] J. L. Feng, K. T. Matchev and F. Wilczek, Phys. Lett. B **482**, 388 (2000) [hep-ph/0004043]. A. G. Cohen, D. B. Kaplan and A. E. Nelson, Phys. Lett. B **388**, 588 (1996) [hep-ph/9607394]. J. D. Wells, Phys. Rev. D **71**, 015013 (2005) [hep-ph/0411041]. N. Arkani-Hamed and S. Dimopoulos, JHEP **0506**, 073 (2005) [hep-th/0405159]. N. Arkani-Hamed, S. Dimopoulos, G. F. Giudice and A. Romanino, Nucl. Phys. B **709**, 3 (2005) [arXiv:hep-ph/0409232 [hep-ph]]. J. L. Feng, K. T. Matchev and D. Sanford, arXiv:1112.3021 [hep-ph].

[74] B. S. Acharya, P. Grajek, G. L. Kane, E. Kuflik, K. Suruliz and L. - T. Wang, arXiv:0901.3367 [hep-ph].

[75] G. L. Kane, E. Kuflik, R. Lu and L. -T. Wang, Phys. Rev. D **84**, 095004 (2011) [arXiv:1101.1963 [hep-ph]].

[76] H. Baer, V. Barger, P. Huang and A. Mustafayev, Phys. Rev. D **84**, 091701 (2011) [arXiv:1109.3197 [hep-ph]]. H. Baer, V. Barger and A. Mustafayev, arXiv:1112.3017 [hep-ph]. M. Carena, S. Gori, N. R. Shah and C. E. M. Wagner, arXiv:1112.3336 [hep-ph]. S. Akula, B. Altunkaynak, D. Feldman, P. Nath and G. Peim, arXiv:1112.3645 [hep-ph].

[77] A. Anandakrishnan and C. S. Hill, arXiv:1403.4294 [hep-ex].

[78] G. Kane, R. Lu and B. Zheng, arXiv:1202.4448 [hep-ph].

[79] G. Kane, P. Kumar,B Nelson and B. Zheng, "To Appear"

[80] B. S. Acharya, G. Kane, S. Watson and P. Kumar, Phys. Rev. D **80**, 083529 (2009) [arXiv:0908.2430 [astro-ph.CO]].

[81] B. S. Acharya, K. Bobkov, G. L. Kane, J. Shao and P. Kumar, Phys. Rev. D **78**, 065038 (2008) [arXiv:0801.0478 [hep-ph]].

[82] J. Fan and M. Reece, JHEP **1310**, 124 (2013) [arXiv:1307.4400 [hep-ph]].

[83] A. Arvanitaki, S. Dimopoulos, S. Dubovsky, N. Kaloper and J. March-Russell, Phys. Rev. D **81**, 123530 (2010) [arXiv:0905.4720 [hep-th]].

[84] P. Fox, A. Pierce and S. D. Thomas, hep-th/0409059.

[85] A. Arvanitaki and S. Dubovsky, Phys. Rev. D **83**, 044026 (2011) [arXiv:1004.3558 [hep-th]].

[86] M. Axenides, R. H. Brandenberger and M. S. Turner, Phys. Lett. B **126**, 178 (1983).

[87] P. A. R. Ade *et al.* [BICEP2 Collaboration], arXiv:1403.3985 [astro-ph.CO].

[88] D. J. E. Marsh, D. Grin, R. Hlozek and P. G. Ferreira, arXiv:1403.4216 [astro-ph.CO].

[89] T. Higaki, K. S. Jeong and F. Takahashi, arXiv:1403.4186 [hep-ph].

[90] G. R. Dvali, hep-ph/9505253.

[91] S. Folkerts, C. Germani and J. Redondo, Phys. Lett. B **728**, 532 (2014) [arXiv:1304.7270 [hep-ph]].

[92] I. Affleck and M. Dine, Nucl. Phys. B **249**, 361 (1985).

[93] M. Dine, L. Randall and S. D. Thomas, Nucl. Phys. B **458**, 291 (1996) [hep-ph/9507453].

[94] G. Kane, J. Shao, S. Watson and H. -B. Yu, JCAP **1111**, 012 (2011) [arXiv:1108.5178 [hep-ph]].

[95] D. H. Lyth and E. D. Stewart, Phys. Rev. D **53**, 1784 (1996) [hep-ph/9510204].

[96] G. G. Ross and O. Vives, Phys. Rev. D **67**, 095013 (2003) [hep-ph/0211279].

11

How Could (Should) We Make Contact Between String/M Theory and Our Four-Dimensional World?

Gordon Kane

Michigan Center for Theoretical Physics,
University of Michigan,
Ann Arbor, MI 48109, USA

String/M theory is an exciting framework within which to try to understand our universe and its properties. Compactified string/M theories address and offer solutions to almost every important question and issue in particle physics and particle cosmology. Earlier goals of finding a top-down "vacuum selection" principle and deriving the 4D theory have not yet been realized. Does that mean we should stop trying, as nearly all string theorists have?

Or can we proceed in the historical way to make a few generic, robust assumptions not closely related to observables, and follow where they lead to testable predictions and explanations? In parallel, there can be efforts to replace assumptions with derivations. Making only very generic assumptions is a significant issue. I discuss how to try to proceed with this approach, particularly in M theory compactified on a 7D manifold of G_2 holonomy. One goal is to understand our universe as a string/M theory vacuum for its own sake, in the long tradition of trying to understand our world, and what that implies. In addition, understanding our vacuum may be a prelude to understanding its connection to the multiverse.

I comment on some successful phenomenological and predictive aspects of this framework, such as moduli stabilization and supersymmetry breaking, the emergence of TeV physics from the Planck scale, Higgs physics, axions, dark matter, LHC and future colliders, rare decays and EDMs, particularly to emphasize that apparent phenomenological successes have reached a level where it is not justifiable to argue that it is too soon to work on relating string/M Theory to the real world, and to aim to identify our string/M Theory vacuum. Which results could be considered as derivations and predictions of the string/M theory framework? Some results, particularly for the Higgs mass and decays, seem to be so generic and robust, and depend so little on the details of the manifold, that they should already be considered as predictions of the compactified theory. Perhaps there is a better language emphasizing correlations among properties of solutions instead of "derivations"?

1. Introduction

It would seem there should be great enthusiasm for following a string/M theory path to achieve the ancient goal of describing, understanding and explaining our world as well as possible. Compactified string theories can contain chiral fermions (thus explaining the parity violation of the weak interactions), can have three families of quarks and leptons, can imply the Standard Model forces, can have N=1 supersymmetry that can be softly broken, can explain Higgs physics, can have axions which solve the strong CP problem, can have stabilized moduli and TeV physics emerging from Planck scale physics, and more, and even are strong candidates for a quantum theory of gravity. These features can all occur in the same theory. Many of these features were recognized early. Enthusiasm was somewhat dampened by strong insistance on a pure top-down vacuum selection principle rather than the traditional physics approach of building up increasingly comprehensive compactified string/M theories, and to some extent the effort to build such theories never recovered, even after the unifying emergence of dualities and M theory. There do not seem to be any physics-based reasons for the less than maximal level of activity toward the traditional goals.

Sometimes people say that the existence of many string vacua, many solutions to the string theory framework, represent a barrier because it could be so difficult to find vacua like ours. Work done in the past decade has clearly shown that is simply not true. Lots of compactified string/M theories look very much like our world, and can obviously be extended to give better and better descriptions and explanations. There is not yet a principle showing what Planck scale theory to compactify, nor what gauge and matter group to compactify to, but there are only a limited number of choices. Some are easily excluded by wrong predictions. Some may be equivalent to others, while some may give distinguishable predictions. The calculations needed to proceed along these lines are interesting and workable for the M Theory case, and perhaps for other corners of string theory. Little is known about the mathematics of G_2 manifolds at present, though it is an area of ongoing research. Some physical results seem to depend little on these mathematical properties, while others do not depend on such details at all. Much work needs to be done to improve the mathematical level of understanding, and to remove currently needed assumptions. The successes of compactified String/M theories emphasize the need for such work, and that it is a good time to pursue such work. In this chapter, I only consider the traditional case of 10/11 dimensional theories with Planck scale size curled up small dimensions.

2. Compactifying M Theory — Assumptions

Let us proceed by listing seven assumptions, after which we simply calculate a long list of observables, explaining and predicting many features of our world. For some assumptions I list motivations, which help select our vacuum out of the multiverse. Some of the assumptions can be separately derived in parallel, and all the assumptions are very reasonable. Along with this list, I show some history of such compactifications with relevant results from the literature. A review [1] as well as an updated version of the review in this volume contain more detailed explanations.

- Compactify M theory on a G_2 manifold in the fluxless sector at the Planck scale. This is motivated because fluxes have mass dimensions and make it hard to find ourselves in vacua with TeV physics. Fluxless compactifications imply a non-perturbative superpotential. Then the resulting theory is a 4D N=1 supergravity relativistic quantum field theory [2], and has non-Abelian gauge matter generically localized on 3-cycles [3]. In addition, chiral fermions are generically supported at points with conical singularities on the manifold [4]. An important result is that supersymmetry breaking is gravity mediated since two 3-cycles generically do not overlap in the 7D G_2 manifold.
- Set the gauge group and matter content at compactification to be that of the MSSM. No principle yet exists to automatically determine the matter and gauge group content. The MSSM will be an allowed choice generically, and larger gauge groups or matter content can be studied later. The MSSM will always be part of the matter and gauge group content.
- Assume no obstacles to constructing compact singular G_2 manifolds exist, though too little of the relevant mathematics is known to be sure of this. The mathematics of G_2 manifolds is an active area of research. The parameters of a G_2 manifold which are relevant for physics are described by a set of constants b_k related

to one-loop beta-function coefficients, a set of superpotential term normalizations A_k, a set of rational numbers a_i that sum to $7/3$, and a set of intergers N_i in the gauge kinetic function that specify the homology class of the three-cycle on which the non-Abelian gauge group is localized. We study the moduli effective potential as a function of the (b_k, A_k, a_i, N_i).

- The actual Kahler potential has not been fully calculated. We assume we can use the generic Kahler potential which satisfies certain homogeneity properties [5].
- The actual gauge kinetic function has not been fully calculated. We assume we can use the generic gauge kinetic function [6].
- Assume the μ parameter is zero in the superpotential via the generic approach of Witten [7], with an accidental geometric discrete symmetry.
- Assume the cosmological constant problem is basically orthogonal to the particle physics goals. This can only be certain after the CC problem is solved, of course, but it is highly plausible. If the CC problem is not solved there does not seem to be any obstacle to stabilizing moduli, breaking supersymmetry, etc, while if the CC problem is solved the solutions generally considered do not seem to help stabilize moduli, break supersymmetry, etc. Care should be taken to ensure that the value of the potential at its minimum can be tuned to be very small, of the order of the observed vacuum energy density.

None of these assumptions have any apparent relation to the physics we hope to study in the compactified theory, such as the Higgs boson mass and decay branching ratios, the LHC spectrum, the size of weak or strong CP violation, dark matter, etc. In order to have a predictive theory, and to understand the solutions that emerge to describe our world, it is important to maintain these generic assumptions. If any extensions are made to, for example, the Kahler potential, this may introduce a dimensionful parameter which in turn means that predictions depend on the value of that parameter, and explanatory power is lost. More thought should be given to maintaining *generic* superpotentials, Kahler potentials, and gauge kinetic functions.

It should be emphasized that historically essentially all tests of theories depend on assumptions, from Galileo's inclined planes to study motion through efficiency calculations for Higgs boson observation. Sometimes people say that string theories are not testable because the associated energy scale is the Planck scale and colliders cannot reach such scales. That is like saying the big bang theory cannot be tested since no one was present when the universe began, inspite of the expanding universe and nucleosynthesis and the CMB radiation, or that we cannot learn how dinosaurs became extinct because (it is usually agreed) no people were present then. Such arguments are nonsense. Obviously the tests of string/M theories should test predictions of the theories compactified to 4D, since we live in a 4D universe. We will list a number of examples next.

A few quantum field theory tests do exist for general theories regardless of the particular force or Lagrangian, such as the requirement that all electrons are identical since they are quanta of the electron field. Perhaps general, non-trivial tests of string/M theories independent of compactifications will exist, but none are known so far, except possibly being a consistent quantum theory of gravity, and qualitative tests such as expecting non-Abelian gauge forces. The lack of such tests of the 10/11 D theories does not in any interesting sense imply that string/M theories are less testable than, say, F=ma, for which all tests require using a particular force (analogous to a particular compactification). Well known tests after compactification include having massless modes or singularities that give Abelian and non-Abelian gauge matter and chiral fermions, and Abelian and non-Abelian gauge symmetries like those of the Standard Model.

3. Results and Predictions of the Compactified M Theory

Given the above assumptions, a remarkable set of results and predictions emerge from calculations with no adjustable parameters. I will list these results below. All the results are from the same theory, using standard supergravity quantum field theory techniques. In addition to the intrinsic interest of obtaining these results and predictions in one underlying theory, the results should encourage readers to think it is not too early for physicists to engage with the compactified string/M Theories.

- The scalar potential is calculated. It depends on all (relevant) moduli, so all moduli interact and are therefore stabilized at the minimum of their potential energy. Their typical vacuum values (vevs) are one to two orders of magnitude below the Planck scale.

- Generic 3-cycles occur, some with larger gauge groups for non-Abelian matter. The associated RGE running leads to gaugino condensation at scales of order 10^{14} GeV, giving non-zero F-terms that break supersymmetry at such scales. Some of course have smaller gauge groups, but they condense at lower scales. Although supersymmetry is broken by gaugino condensation at scales Λ of order 10^{14} GeV, the gravitino mass comes out to be of order 50 TeV from dimensional transmutation (see below). These non-Abelian gauge groups may also play a role in understanding dark matter.

- Conical singularities lead to chiral fermions and thus meson condensates that also generically occur at scales Λ of order 10^{14} GeV, giving non-zero F-terms that also break supersymmetry at such scales. The gaugino and meson condensates give a unique deSitter vacuum without needing any extra "uplifting".

- The full supersymmetry soft-breaking Lagrangian is calculated. Tree level scalar masses are degenerate and equal to the gravitino mass to a good approximation. Tree level trilinear couplings have contributions proportional to Yukawa couplings, and 3rd family ones are about 1.5 times tree level scalar masses. Gaugino masses are suppressed because they are proportional to the derivative of the Standard Model gauge kinetic function f_{SM} with respect to the moduli, but the SM gauge kinetic function does not depend on the meson condensate fields. Therefore its derivative with respect to the meson F-terms vanishes, since $M_{1/2} \sim F_m \partial_n f_{SM}$. This gaugino mass suppression is a generic and robust prediction. Tree level gaugino masses are also degenerate, but there is an anomaly mediation contribution to gaugino masses that is similar in size to the suppressed tree level ones, so the resulting gaugino masses are not degenerate, even at the high scale. We will review the derivation of these results in the next section.

- All terms in the soft-breaking Lagrangian turn out to have the same phase, so their phase can be rotated away. Since the $\mu-$term from the superpotential has been guaranteed to be zero by the Witten mechanism, μ arises from the Kahler potential and its phase (along with that of $B\mu$) can also be rotated away by a PQ rotation, and thus at tree level the "susy CP problem" is solved. Since the scalars are of order the gravitino mass, tens of TeV, there are no CP related phenomenological issues. EDMs are predicted to be non-zero only because the phase(s) from the Yukawa couplings (that leads to the CKM phase) also appears in the trilinear couplings, and gets rotated by the RGE running into the scalar masses at the weak scale. This generates interesting EDMs, smaller than current limits but not a great deal smaller [8, 9]. Thus EDMs are calculable because they only depend on the CKM phase. Since scalars are heavier than about 25 TeV, squarks will not be observable at LHC, but some may be at a 100 TeV pp collider. Gluinos and some lighter electroweakinos will be observable at LHC if it reaches design energy and luminosity. Since scalars are heavy, the rare decay $B_s \to \mu\mu$ will not be measurably different from its SM value (as was predicted in the compactified theory well before the LHC data), and $(g - 2)_\mu$ should also not deviate more than a few per cent (from loop effects) from its SM value (so the current $\sim 3\sigma$ effect should disappear).

- The moduli mass matrix is calculable. Moduli have only gravitational interactions, so their lifetimes for decay to all SM particles, superpartners and axions are calculable.
- The gravitino mass $M_{3/2}$ can be approximately calculated, by the traditional dimensional transmutation top down method. The superpotential W dimensionally is of order $(\Lambda/M_{PL})^3$ and has a further factor $e^{K/2}$ that is approximately a volume suppression. The gravitino mass comes out to be about 50 TeV, to about a factor of two. This result is crucial for this framework, and we will summarize its derivation in the next section.
- Remarkably, the gravitino mass and the lightest eigenvalue of the moduli mass matrix can be related [10–12]. Qualitatively, the (complex) scalar gravitino superpartner is degenerate with the gravitino, and mixes with moduli. The scalar gravitino 2 × 2 mass matrix is somewhere inside the positive definite moduli mass matrix and can be moved to the diagonal, and the theorem that the lightest eigenvalue of the full matrix is less than the eigenvalues of any of the 2 × 2 submatrices gives the stated result. The full result depends on the Kahler curvature of the full matrix, but for the compactified theory of interest to us that should not matter, and in any case one can prove a relevant result.
- Then the lightest moduli mass should not be lighter than about the gravitino mass, about 25 TeV from the top-down calculation just described, so the moduli decay before BBN and there is no cosmological moduli (or gravitino) problem.

- String axions are present [13]. One combination is massless down to the QCD scale and provides a solution to the strong CP problem. Upper bounds on the axion decay constants increase because the universe is matter dominated instead of radiation dominated after inflation.
- When moduli are stabilized, the discrete symmetry Witten proposed to solve doublet-triplet splitting and enforce $\mu = 0$ is broken to generate nonzero μ. Calculating the resulting value of μ precisely requires knowledge of G_2 manifolds which is currently unavailable. But the resulting value of μ should vanish if supersymmetry became unbroken, so it should be proportional to $M_{3/2}$, and μ should also vanish if the moduli were not stabilized, so it should be proportional to typical moduli vevs divided by the Planck mass. As a result, μ should be suppressed from $M_{3/2}$ by an order of magnitude or so.
- The moduli make the universe matter dominated instead of radiation dominated until they decay near BBN time, so the cosmological history is non-thermal. Large entropy from moduli decay, proportional to the temperature cubed so of order 10^9, wash out any freezeout dark matter, and any early matter asymmetry. Dark matter is regenerated from moduli decay into superpartners, axions and LSPs, and a matter asymmetry is generated. An initial matter asymmetry of order unity at the end of inflation from Afflek-Dine baryogenesis can lead to the observed matter asymmetry and can explain the ratio of matter to dark matter [14].
- R-Parity conservation. Because the coefficient μ of $H_u H_d$ and the coefficient κ of the bilinear R-parity violating term come from similar Kahler operators and are therefore about the same size in the compactified theory, one can show that all R-parity operators must be absent or very large neutrino masses would be generated [15].

Conservatively, one can say that the M theory vacuum we are studying can contain all these results in a correlated way. It is very encouraging that so far nothing has gone wrong. The previously stated assumptions still need to be satisfied, though hopefully most will be derived in the near future.

4. Derivation of $M_{3/2} \sim 50$ TeV from Moduli Stabilization and Supersymmetry Breaking

A key result which is crucial for the connection between compactified M theory and TeV scale physics is the fact that moduli stabilization fixes the gravitino mass to be 50 TeV, within a factor of 2 or so. In order to keep this review as self contained as possible, we provide a summary of this derivation below, though the relevant results were first obtained in [16–19].

In G_2 compactifications of M theory, the moduli fields s_i in the effective 4-D theory arise from Kaluza-Klein zero modes of the covariantly constant 3-form Φ, which is uniquely determined by the metric of the G_2 manifold X. Note that the 3-form Φ is real, such that the moduli fields s_i are real scalar fields in 4-D. The moduli Kahler potential can be inferred from the classical moduli space metric to be [5]:

$$\hat{K} = -3 \log 4\pi^{1/3} V_X \tag{1}$$

where V_X is the volume of the G_2-manifold in 11-d Planck units:

$$V_X = \frac{1}{7} \int_X \Phi \wedge *\Phi. \tag{2}$$

Without explicit knowledge of the metric for a particular G_2 manifold, the functional form for V_X in terms of the moduli fields s_i can not be determined. However, it is possible to argue on general grounds that V_X is a homogeneous function of degree 7/3 in the moduli s_i, in other words $V_X \to A^{7/3} V_X$ if the moduli are scaled by a common factor $s_i \to A s_i$. Using this homogeneity property, one can derive the identity:

$$\sum_{i=1}^{N} s_i \hat{K}_i = -7 \tag{3}$$

where $\hat{K}_i \equiv \partial \hat{K}/\partial s_i$. We will use this homogeneity property in Section 2 to derive general results from moduli stabilization without needing to specify a particular form for V_X.

The Kahler potential for the matter fields is given by:

$$K = \kappa(s_i) \frac{QQ^\dagger}{V_X} \tag{4}$$

where Q represents some chiral matter supermuliplet and $\kappa(s_i)$ is a scale-invariant function of the moduli s_i. This form for the Kahler potential was motivated in Section III of [19] by three independent arguments: (i)

dimensional reduction, (ii) locality of the physical Yukawa couplings and
(iii) matching KK threshold corrections to the 4-D gauge coupling [20] in
the effective $\mathcal{N} = 1$ SUGRA theory. For brevity we will not reproduce these
arguments here.

Moduli Superpotential

In order to preserve $\mathcal{N} = 1$ SUSY in the effective Lagrangian, the real
moduli fields s_i must combine with some other scalar degree of freedom in
order to form the complex scalar component of some chiral mutliplet. Such
scalar degrees of freedom arise from the KK zero-modes of the real 3-form
C-field present in 11-dimensional supergravity. The complexified moduli z_i
can be expressed as the zero-modes of the "complexified" 3-form $\Phi + iC$:

$$C + i\Phi = \sum_{i=1}^{b^3(X)} (a_i + is_i)\,\phi_i = \sum_{i=1}^{b^3(X)} z^i \phi_i \tag{5}$$

where ϕ_i represent harmonic 3-forms $\phi_i \in H^3(X, \mathbb{Z})$ and $b^3(X)$ is the 3rd
Betti number of the G_2 manifold. From explicit constructions of smooth
G_2-manifolds we expect $b^3(X) \sim \mathcal{O}(100)$.

The a_i, which are the scalar zero-modes of C, are axionic as they inherit
a shift symmetry in the 4-D effective theory from the underlying higher-
dimensional gauge symmetry of the C-field. This provides a very important
M-theoretic input for the 4-D effective lagrangian, namely that *polynomials
of the complexified moduli fields can not appear in the superpotential*, assuming that the axionic shift symmetry is violated only by non-perturbative
effects. In order to stabilize the moduli s_i, we must assume the presence
of non-Abelian "hidden sector" gauge groups in addition to the Standard
Model one. In G_2 compactifications of M theory, non-Abelian gauge fields
are localized along 3-dimensional submanifolds which parameterize families of ADE orbifold singularities [3, 21]. Thus requiring the presence of
hidden sectors is equivalent to assuming that there are some other 3-cycles
in the G_2 manifold which support non-Abelian gauge fields, in addition to
the visible sector 3-cycle which supports the SM gauge fields. Because two
3-cycles will generically not intersect in a 7-dimensional space, we assume
no light matter charged under both the visible and hidden sector gauge
groups and thus SUSY breaking will be gravity mediated.

In the presence of a pure $SU(Q)$ SYM hidden sector, non-perturbative
dynamics generate an effective moduli superpotential of the form $W = Am_{pl}^3 e^{i2\pi bf}$ where f is the hidden sector gauge kinetic function $f = \sum_i N_i z_i$

and $b = 1/Q$. The integers N_i are determined by the homology class of the 3-cycle. Such a superpotential will stabilize all moduli; however with the presence of only a single hidden sector may not yield vacuaa which are within the supergravity approximation ($V_X \gg 1$ in 11-D Planck units). Introducing another pure SYM hidden sector such that $W = m_{pl}^3 \left(A_1 e^{i2\pi b_1 f_1} + A_2 e^{i2\pi b_2 f_2} \right)$ will stabilize moduli within the supergravity approximation, but gives only AdS vacuaa. In order to obtain de Sitter vacua with moduli stabilized in a region where the supergravity approximation is valid, at least one hidden sector $SU(P)$ gauge group with charged matter is required. It was discussed in [4] how chiral matter charged under a particular gauge group naturally arises from isolated conical singularities in the G_2 manifold.

In the minimal setup, we assume that there are two hidden sectors; an $SU(Q)$ pure SYM hidden sector, and a $SU(P+1)$ hidden sector with $N_f = 1$ flavor of fundamental + antifundamental chiral multiplets. The Affleck-Dine-Seiberg effective superpotential [22] is then given by:

$$W = A_1 \phi^{-2b_1} m_{pl}^3 \, e^{i2\pi b_1 f_1} + A_2 m_{pl}^3 \, e^{i2\pi b_2 f_2} \tag{6}$$

where $b_1 = 1/P$, $b_2 = 1/Q$, f_1, f_2 are the gauge kinetic functions of the $SU(P+1)$ and $SU(Q)$ hidden sectors, and A_1 and A_2 represent instanton prefactors. ϕ represents the $Q\overline{Q}$ meson condensate in $SU(P+1)$ sector. There in principal will be other gauge groups which also contribute nonperturbative superpotential terms. However their contribution to W will scale like $W \propto e^{i2\pi f/(N-N_f)}$, so hidden sectors with large $N - N_f$ will provide the dominant contribution to $\langle W \rangle$ while other smaller rank hidden sectors will provide a subdominant contribution to $\langle W \rangle$ and can be neglected in the moduli stabilization analysis.

Moduli Stabilization and Determining $M_{3/2}$

From the discussion in the preceding section, we have argued that in G_2 compactifications of M theory motivate the following forms for the Kahler potential and moduli superpotential in the effective theory:

$$K = -3 \log 4\pi^{1/3} V_X + \kappa_h(s_i) \frac{\phi\phi^{\dagger}}{V_X}$$
$$W = A_1 \phi^{-2b_1} m_{pl}^3 \, e^{i2\pi b_1 f_1} + A_2 m_{pl}^3 \, e^{i2\pi b_2 f_2} \tag{7}$$

where again we take ϕ to be the $SU(P+1)$ hidden sector meson condensate. We have neglected visible sector fields, as they do not develop *vev*'s at this

stage and are thus irrelevant for the moduli stabilization analysis. In the
following analysis we assume that 3-cycles which support the $SU(Q)$ and
$SU(P+1)$ gauge groups are equivalent in cohomology so that $f_1 = f_2 \equiv f$.
This simplifies the moduli stabilization analysis; the more general case is
considered in [23]. Given W and K in (7), we can now compute the scalar
potential for the effective 4-D $\mathcal{N} = 1$ SUGRA Lagrangian:

$$V = e^K \left(g^{n\overline{m}} F_n \overline{F}_{\overline{m}} - 3 \, |W|^2 \right) \tag{8}$$

where the Kahler metric $g_{n\overline{m}}$ is obtained in the usual way by differentiating
K with respect to the moduli and meson fields. The resulting form of the
Kahler potential in terms of the moduli fields along with the minimization
conditions are rather cumbersome, so we will not reproduce them here. The
details are provided in pages 21–25 of [19]. Here we simply recapitulate the
main points of the analysis, which in particular demonstrates that solutions
with stabilized moduli exist.

The stabilized moduli vev's are given by the following ansatz:

$$s_i = \frac{a_i}{N_i} \frac{3}{7} V_Q \tag{9}$$

where V_Q is the volume of the hidden sector gauge kinetic functions, $V_Q \equiv
Im(f) = \sum_i N_i s_i$. The a_i are defined as $3a_i \equiv -s_i \, \partial \hat{K}/\partial s_i$, and given an
explicit form for V_X their value at the de Sitter minimum can be obtained
by solving the following transcendental equation:

$$\left. \frac{\partial \hat{K}}{\partial s_i} \right|_{s_i = a_i/N_i} = -3N_i. \tag{10}$$

Regardless of the form for V_X, the homogeneity property (3) implies
$\sum a_i = 7/3$. Given the ansatz (9) for s_i, one can numerically solve for
the value of V_Q which satisfies the moduli stabilization equations. A good
approximation for V_Q is given by:

$$V_Q = \frac{1}{2\pi} \frac{PQ}{Q - P} \log \left(\frac{Q A_1 \phi_0^{-2/P}}{P A_2} \right) \tag{11}$$

in the limit where $V_Q \gg 1$. In a regime where the supergravity approximation is valid $V_X \gg 1$; since the homogeneity properties of V_X imply
$V_X \propto s^{7/3}$, we expect $s_i \gg 1$ and thus $V_Q = \sum N_i s_i \gg 1$ if the supergravity approximation is valid.

Given the ansatz (9) along with the homogeneity property $\sum_i a_i = 7/3$,
· the value of the scalar potential (8) at the minimum to lowest order in

$\mathcal{O}(1/P_{eff}{}^2)$ is given by:

$$\frac{V_0}{m_{pl}^2 m_{3/2}^2} = \left[\left(\frac{2}{Q-P} + \frac{\phi_0^2}{V_X} \right)^2 \right.$$
$$\left. + \frac{14}{P_{eff}} \left(1 - \frac{2}{3(Q-P)} \right) \left(\frac{2}{Q-P} + \frac{\phi_0^2}{V_X} \right) - 3\frac{\phi_0^2}{V_X} \right] \frac{V_X}{\phi_0^2} \quad (12)$$

where we have defined:

$$P_{eff} = P \log \left(\frac{QA_1 \phi_0^{-2/P}}{PA_2} \right). \quad (13)$$

Thus the superpotential coefficients A_1 and A_2 have been absorbed into our definition of P_{eff}. Note from (11) that $V_Q \approx P_{eff}Q/2\pi(Q-P)$, so $V_Q \gg 1$ ensures $P_{eff} \gg 1$.

We now require that the minimization condition for the meson condensate ϕ_0 causes the tree level scalar potential (12) to vanish. To leading order in $1/P_{eff}^2$, this imposes the requirement that[a]:

$$P_{eff} = \frac{14(3(Q-P)-2)}{3(3(Q-P)) - 2\sqrt{6(Q-P)}}. \quad (14)$$

Thus given particular values of Q and P, imposing a vanishing cosmological constant actually fixes the ratio A_1/A_2. Put another way, tuning the vacuum energy to zero at tree level upon moduli stabilization is only possible for a particular value of A_1/A_2. Requiring the s_i to be stabilized at positive values requires $P_{eff} > 0$ as can be seen from (9) and (11), which imposes the constraint $Q - P \geq 3$. If (14) is satisfied, then the meson vev is obtained by determining the value of ϕ_0 which causes (12) to vanish:

$$\frac{\phi_0^2}{V_X} = \frac{2}{Q-P} + \frac{7}{P_{eff}} \left(1 - \frac{2}{3(Q-P)} \right) + \mathcal{O}(1/P_{eff}^2). \quad (15)$$

Thus we have shown that ϕ_0 and s_i are all stabilized by non-perturbative effects from hidden sector strong dynamics. We can now compute the value of the bare gravitino mass from (7)-(14):

$$m_{3/2} = m_{pl} e^{K/2} |W| = m_{pl} \frac{e^{\phi_0^2/2V_X}}{8\sqrt{\pi}V_X^{3/2}} |P - Q| \frac{A_2}{Q} e^{-P_{eff}/(Q-P)}. \quad (16)$$

Thus given the ranks of the hidden sector gauge groups $SU(Q)$ and $SU(P+1)$, the value of $m_{3/2}$ is completely determined up to the unknown instanton

[a]Note that we are neglecting the ϕ_0 dependence of P_{eff}, due to the smallness of $2/P$ along with the fact that the dependence is logarithmic.

prefactor A_2. Field theoretic computations [24] indicate that $A_2 = Q$ up to RGE-scheme dependent and threshold corrections, so we expect $A/Q \lesssim 1$.

Now we examine constraints on the possible values of $Q - P$, which fixes the exponential factor in (16). As discussed, we require $Q - P \geq 3$ so that $V_Q \equiv Im(f) = \sum N_i s_i$ stays positive. From (11), (13) and (14) we see that for $Q - P = 3$, $V_Q \approx 3.37\, Q$ while for $Q - P > 3$, $V_Q < 0.84\, Q$. Recall that for the supergravity approximation to be valid we require $V_X \gg 1$ from which we expect $s_i > 1$. Thus we expect $V_Q = \sum_i N_i s_i \gtrsim N_{mod}$, where N_{mod} is the number of moduli fields. As mentioned in Section 1, from explicit constructions we expect $N_{mod} = b^3(X) \sim \mathcal{O}(100)$, so we roughly expect $V_Q \sim \mathcal{O}(100)$ if the supergravity approximation is valid. Thus the larger the value of $Q - P$, the more difficult it is to stabilize moduli in a region where the supergravity approximation is valid. To ensure the validity of the SUGRA approximation we focus on the case where $Q - P = 3$; the general case will be discussed at the end of this section. This fixes the gravitino mass to be:

$$m_{3/2} \approx 10^6 \,(\text{TeV})\, \frac{A_2}{Q V_X^{3/2}}. \tag{17}$$

The value of V_X in (17) can be fixed from dimensional reduction arguments by requiring that the 11-D supergravity theory gives the correct value for the 4-D visible sector gauge coupling α_{GUT} at the GUT scale [20]. This constrains V_X to be:

$$V_X \approx 137 L(\mathcal{Q})^{2/3} \tag{18}$$

where $L(\mathcal{Q})$ is a topological invariant, related to the analytic torsion of the 3-cycle $\mathcal{Q}$ on which visible sector gauge fields are localized. The dependence of V_X on $L(\mathcal{Q})$ arises from computing KK-threshold corrections to the visible sector gauge coupling. Motivated by triplet-doublet splitting [7], the reference [20] assumes $\mathcal{Q} \cong S_3/Z_q$ in which case $L(\mathcal{Q}) = 4q \sin^2(5\pi\omega/q)$. ω is an integer determined by the geometry of $\mathcal{Q}$ such that $Mod(5\omega, q) \neq 0$. The Poincare conjecture seems to imply that this form for $L(\mathcal{Q})$ is fairly general, but we are still currently working on understanding this issue in more detail. It is also straightforward to compute the scale of gaugino condensation in the $SU(Q)$ SYM hidden sector:

$$\Lambda \sim m_{pl} \frac{e^{-\frac{2\pi}{3Q} V_Q}}{2\pi^{1/6} V_X^{1/2}} \approx \frac{1.1 \times 10^{14}\,\text{GeV}}{L(\mathcal{Q})^{1/3}}. \tag{19}$$

Assuming $Q - P = 3$ and $L(\mathcal{Q}) = 4q \sin^2(5\pi\omega/q)$, we can combine (17) and (18) to obtain some representative values for $m_{3/2}$, given in Table 1 of [19]. Depending on the values of p and q, $20\,\mathrm{TeV} \lesssim m_{3/2}\,(A_2/Q) \lesssim 100\,\mathrm{TeV}$. Thus up to the ratio A_2/Q which we expect to be $\lesssim 1$, we have shown with a pure $SU(Q)$ gauge group as well as an $SU(P+1)$ gauge group with $N_f = 1$, setting $Q - P = 3$ and stabilizing moduli naturally gives $m_{3/2} \sim \mathcal{O}(50)$ TeV within about a factor of 2.

We have shown that there exists a generic class of compactified M theories which naturally give $m_{3/2} \sim 50$ TeV upon moduli stabilization; however a crucial assumption we have made is $Q - P = 3$ to ensure the validity of the supergravity approximation $V_X \gg 1$. It may be in principal possible to find valid solutions with $Q - P > 3$, which will greatly change $m_{3/2}$ due to the exponential sensitivity in (16). However, it is straightforward to show from (14) and (16) that for $Q - P \geq 4$, $P_{eff} \geq 20$, which results in $m_{3/2} \gtrsim 10^{-2} M_{pl}$. Therefore $Q - P = 3$ is the only possibility which allows a solution to the hierarchy problem between $m_{3/2}$ and M_{pl}. In other words, if one demands a solution the hierarchy problem from moduli stabilization in this framework, $Q - P = 3$ and $m_{3/2} \sim 50$ TeV is a robust prediction. If one takes the appropriate attitude for string phenomenology of looking for a generic set of solutions that could describe our world, with stabilized moduli, the set of solutions that emerges and solves the hierarchy problem has $m_{3/2} \sim 50$ TeV.

Hierarchy between Gaugino Masses and $M_{3/2}$

We now briefly discuss the hierarchy between $m_{3/2}$ and the soft SUSY breaking gaugino masses, namely $M_{1/2}/m_{3/2} \sim 10^{-2}$, which results from the moduli stabilization procedure mentioned in Section 2. The universal tree-level contribution to the gaugino masses from the supergravity Lagrangian is given by [25]:

$$M_{1/2} = \frac{e^{K/2} F^i \partial_i f_{vis}}{2i\,Im(f_{vis})} \tag{20}$$

where F_i are the F-terms for the moduli fields s_i and f_{vis} is now the visible sector gauge kinetic function, $f_{vis} = \sum_i N_i^{vis} z_i$. The suppression of $M_{1/2}$ with respect to $m_{3/2}$ arises from a suppression of the moduli F-terms F_i with respect to the hidden sector meson condensate F-term F_ϕ. In particular, the moduli stabilization procedure discussed in the previous section yields the following F-term *vev*'s:

$$\left| e^{K/2} F^i \right| \approx \frac{2s_i}{P_{eff}} m_{3/2}, \quad \left| e^{K/2} F^\phi \right| \approx \phi\, m_{3/2} \tag{21}$$

The moduli stabilization procedure yields $s_i \sim \phi$, and therefore the moduli F-terms are suppressed by $1/P_{eff}$ with respect to the meson condensate F-terms; therefore the meson condensate F-terms dominate the vacuum energy. Since f_{vis} does not depend on the meson condensate ϕ, $\partial_\phi f_{vis} = 0$ and only the smaller moduli F-terms contribute to (20). Consequently, (12) yields:

$$|M_{1/2}| \approx \frac{1}{P_{eff}} \left(1 + \frac{2V_X}{(Q-P)\phi_0^2}\right) m_{3/2}. \tag{22}$$

As discussed in the previous section, for solutions with $Q-P=3$, $P_{eff} \approx 61$ and $\phi_0^2/V_X \sim 0.5$, and thus $M_{1/2} \approx 0.03\, m_{3/2}$. Thus the universal tree level contribution to gaugino masses from gravity mediation is suppressed relative to $m_{3/2}$. There are also the usual non-universal anomaly-mediated contributions to the gaugino masses, as well as non-universal effects which arise from renormalizing parameters down to the electroweak scale, but they are of the same order of magnitude as the tree level contributions from (20).

This result is significant for LHC phenomenology, as in gravity mediation the soft SUSY breaking scalar masses are all $\mathcal{O}(m_{3/2})$ assuming no sequestering, as is expected to be generic in M theory compactifications. This implies that all scalar superpartners will have masses $\tilde{m} \sim m_{3/2} \sim 10$'s of TeV, while gauginos will have masses in the $200\,\text{GeV} \lesssim M \lesssim 1.5$ TeV range (depending on the particular gaugino in question). Thus the compactified M theory framework yields a robust prediction that scalar superpartners will be out of reach of the LHC, while the gauginos (the gluino in particular) should be light enough to be within kinematic reach. As we have discussed, these predictions are largely insensitive to details regarding the precise mathematical structure of compact G_2 manifolds.

5. Higgs Mass "Prediction"

Next we want to ask about Higgs physics in this compactified M theory vacuum. How can one approach the issue of Higgs physics in string/M Theory? Vacua that do not allow Electroweak Symmetry Breaking (EWSB) could not describe our world. Presumably we do not expect all vacua to break the electroweak symmetry. It's not clear what a pure top-down approach might be. We think an approach where we consider the subset of all vacua that can allow EWSB and study the resulting Higgs sector and calculate observables in such vacua is appropriate. Various vacua might predict a range of Higgs masses and properties.

In the N=1 supersymmetric theory (already derived for the compactified M Theory) there are two Higgs doublets. The lightest mass eigenstate is identified with "the Higgs boson". Given the calculated Lagrangian soft-breaking terms M_{H_u} and M_{H_d} one calculates the Higgs sector scalar potential and identifies the coefficient λ_{eff} of the h^4 term. The EWSB conditions are normally expressed in terms of the ratio of up- and down-type Higgs vevs, $\tan\beta$. For values of $\tan\beta$ not too small, the EWSB conditions become [26] approximately $\tan\beta \approx M_{3/2}/1.7\mu$. The value 1.7 would have been 2 at the unification scale before RGE running, but since $\tan\beta$ does not exist at the high scale until the RGE running generates the vevs, this form is more useful.

Amazingly it turns out [27] that all solutions that satisfy the EWSB conditions have the same Higgs boson mass, and it is 126 GeV, with an uncertainty of about ±1.5 GeV. The uncertainty is largely due to the experimental errors in the top quark mass, since the (not precisely known) top Yukawa couplings give large 1-loop corrections to the Higgs mass. The resulting uncertainty in the Higgs mass to be compared to experimental data is currently somewhat larger than the reported experimental error. Both will improve with time. The Higgs mass is probably the most precise quantity that can be predicted from a compactified string/M Theory, because the combination of the large gravitino mass, the connection of scalar masses to the gravitino mass, and the EWSB conditions force the outcome of a precise value for M_h. It is worth remarking that the central value of the predicted M_h was reported in summer 2011, well before any experimental information. The subtle evaluation of the uncertainty in the prediction from connecting to the low scale, and including radiative corrections and errors in input informatioin on the top mass etc was completed in November 2011, before the LHC reports.

Recently, three-loop contributions to the Higgs mass prediction have become available [28–30]. They have been incorporated into our calculation [31]. For heavy scalars at the few tens of TeV level that are expected from the compactified M Theory, the three loop corrections can have large logs and could have had large contributions (of the order of several GeV) to the Higgs mass. When the full calculation is carried out however, the effects are not large, increasing the Higgs mass by about one GeV. There is a robust set of solutions with correlated values of M_h, $\tan\beta$, μ, and $M_{3/2}$. This need not have happened, and that it did is strong endorsement of the compactified M Theory as a good candidate for describing our vacuum. The required values of μ and $\tan\beta$ are consistent with compactified M theory predictions, but are more tightly contrained than what can at present calculated from the theory. Therefore the result is predictive without adjustable parameters, but our ability to calculate μ, $\tan\beta$, M_h and $M_{3/2}$ separately from the generic theory is not yet good enough to capture all compactified M theory solutions.

6. Standard Model Higgs? Metastability? Naturalness?

The Higgs boson discovered at LHC had several apparently surprising properties. From the point of view of the compactified M Theory these properties are not surprising, and in fact are largely not only expected but are generic.

- The Higgs mass value was larger than naively predicted in non-string/M Theories. This follows automatically in the compactified M Theory since the gravitino is calculated to be tens of TeV, and the scalars masses are approximately equal to the gravitino mass. Then one is in the decoupling limit of the MSSM Higgs sector, and one needs to include supersymmetric loop effects which then leads to the 126 GeV result. The inputs to this calculation are fixed by the compactified M theory. From the theory point of view all superpartners should naively be of order the gravitino mass, that is tens of TeV. Luckily the largest F-term from the meson condensate does not contribute to gaugino masses so the gluinos and electroweakinos should have suppressed masses, some in the one TeV or fraction of a TeV range. These should be observable at the upgraded LHC if it reaches design energy and luminosity. The spectrum is shown in Figure 1. Some of the squarks and electroweakinos will be observable at a 100 TeV pp collider.
- The Higgs decay branching ratios were those of a SM Higgs boson to good accuracy. This is standard and unavoidable in the decoupling limit of the MSSM Higgs sector, a successful prediction.
- The Higgs potential has a shape that is largely determined by the values of M_{top} and M_h, and ends up in a region that would be on the edge of instability if the theory were the SM theory. In the supersymmetric case the vacuum is automatically guaranteed

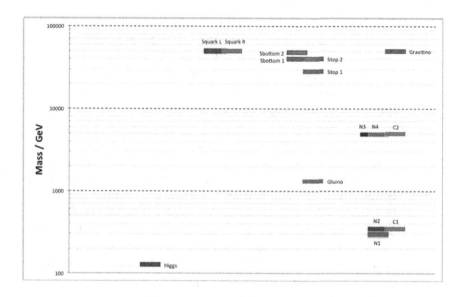

to be stable – the value of λ_{eff} is never below about 0.1, so the metastability issue never arises. So the "apparent" SM metastability must be accidental. That is not an interesting coincidence, because if one computes the allowed region on the $M_{top} - M_h$ plane in the supersymmetric case one finds that region always includes the SM metastability point (see for example the talk of G. Isidori at Supersymmetry 2013).

• One can think of the breaking of the electroweak symmetry as the mechanism to explain the gauge boson masses M_Z and M_W, or equivalently the Higgs vev v. If one just takes generic values for all quantities in a top-down calculation of v, one finds a value of a few TeV, rather than the actual value of a quarter of a TeV. This is called the "little hierarchy problem". The usual equations are for the squares of the masses, which then come out about two orders of magnitude too large. The compactified M Theory suggests a possible solution for this. The soft-breaking trilinear couplings are calculable and come out about 1.5 times the gravitino mass.

Then the RGE running produces values of $M^2_{H_u}$ and $M^2_{H_d}$ that do allow correct values of v or M^2_Z. Basically $M^2_{H_u}$ always runs to nearly zero at or below the TeV scale, because the trilinears are generically large, and do affect the running in the right way. This is the first time a theoretical approach led to such large trilinears. But the relevant quantities need to be strongly correlated to give the small values of v, etc., so although this approach is promising it is not yet a robust solution.

- There has been much talk about the failure of naive naturalness, since superpartners have not been observed yet, and the Higgs boson is heavier than naively expected. Many talks and papers by experts and non-experts have speculated on the implications. Speakers at meetings in the past year often list alternatives. Surprisingly, the most sensible alternative is seldom mentioned. The opposite of unnatural is having a theory! Naturalness is basically the argument that dimensional analysis in an effective theory will allow masses and other properties to be estimated with some success. When naturalness fails I think the obvious implication is that there is an underlying theory, and the theory has outcomes that are more complicated. That is just what happens in the compactified M Theory case, where the gravitino mass sets the scale for electroweak scale physics, but solutions having electroweak symmetry breaking are then in the decoupling limit and have a suppressed Higgs mass compared to the gravitino mass while the Higgs mass is enhanced compared to the Z mass, and gaugino masses are suppressed because the dominant F-term does not contribute to the gaugino masses. The compactified theory naturally predicts the observed Higgs mass, but the word "naturalness" has been abused to mean something different, so we should probably not use it in this context.

7. Some Predictions Robust, Others Not

The Higgs mass and branching ratios are calculable in the compactified theory to very good accuracy, because of the physics of how they arise, via heavy scalars. The gluino mass is suppressed, so it should be around one TeV, but it's mass depends on quantities that are not easily calculable, such as the gaugino condensate F-term, and the derivative of the SM gauge kinetic function. One can try to use precision unification conditions to constrain the gluino mass, but because the tree level mass terms and the anomaly mediation contribution to the gaugino masses are comparable, and because there are threshold corrections to the unification calculations [20], an accurate calculation is again difficult. So gluinos should be a TeV to about a factor of two. It is important to do a more accurate calculation before LHC data is available. The cosmological history and solution of the moduli problem depends only on moduli being heavier than about 25 TeV, so that is robust.

8. What Does the String/M Theory Give Us Compared to Models and Effective Theories?

Lots of authors like to make arguments about what kinds of models and model parameters give a Higgs mass like the experimental one. What do we gain if we have the compactified string/M Theory rather than just models? Below is a table comparing the approaches, and one can see the major gains in understanding resulting from the compactified M Theory.

One way to summarize why the compactified theory is far better than purely effective theory approaches is to realize that in effective theories, undetermined coefficients of effective operators are independent free parameters. In contrast, in the compactified theory all such coefficients are determined by the theory and explicitly related to one another, which allows one to make strong correlations between different physical observables.

9. More To Do

I have argued that the M Theory compactification, and perhaps others, are successful enough so that the point of view that it is too early to try to use compactified string/M theories to improve our understanding of our world is no longer valid. There is evidence that if people work on it, success will follow. I have focused on the M Theory approach where my own work has been. In this framework one can identify several directions that are ripe for improvement, and have not been worked on only because of lack of time and people. In the following, I list some of these open problems.

- Any satisfactory theory must be able to derive the existence of a top quark with Yukawa coupling of order unity, and no other quarks or leptons with Yukawa couplings of order unity. There

COMPACTIFIED(STRING)M THEORY	SPLIT SUSY (ETC) MODELS
• **Derive** solution to large hierarchy problem	• Assumes **no solution (possible) for large hierarchy problem**
• Generic solutions with **EWSB derived**	• **EWSB assumed**, not derived
• main F term drops out of **gaugino masses so dynamically suppressed**	• **Gauginos suppressed by assumed** R-symmetry, suppression arbitrary
• Trilinears > $M_{3/2}$ necessarily	• Trilinears small, suppressed compared to scalars
• **μ incorporated in theory (M-theory)**	• **μ not in theory** at all; guessed to be $\mu \sim M_{3/2}$
• Little hierarchy significantly reduced	• No solution to little hierarchy
• **Scalars = $M_{3/2} \sim$ 50 TeV necessarily**, scalars not very heavy	• Scalars **assumed** very heavy, whatever you want, e.g. 10^{10} GeV
• Gluino lifetime about 10^{-19} sec, decay in beam pipe	• Long lived gluino, perhaps meters or more
• $M_h \approx$**126 GeV unavoidable**, predicted	• Any M_h allowed

has been some question as to whether this derivation works in M Theory. Such a derivation is underway, and we think results will be satisfactory [32].

- Any satisfactory theory must at least provide a model of hierarchical Yukawa couplings for quark, lepton, and neutrino masses, with at most a few parameters that are calculable in principle but perhaps not in practice. The theory must also satisfy constraints from the CKM and PMNS matrices.
- Because of the coming LHC data, a fairly precise calculation of the gluino mass is very important.
- The theory should explain the dark matter. The simplest possibility is that in a uiniverse with a non-thermal history, some or all of the dark matter is a visible sector wino-like lightest superpartner, but this seems to be excluded by indirect detection satellite data [33]. M Theory also has axionic dark matter with a relic density of order unity. The visible sector LSP can decay into lighter

hidden sector gauge or chiral fermion matter, so it is necessary to work out the associated relic density and observable effects.

- The compactified M Theory approach with its non-thermal cosmological history suggests that the matter asymmetry is of order unity, from Affleck-Dine baryogenesis just after inflation. Then when the lightest modulus decays somewhat before BBN the introduced entropy dilutes the matter asymmetry to its present value, and the dark matter (WIMPs plus axions) emerges in the modulus decay [14]. This model also could explain the matter to dark matter ratio since both emerge from moduli decay. Much better calculations are needed of the Affleck-Dine mechanism and moduli vevs and the amounts of dark matter.

- Mechanisms that could provide for the "little hierarchy" do exist in the theory, but it would be good to sharpen them technically.

- Kahler potential corrections to gauge coupling unification, sizes of trilinears, the gluino mass, and more should be brought under better control.

- Although all the physics we consider begins as moduli oscillate in their potentials at the end of inflation, it would be good to identify inflatons with actual physical states in the theory, and connect them and their properties to the other aspects of the theory.

10. Concluding Remarks

The main point of this paper is to argue that understanding of compactified string/M Theories has reached the level where work in such areas, physics and mathematics and removal of assumptions, should become mainstream. Successful correlations of particular vacua and phenomenological observables exist. Some calculations of observables (such as the Higgs mass) turn out to depend very little on missing knowledge of the manifolds. The traditional goal to understand *our* universe better looks promising via compactified string/M Theories.

The cover image would be good to keep in mind. Top-down approaches (just the upper handle of the nutcracker) will not let us open up the content of the curled up manifolds, nor will bottom-up ones such as model building or data alone (the bottom handle of the nutcracker). Together they have a greatly increased chance of leading to important progress. Using both handles of the nutcracker is the approach of string phenomenology. The background is LHC data, and in particular a Higgs boson event – understanding and calculating the Higgs physics is a crucial challenge for any approach to understanding our world.

Acknowledgments

I am particularly grateful to Bobby Acharya for a long and fruitful collaboration and all that he has taught me. I am very grateful to Piyush Kumar for our collaboration and discussions, and lessons he has taught me since he began as a PhD student through his now mature leadership with our work. I also thank Konstantin Bobkov, Eric Kuflik, Jing Shao, Ran Lu, and Bob Zheng for many informative discussions and collaboration on our M Theory work. I thank Sebastian Ellis for recent calculations regarding the superpartner spectrum. I also thank David Gross and Joe Polchinski for illuminating discussions.

References

[1] B. S. Acharya, G. Kane, and P. Kumar, Compactified String Theories – Generic Predictions for Particle Physics, *Int.J.Mod.Phys.* **A27**, 1230012 (2012). doi: 10.1142/S0217751X12300128.

[2] G. Papadopoulos and P. Townsend, Compactification of D = 11 supergravity on spaces of exceptional holonomy, *Phys.Lett.* **B357**, 300–306 (1995). doi: 10.1016/0370-2693(95)00929-F.

[3] B. S. Acharya, M theory, Joyce orbifolds and superYang-Mills, *Adv.Theor.Math.Phys.* **3**, 227–248 (1999).

[4] B. S. Acharya and E. Witten, Chiral fermions from manifolds of G(2) holonomy (2001).

[5] C. Beasley and E. Witten, A Note on fluxes and superpotentials in M theory compactifications on manifolds of G(2) holonomy, *JHEP.* **0207**, 046 (2002).

[6] A. Lukas and S. Morris, Moduli Kahler potential for M theory on a G(2) manifold, *Phys.Rev.* **D69**, 066003 (2004). doi: 10.1103/PhysRevD.69.066003.

[7] E. Witten, Deconstruction, G(2) holonomy, and doublet triplet splitting. pp. 472–491 (2001).

[8] G. Kane, P. Kumar, and J. Shao, CP-violating Phases in M theory and Implications for EDMs, *Phys.Rev.* **D82**, 055005 (2010). doi: 10.1103/PhysRevD.82.055005.

[9] Ellis and Kane, EDM's in the CCSSSM (In Preparation).

[10] F. Denef and M. R. Douglas, Distributions of nonsupersymmetric flux vacua, *JHEP.* **0503**, 061 (2005). doi: 10.1088/1126-6708/2005/03/061.

[11] M. Gomez-Reino and C. A. Scrucca, Locally stable non-supersymmetric Minkowski vacua in supergravity, *JHEP.* **0605**, 015 (2006). doi: 10.1088/1126-6708/2006/05/015.

[12] B. S. Acharya, G. Kane, and E. Kuflik, String Theories with Moduli Stabilization Imply Non-Thermal Cosmological History, and Particular Dark Matter (2010).

[13] B. S. Acharya, K. Bobkov, and P. Kumar, An M Theory Solution to the Strong CP Problem and Constraints on the Axiverse, *JHEP.* **1011**, 105 (2010). doi: 10.1007/JHEP11(2010)105.

[14] G. Kane, J. Shao, S. Watson, and H.-B. Yu, The Baryon-Dark Matter Ratio Via Moduli Decay After Affleck-Dine Baryogenesis, *JCAP.* **1111**, 012 (2011). doi: 10.1088/1475-7516/2011/11/012.

[15] B. S. Acharya, G. L. Kane, P. Kumar, R. Lu, and B. Zheng, R-Parity Conservation from a Top Down Perspective (2014).

[16] B. S. Acharya, K. Bobkov, G. Kane, P. Kumar, and D. Vaman, An M theory Solution to the Hierarchy Problem, *Phys.Rev.Lett.* **97**, 191601 (2006). doi: 10.1103/PhysRevLett.97.191601.

[17] B. S. Acharya, K. Bobkov, G. L. Kane, P. Kumar, and J. Shao, Explaining the Electroweak Scale and Stabilizing Moduli in M Theory, *Phys.Rev.* **D76**, 126010 (2007). doi: 10.1103/PhysRevD.76.126010.

[18] B. S. Acharya, K. Bobkov, G. L. Kane, J. Shao, and P. Kumar, The G(2)-MSSM: An M Theory motivated model of Particle Physics, *Phys.Rev.* **D78**, 065038 (2008). doi: 10.1103/PhysRevD.78.065038.

[19] B. S. Acharya and K. Bobkov, Kahler Independence of the G(2)-MSSM, *JHEP.* **1009**, 001 (2010). doi: 10.1007/JHEP09(2010)001.

[20] T. Friedmann and E. Witten, Unification scale, proton decay, and manifolds of G(2) holonomy, *Adv.Theor.Math.Phys.* **7**, 577–617 (2003).

[21] B. S. Acharya, On Realizing N=1 superYang-Mills in M theory (2000).

[22] I. Affleck, M. Dine, and N. Seiberg, Dynamical Supersymmetry Breaking in Supersymmetric QCD, *Nucl.Phys.* **B241**, 493–534 (1984). doi: 10.1016/0550-3213(84)90058-0.

[23] B. S. Acharya and M. Torabian, Supersymmetry Breaking, Moduli Stabilization and Hidden U(1) Breaking in M Theory, *Phys.Rev.* **D83**, 126001 (2011). doi: 10.1103/PhysRevD.83.126001.

[24] D. Finnell and P. Pouliot, Instanton calculations versus exact results in four-dimensional SUSY gauge theories, *Nucl.Phys.* **B453**, 225–239 (1995). doi: 10.1016/0550-3213(95)00318-M.

[25] A. Brignole, L. E. Ibanez, and C. Munoz, Soft supersymmetry breaking terms from supergravity and superstring models (1997).

[26] D. Feldman, G. Kane, E. Kuflik, and R. Lu, A new (string motivated) approach to the little hierarchy problem, *Phys.Lett.* **B704**, 56–61 (2011). doi: 10.1016/j.physletb.2011.08.063.

[27] G. Kane, P. Kumar, R. Lu, and B. Zheng, Higgs Mass Prediction for Realistic String/M Theory Vacua, *Phys.Rev.* **D85**, 075026 (2012). doi: 10.1103/PhysRevD.85.075026.

[28] P. Kant, R. Harlander, L. Mihaila, and M. Steinhauser, Light MSSM Higgs boson mass to three-loop accuracy, *JHEP.* **1008**, 104 (2010). doi: 10.1007/JHEP08(2010)104.

[29] T. Hahn, S. Heinemeyer, W. Hollik, H. Rzehak, and G. Weiglein, High-

precision predictions for the light CP-even Higgs Boson Mass of the MSSM (2013).

[30] P. Draper, G. Lee, and C. E. M. Wagner, Precise Estimates of the Higgs Mass in Heavy SUSY (2013).

[31] Acharya, Kane, Kumar, Lu, and Zheng, Generating a Viable μ in M theory; A Discrete Mechanism (In Preparation) .

[32] Bourjaily, Kane, Kumar, Lu, Perry, and Zheng, The Top Yukawa from M Theory (In Preparation) .

[33] J. Fan and M. Reece, In Wino Veritas? Indirect Searches Shed Light on Neutralino Dark Matter, *JHEP.* **1310**, 124 (2013). doi: 10.1007/JHEP10(2013) 124.

12

String Cosmology — Large-Field Inflation in String Theory

Alexander Westphal

Deutsches Elektronen-Synchrotron DESY, Theory Group,
D-22603 Hamburg, Germany
alexander.westphal@desy.de

This is a short review of string cosmology. We wish to connect string-scale physics as closely as possible to observables accessible to current or near-future experiments. Our possible best hope to do so is a description of inflation in string theory. The energy scale of inflation can be as high as that of Grand Unification (GUT). If this is the case, this is the closest we can possibly get in energy scales to string-scale physics. Hence, GUT-scale inflation may be our best candidate phenomenon to preserve traces of string-scale dynamics. Our chance to look for such traces is the primordial gravitational wave, or tensor mode signal produced during inflation. For GUT-scale inflation this is strong enough to be potentially visible as a B-mode polarization of the cosmic microwave background (CMB). Moreover, a GUT-scale inflation model has a trans-Planckian excursion of the inflaton scalar field during the observable amount of inflation. Such large-field models of inflation have a clear need for symmetry protection against quantum corrections. This makes them ideal candidates for a description in a candidate fundamental theory like string theory. At the same time the need of large-field inflation models for UV completion makes them particularly susceptible to preserve imprints of their string-scale dynamics in the inflationary observables, the spectral index n_s and the fractional tensor mode power r. Hence, we will focus this review on axion monodromy inflation as a mechanism of large-field inflation in string theory.

1. Introduction

Cosmology has long been considered as a speculative field. Envisioning now a combination of a candidate theory of quantum gravity and unification yet to be fully understood, string theory, with early universe cosmology may at

first warrant even more caution. However, before rushing to this conclusion
we shall start our discussion employing judicious use of the structure of effective field theory as a bottom-up tool and controlled approximations or
constructions in string theory. This will lead us to realize that certain phenomena of early and late-time cosmology can be reasonably well embedded
and described by certain classes of solutions of string theory.

We will start making the following observation. String theory is a candidate theory for a fundamental unification of quantum mechanics with general relativity. String theory solutions are typically subject to a requirement
of 10 dimensions of space-time. To make contact with our four dimensional
large-scale space-time, we need to compactify the six extra-dimensions on
an undetectably small internal manifold. However, at energies small compared to the inverse string and compactification length scales string theory
reduces to an effective quantum field theory (EFT). This EFT contains
gauge fields coupled to fermionic matter and a number of scalar fields as
well as general relativity extended by a series of higher-order curvature
corrections from the string α'-expansion.

Hence, below string and compactification scales the usual decoupling
theorem of QFT ensures the absence of strong string-theoretic effects from,
e.g., weak-scale physics – except if weakly-broken symmetries protect a
hierarchical suppression of scales for some of the effects of string theory.
Examples for the latter are low-scale broken supersymmetry, and axionic
shift symmetries originating from higher-dimensional gauge symmetries of
string theory which usually enjoy exponentially suppressed scales of symmetry breaking due to its non-perturbative nature. Without the presence
of symmetries surviving unbroken to low scales, we conclude that the only
phenomena in cosmology which may warrant or even necessitate a description in a theory of quantum gravity must either live by themselves close
to the string scale or display explicit ultra-violet (UV) sensitivity to UV-
divergent quantum corrections.

From this argument we discern three such phenomena in need of a description beyond EFT coupled to general relativity:

- The initial cosmological singularity.[a]
- A very early period of cosmological inflation. As discussed further
  below, there is increasingly strong evidence that the CMB temperature fluctuations are down to inflationary scalar quantum fluctua-

---

[a]Note, that even in presence of eternal inflation space-time is geodesically incomplete to
the past, so even then there was a cosmological singularity in our past.[1]

tions with almost scale-invariant 2-point function power spectrum $\Delta_{\mathcal{R}}^2 \sim H^2/\epsilon$. Barring suppression of the first slow-roll parameter ϵ by tuning, we expect $\epsilon \lesssim 1/N_* \sim 0.01$ where $N_* \simeq 60$ denotes the observable span of slow-roll inflation. Hence, from $\Delta_{\mathcal{R}}^2 \sim 10^{-9}$ we expect slow-roll inflation to happen near the GUT scale. Moreover, we will see below that slow-roll inflation is inherently UV-sensitive to quantum corrections of the effective potential. This sensitivity is only enhanced by the closeness of the inflationary energy scales to the string scale.

- The late-time form of dark energy driving the observed contemporary accelerated expansion of our universe. This phenomenon has two aspects of needing a UV completion in some theory of quantum gravity. Firstly, the contribution to dark energy in form of a cosmological constant from quantum field theory is manifestly power law UV-divergent. As such, EFT coupled to general relativity can only explain the present-day amout dark energy $\sim 10^{-122} M_P^4$ by manifest and insane amounts of fine-tuning. Hence the need for UV completion. Secondly, a pure cosmological constant generates eternal de Sitter (dS) space-time in general relativity. QFT on eternal dS space-time is not well defined as the S-matrix does not exist. Hence de Sitter space is in need of a more fundamental description, presumably a UV complete theory of quantum gravity.

While we do have in recent years some important progress in the description of cosmological singularities in string theory, it is probably still fair to say that a full controlled description of space-time near a realistic cosmological singularity (i.e. the one in the past of null-energy condition satisfying FRW space-times) remains an open question (for reviews see e.g. Refs. 2–4).

Given the stage set by this general discussion, it is important to recall that it was the advent of several waves of observational progress during the last two decades which have required string theory to confront de Sitter space and the description of cosmological inflation. Moreover, the accuracy of these new experimental results fundamentally changed the speculative status of cosmology, leading to an era of precision cosmology.

At first, highly reliable and increasingly precise measurements of the Hubble Space Telescope (HST) yielded a determination of present-day Hubble parameter $H_0 = 73.8 \pm 2.4 \, \mathrm{km \, s^{-1} Mpc^{-1}}$.[5,6] This milestone enabled the construction of an ultra deep-space distance ladder by type IA supernovae.

Systematic observations of such type IA supernovae at high red-shift culminated in the detection of a form of dark energy consistent with extremely small ($\sim 10^{-122} M_P^4$) and positive cosmological constant, which drives a late-time accelerated expansion of our Universe.[7,8]

In the next step, the space-based satellite missions WMAP[9] and PLANCK[10,11] as well as the ground-based telescopes ACT[12] and SPT[13,14] probed the cosmic microwave background (CMB) radiation with unprecedented precision and resolution. Their combined results together with the HST and type IA supernova data led to the concordance model of observational cosmology. In its 'essence' this new standard model is consistent with certain simple features of our observed Universe. We find ourselves to a good approximation (sub-%-level) in a FLRW universe which is spatially flat and undergoes accelerated late-time expansion driven by dark energy. Moreover, the recent CMB missions provided measurements of the two-point function power-spectrum of the 10^{-5}-level thermal fluctuations with unprecedented precision. These results now bear increasingly strong evidence that the very early Universe went through a much earlier and extremely rapid phase of accelerated expansion driven by the vacuum energy of a slowly-rolling scalar field, called inflation[15-18] (see e.g. Ref. 19 for a recent review).

The recent high-precision data from PLANCK[10,11] as well as the ground-based telescopes ACT[12] and SPT,[13,14] and in particularly the strong limits on the presence of non-Gaussianity in form of a non-vanishing three-point function from the PLANCK mission,[20] are consistent with a picture of simple slow-roll inflation driven by the scalar potential of a single canonically normalized scalar field.

Finally, the recent report from the BICEP2 telescope[21] announced the detection of degree angular scale B-mode polarization in the CMB. B-mode polarization in the CMB at such large angular scales can have a primordial origin only from inflationary gravitational waves, so-called tensor modes. However, polarized emission from galactic dust may provide the same type polarization pattern as a foreground contaminatio.[22,23] Discrimination may be possible due to different angular power spectra and radio frequency spectra of such dust induced B-modes versus the ones from primordial tensor modes. If the B-mode detection by BICEP2 turns out to be of at least partial primordial original, then the inflationary energy scale would be at the GUT scale, and inflation would have to proceed as a large-field model with trans-Planckian field excursion.[b] Both features would spell a direct need for embedding inflation into a fundamental candidate theory of quantum gravity such as string theory as we will discuss below.

This progress of the last 15 years forced candidate theories for a fundamental unification of quantum mechanics with general relativity, such as string theory, to accommodate both an extremely tiny positive cosmological constant and the dynamics of slow-roll inflation for the very early Universe. In particular, string theory solutions are typically subject to a requirement of 10 dimensions of space-time. To make contact with our four dimensional large-scale space-time, we need to compactify the six extra-dimensions on an undetectably small internal manifold. This process of compactification produces a huge set of possible suitable manifolds, each of which is accompanied by set of massless 4d scalar moduli fields describing the allowed deformation modes of each manifold. Crucial progress towards describing late-time and early-time (quasi) de Sitter (dS) stages of positive vacuum energy in string theory involved the construction of string vacua with complete stabilization of all geometrical moduli fields and controllable supersymmetry breaking, largely in type IIB string theory (see e.g. Refs. 26–28 for reviews), and more recently also in heterotic string theory.[29]

Hence, the outline of our discussions proceeds as follows. We will start with a discussion of the basic ingredients of inflation and string theory in Sections 2 and 3.

Section 4 will discuss inflation in string theory, emphasizing the necessity of having controlled string vacua with full moduli stabilization before even beginning to search for inflationary regimes. Some of the major progress in recent years involved the construction[30–33] of controlled type II string flux compactifications (see e.g. Refs. 26, 27 for a review) with full moduli stabilisation and positive vacuum energy necessary to make contact with cosmologically viable 4d space-time descriptions incorporating the observed late-time acceleration. A typical compactification has $\mathcal{O}(100)$ moduli fields allowing for an exponentially large number of possible combinations of fluxes. This led to the discovery of an exponentially large landscape of isolated dS vacua in string theory.[31]

---

[b] A very recent analysis of the 2D genus topology statistics and the cross-correlation between the BICEP2 150 GHz data and PLANCK polarization data in the BICEP2 region at 353 GHz [24] by Colley and Gott III allowed for a first more dust-model independent estimate of the dust emission fraction in the BICEP2 signal. This analysis yields the dust part to be a bit less than 50% and results in $r = 0.11 \pm 0.04$ $(1 - \sigma)$ providing $2.5 - \sigma$ evidence for $r > 0$.[25]

Within the extant constructions of string theory dS vacua inflation must then arise from a flat region of the moduli scalar potential supporting the criteria of slow-roll by a flat potential, or higher-derivative kinetic terms originating from mobile D-branes. Since we wish to focus on connecting string-scale physics to the observables of inflation as closely as possible, this motivates us to concentrate our discussion on large-field inflation in the form of axion monodromy[34,35] or aligned natural inflation.[36] The book by Baumann and McAllister[37] contains an exhaustive discussion of the many models of small-field inflation in string theory (e.g. warped D3-brane inflation, D3-D7 inflation, Kähler moduli inflation, fibre inflation, DBI inflation, etc.) following the seminal work of Kallosh *et al.*,[38] which we consequently omit here.

Our discussion will need to include corrections to a scalar potential driving slow-roll inflation. They will arise generically due to dimension-six operators from radiative corrections or integrating out massive states, rendering inflation UV sensitive. Hence, candidate fundamental theories of quantum gravity such as string theory are necessary for a full description of inflation beyond the limits of effective field theory.

Using this stage, we will try explain the general mechanism and structure of axion monodromy which underlies the various models of large-field inflation with axions arising from higher-dimensional gauge-fields in string theory (see e.g. Refs. 37, 39 for very recent reviews). I will openly say at this point, that you will find the most probably definitive reference for inflation in string theory in the recent book by Baumann and McAllister.[37]

2. Basics of Inflation

Inflation is a period of quasi-exponential expansion of the very early Universe, invented originally by Guth[15] to overcome several initial condition problems of the conventional hot big bang cosmology. A short discussion of the two most pertinent of these, the horizon and flatness problems, will serve us well to illuminate the essence of and need for inflation.

The horizon problem of the hot big bang has its root in the differing behavior of the causal horizon distance and the stretching of physical length scales in an expanding universe driven by radiation or matter.

2.1. *Classical theory*

To begin, we recall a few basic facts about the description of spatially homogeneous and isotropic expanding space-times in General Relativity. In an FRW universe driven by any type of matter or energy in perfect fluid form except vacuum energy, we have the energy density diluting as $\rho \sim a^{-3(1+w)}$. Here w denotes the equation of state of the fluid dominating the energy budget. The 1st Friedmann equation governing an expanding FRW universe, possibly with spatial curvature denoted by a parameter $k = 0, \pm 1$ reads

$$H^2 \equiv \left(\frac{\dot{a}}{a}\right)^2 = \frac{1}{3}\rho - \frac{k}{a^2} \quad . \tag{1}$$

In a spatially flat universe $k = 0$ this describes an expansion following

$$a \sim t^p \quad , \quad p = \frac{2}{3(1+w)} \quad \text{and} \quad H = \frac{p}{t} \quad . \tag{2}$$

At $t = 0$ we have $a = 0$, implying a curvature singularity, that is, the initial big bang singularity. The exception to the above is pure vacuum energy, e.g. the potential energy of a scalar field, which has $w = -1$ and thus $H = const.$ and $a \sim e^{Ht}$ (i.e. formally $p \to \infty$).

A given physical length scale, e.g. radiation of a given wavelength λ_{ph}, will stretch with the expansion of an FRW universe as

$$\lambda_{ph}(t) = \frac{\lambda_{ph}(t_0)}{a(t_0)} a(t) \quad . \tag{3}$$

Hence, we can think about the continually stretching physical length scales in terms of associated 'comoving' length scales $\lambda \equiv \lambda_{ph}/a$, that is with reference length scales obtained by scaling out the 'stretching' with the scale factor a.

We now compare a comoving length scale with the comoving light-travel distance since the initial singularity, the comoving horizon distance τ

$$\tau = \int \frac{dt}{a} = \int \frac{d\ln a}{aH} \quad . \tag{4}$$

We see from the preceding discussion that for any fluid with $w > -1$ driving the expansion we get

$$(aH)^{-1} \sim \frac{1}{p} a^{\frac{1-p}{p}} \quad \Rightarrow \quad \tau = \frac{1}{1-p} a^{\frac{1-p}{p}} \sim \frac{1}{aH} \tag{5}$$

while for $w = -1$ directly $\tau = 1/aH$.

The last result displays the horizon problem: An expanding universe driven by a fluid with $w > -1/3$ or equivalently $p < 1$ has a *growing* comoving horizon $(aH)^{-1} \sim a^{(1-p)/p}$ with $(1-p)/p > 0$ for $p < 1$. All comoving scales λ enter the horizon from outside from the past towards the future. Therefore, all observable length scales were outside the horizon and thus out of causal contact at sufficiently early times. For the CMB sky this corresponds to patches separated by more than about 1 degree. Yet these patches are all at the same temperature to better than 1 part in 10^4. Why?

The problem arises from the integral in τ and the fact that $1/aH$ is growing with a for $w > -1/3$ or $p < 1$. The integral gets its main contribution from late times. The problem was solved if at an earlier time we can arrange for the comoving horizon $1/aH$ to *decrease* with increasing a. Comoving length scales would then *leave* the horizon at an early time, and then re-enter later on, after the expansion has changed from decreasing comoving horizon to the increasing comoving horizon of matter or radiation domination. Therefore, at a very early time all observable comoving length scales would have been inside the horizon and in causal contact, despite leaving the horizon later on for quite a while.

Since in a expanding universe we have $\dot{a} > 0$ always, this means we need arrange for an early period where

$$\frac{d}{dt}\left(\frac{1}{aH}\right) < 0 \quad \text{for} \quad t < t_e \tag{6}$$

and positive thereafter. We denote with t_e the time when this early phase of a decreasing comoving horizon ended.

What does this early phase mean? We evaluate $\frac{d}{dt}(aH)^{-1}$ and by differentiating in turn the Friedmann equation we arrive at

$$\frac{d}{dt}\left(\frac{1}{aH}\right) = -\frac{1}{a}\left(1 + \frac{\dot{H}}{H^2}\right) = -\frac{1}{(aH)^2}\ddot{a} \; . \tag{7}$$

An early phase of *decreasing* comoving horizon requires an interval of *accelerated* expansion $\ddot{a} > 0$ ending at $t = t_e$, and this is inflation in its most general sense.

By inspecting the last result, we conclude that

$$\frac{\ddot{a}}{a} = H^2\left(1 + \frac{\dot{H}}{H^2}\right) = H^2(1 - \epsilon_H) \quad \text{with} \quad \epsilon_H \equiv -\frac{\dot{H}}{H^2} \; . \tag{8}$$

Acceleration implies a condition on the *first slow-roll parameter* ϵ_H, namely we need $\epsilon_H < 1$. While this is the general condition for inflation, we see a very simple means of realizing this regime: We look for an energy source yielding $\dot{H} \approx 0$ in the sense of $-\dot{H} \ll H^2$ (since inflation must end at $t = t_e$, we cannot have $\dot{H} = 0$ strictly). This implies *Hubble slow-roll* $H \approx const.$ and leads to a quasi-exponential phase of expansion $a \sim e^{Ht}$. Thus, building inflation in the more narrow sense means constructing a fluid which behaves almost like vacuum energy with $w \approx -1$ for a while and then quickly changes towards $w \geq 0$ at $t = t_e$.

We now see that inflation driven by a source similar to vacuum energy with $\rho \approx const.$ solves the flatness problem as well. The contribution of spatial curvature $-k/a^2$ to the Friedmann equation shrinks exponentially during inflation compared to the source $\rho \approx const. > 0$ driving inflation. Hence, if the exponential expansion latest long enough, inflation can render the universe spatially flat enough at the beginning t_e of the matter or radiation driven expansion to avoid spatial curvature growing to more than about a percent fraction now.

Plugging in the ratio of scales between, in the most extreme case, the GUT scale and the largest cosmological scales visible today, we see that inflation must grow the scale factor by at least

$$a(t_e) \simeq a(t_*)e^{60} \tag{9}$$

60 efolds, to suppress spatial curvature to less than a percent in our present late-time universe. Solving the horizon problem, i.e. requiring that inflation lasted long enough to have all the scales between the GUT scale and cosmological scales today inside the horizon at the beginning t_* of the needed amount of inflation, needs again about 60 efolds of inflation.

A very large class of inflation models realizing this path uses the dynamics of one or several scalar fields. A scalar field minimally coupled to Einstein gravity is described in 4d at the 2-derivative level by an action

$$S = \int d^4x \sqrt{-g} \left[\frac{1}{2}R + \frac{1}{2}(\partial_\mu \phi)^2 - V(\phi) \right] \quad . \tag{10}$$

If we can arrange for a regime where $V > 0$ and $\dot{\phi}^2 \ll V$, then the scalar potential will act effectively as a positive vacuum energy and drive exponential expansion. The condition

$$\dot{\phi}^2 \ll V \tag{11}$$

guarantees

$$\epsilon_H \ll 1 \quad , \tag{12}$$

the first Hubble slow-roll condition for inflation. In this case, we can also further express $\epsilon_H \ll 1$ by a 1st slow-roll condition on the scalar potential itself

$$\epsilon_H \simeq \epsilon \equiv \frac{1}{2} \left(\frac{V'}{V} \right)^2 \ll 1 \tag{13}$$

where we denote $()' = \frac{d}{d\phi}()$. Maintaing this condition for a long time to generate at least about 60 efolds of exponential expansion, then requires a 2nd slow-roll condition on the scalar potential to hold

$$\eta \equiv \frac{V''}{V} \ll 1 \quad . \tag{14}$$

Scalar field models of inflation realizing Hubble slow-roll this way are called slow-roll models of inflation.[17,18]

Alternatively, higher-derivative kinetic terms can generate a phase $\epsilon_H \ll 1$ even if $V = 0$ or if $\epsilon, \eta > 1$ for the potential itself. One example of this in field theory is k-inflation,[40] and a simple realization of the same higher-derivative mechanism to generate Hubble slow-roll in string theory is DBI-inflation.[41]

2.2. Quantum fluctuations during inflation

The description of inflation using the dynamics of a scalar field has a very far-reaching consequence. We have seen, that successful slow-roll inflation entails the scalar inflaton being light $\eta \ll 1 \leftrightarrow m_\phi^2 \ll H^2$. However, light scalar degrees of freedom are subject to quantum mechanical vacuum fluctuations. Inflation takes these fluctuations and stretches their wavelength so rapidly, that they become larger than the Hubble horizon ('super-horizon') after a finite time. The finite average amplitude, which a given fluctuation has at that point, then 'freezes' since super-horizon wave-length fluctuations cease to evolve during inflation. They have become an essentially classical field profile at this stage. Once inflation ends, these frozen large wave-length modes re-enter the horizon at a certain time given as a function of their comoving wavenumber. The modes who left the horizon during inflation first, re-enter last after inflation. The scalar field distribution described by these long wave-length modes causes a variation of the gravitational potential. This 'curvature perturbation' generates an initial field of density perturbations in the matter distribution after inflation. Hence, in this picture the inflationary quantum fluctuations are the ultimate cause of the primordial density perturbations which are the seed of structure formation in the observable universe.[42]

We describe the curvature perturbation ζ by expanding around the FRW metric

$$ds^2 = N^2 dt^2 - h_{ij} \left(dx^i + N^i dt \right)(dx^j + N^j dt) \tag{15}$$

where N, N^i denote the lapse function and shift vector, respectively, which enforce constraints containing the gauge invariance of GR. The spatial metric reads

$$h_{ij} = a^2(t) \left(e^{2\zeta} \delta_{ij} + \gamma_{ij} \right) \quad . \tag{16}$$

During inflation we have $a(t) \approx e^{Ht}$. Plugging this ansatz into the action eq. (10) and expanding in ζ and the inflaton fluctuation $\delta\phi$ leads at leading order to a 2-derivative action for ζ and $\delta\phi$ without self-interactions. Time reparametrization invariance relates ζ and $\delta\phi$. This gives us the freedom to trade between ζ and $\delta\phi$ and implies the relation

$$\zeta = \frac{H}{\dot\phi} \delta\phi \quad . \tag{17}$$

We usually characterize the fluctuation spectrum of ζ in terms of the power spectrum Δ_ζ^2, which is the Fourier transform of the 2-point function $\langle \zeta(\mathbf{x})\zeta(\mathbf{y}) \rangle$. We define the power spectrum via

$$\langle A_k A_{k'} \rangle \equiv \frac{\Delta_A^2}{k^3} \delta(\mathbf{k} + \mathbf{k}') \tag{18}$$

for a fluctuation field $A(\mathbf{x})$. The relation between ζ and $\delta\phi$ implies

$$\langle \zeta_k \zeta_{k'} \rangle = \frac{H^2}{\dot\phi^2} \langle \delta\phi_k \delta\phi_{k'} \rangle \quad . \tag{19}$$

In slow-roll inflation we have

$$\langle \delta\phi_k \delta\phi_{k'} \rangle = \frac{1}{k^3} \left(\frac{H}{2\pi} \right)^2 \delta(\mathbf{k} + \mathbf{k}') \quad , \tag{20}$$

and so we get

$$\langle \zeta_k \zeta_{k'} \rangle = \frac{H^4}{4\pi^2 \dot\phi^2 k^3} \left(\frac{k}{k_\star} \right)^{n_s - 1} \delta(\mathbf{k} + \mathbf{k}') \quad . \tag{21}$$

Here $k_\star$ defines a reference wavenumber corresponding to large-scale CMB fluctuations (e.g. corresponding to $\ell_\star = 200$ after decomposition of the wave field into spherical harmonics). After the magnitude Δ_ζ^2 of the curvature perturbation power spectrum, the spectral tilt n_s constitutes our 2nd inflationary observable. n_s captures the 1st-order variations of the slowly rolling scalar field and is calculable for any given model of inflation.

Generically, interactions of the inflaton will generate higher-point functions $\langle \zeta_{k_1} \cdots \zeta_{k_n} \rangle$ from expanding out Eq. (10) beyond 2nd order. Non-vanishing odd-point functions, such as the 3-point function $\langle \zeta_k \zeta_{k'} \zeta_{k''} \rangle$, constitute 'non-Gaussianity'. Two central results establish that non-Gaussianity is small for single-field slow-roll inflation, while the magnitude of non-Gaussianity is linked to the 'speed of sound' c_s of the curvature perturbation. Furthermore, there are certain relations between the 'shapes' in momentum space which the 3-point function can take.[20,43,44] For instance, DBI inflation[41] produces a distinctive pattern of non-Gaussianity, where the 3-point function peaks in the 'equilateral' configuration, where all three momenta have roughly equal magnitude. This equilateral shape is not yet that strongly constrained from the PLANCK data $f_{NL}^{equil.} = 42 \pm 75$, contrary to the local shape from multi-field inflation $f_{NL}^{loc.} = 2.7 \pm 5.8$.[20]

Finally, there are gravitational waves in general relativity. These appear as the fluctuations γ_{ij}, hence called 'tensor modes'. Each polarization s of a gravitational wave formally constitutes a massless scalar degree of freedom. Hence, inflation generates a long wave-length spectrum of gravitational waves with a power spectrum

$$\sum_{s,s'=1,2} \langle \delta\gamma_{k,s} \delta\gamma_{k',s'} \rangle \equiv \frac{\Delta_T^2}{k^3} \delta(\mathbf{k} + \mathbf{k}') = \sum_{s,s'} \frac{1}{k^3} \left(\frac{H}{\pi} \right)^2 \delta ss' \, \delta(\mathbf{k} + \mathbf{k}'). \quad (22)$$

Observationally, we often refer instead to the 'tensor-to-scalar ratio' r of gravitational wave power to curvature perturbation power

$$r \equiv \frac{\Delta_T^2}{\Delta_\zeta^2} = 8 \frac{\dot{\phi}^2}{H^2} \quad . \quad (23)$$

From the PLANCK[10,11,20] and WMAP[9] satellite data, we know $\Delta_\zeta^2 \simeq 2.2 \times 10^{-9}$. Hence, any detection of $r \gtrsim 0.01$ implies a GUT-scale inflaton potential $V_{inf.}^{1/4} \sim M_{GUT} \sim 10^{16}$ GeV in the context of single-field slow-roll inflation. This is one of the main reasons driving the search for tensor modes.

Tensor modes are detectable in the CMB as a B-mode polarization pattern. Recently, the BICEP2[21] experiment reported on the detection of B-mode signal in the CMB on large angular scales $\ell \sim 100$. If future analyses along the line of Refs. 22, 23 using e.g. PLANCK polarization data confirm a part of this B-mode signal to be primordial as opposed to coming from polarized galactic dust, then we would know that $r \sim 0.1$.

In models of single-field slow-roll inflation we can compute n_s and r as functions of the slow-roll parameters. To leading order we get

$$n_s = 1 - 6\epsilon + 2\eta \quad , \quad r = 16\epsilon \quad . \tag{24}$$

We can express these results in terms of the number of e-folds $N_e(\phi)$ that inflation lasts from the field value ϕ until the end of inflation at ϕ_e. We have

$$N_e = \int_{t(\phi_e)}^{t(\phi(N_e))} H \, dt = \int_{\phi_e}^{\phi(N_e)} \frac{d\phi}{\sqrt{2\epsilon}} \quad . \tag{25}$$

CMB scales correspond to about $N_e = 50 \ldots 60$ e-folds before the end of inflation. If we assume ϵ to increase monotonically with ϕ then we can bound N_e by

$$N_e < \frac{\Delta\phi}{\sqrt{2\epsilon}} \quad . \tag{26}$$

This implies the Lyth bound[45]

$$\Delta\phi > \sqrt{\frac{r}{8}} N_e \quad . \tag{27}$$

For $N_e \simeq 50 \ldots 60$ we see that $r > 0.01$ implies $\Delta\phi > M_{\rm P}$. This ties a large B-mode signal from primordial tensor modes to having a so-called 'large-field model' of inflation, where sufficient inflation requires a trans-Planckian initial field displacement.

2.3. *Effective field theory and the role of symmetries*

We now need to discuss inflation in effective field theory, the role which the amount of field displacement $\Delta\phi$ plays for the ultraviolet (UV) sensitivity of inflation. In effective field theory in the Wilsonian sense we Taylor expand the effective action of the inflaton scalar field around a given point ϕ_0 in field space. We can organize this expansion in terms of the scaling of the various operators under dilatations of space and time $x^\mu \to a x^\mu$. The operator dimension of ϕ follows from the requirements of an invariant kinetic term, so $\phi \to \phi/a$. This is necessary for the action to be dimensionless as it is the phase in the path integral. Expanding the non-kinetic part of the action we get (assuming a Z_2 symmetry $\phi \to -\phi$ for slight simplification)

$$S = \int d^4 \sqrt{-g} \left[\frac{1}{2} (\partial_\mu \phi)^2 - \frac{m^2}{2} (\phi - \phi_0)^2 + \sum_{n \geq 0} \frac{\lambda_{4+2n}}{M_\star^{2n}} (\phi - \phi_0)^{4+2n} \right.$$
$$\left. + \frac{\lambda_{4,4}}{M_\star^4} (\partial_\mu \phi)^4 + \ldots \right] . \tag{28}$$

Except for the mass all couplings are dimensionless as the relevant mass scale $M_\star$ suppressing the higher-dimension contributions appears everywhere. The higher-dimension operators scale under dilatation as $\mathcal{O} \to \frac{1}{a^\Delta} \mathcal{O}$ where e.g. for the non-derivative operators above we have $\Delta_{4+2n} = 4 + 2n$ at weak coupling. We see that these operators induce scattering amplitudes with effective couplings $\lambda_{4+2n,eff.} = \lambda_{4+2n} (E/M_\star)^{\Delta_{4+2n}-4}$. Hence, operators with $n < 0$, such as the mass term, are 'relevant' at low energies. The quartic self-coupling of a scalar ($n = 0$) is an example of a 'marginal' operator. The great majority of operators have $n > 0$. These are 'irrelevant' as they die out in the infrared. In the following discussion we will typically have the UV mass scale $M_\star = M_P$

However, the viability of inflation depends on the slow-roll parameters being small $\epsilon, \eta \ll 1$. As they contain ϕ-derivatives, we see that in particular η which pulls down two powers of ϕ can receive large contributions from 'dangerously irrelevant' operators $\mathcal{O}_{4+2n}, n > 0$. In particular, an operator of the form $\mathcal{O}_6 = \lambda_6 V_0(\phi)(\phi - \phi_0)^2/M_P^2$ corrects η by an $\mathcal{O}(1)$ shift which destroys inflation. This happens even at low energy densities during inflation $H/M_P \ll 1$ and for any field displacement $\Delta\phi = \phi - \phi_0$.

Moreover, we see that for $\Delta\phi \gg M_P$ *all* higher-dimension operators of type $\mathcal{O}_{4n+2}$ above correct η by $\mathcal{O}(1)$ values, while for $n \geq 2$ their contribution is suppressed in $(\Delta\phi/M_P)^{2n-2} \ll 1$ at small $\Delta\phi \ll M_P$. This is the true significance of the split into 'small-field' $\Delta\phi \ll 1$ and 'large-field' $\Delta\phi$ models of inflation.

We see that viable large-field inflation requires extra suppression of all higher-dimension corrections $\lambda_{4-2n} \ll 1$. This amounts to the presence of a symmetry effectively forbidding these terms at the high scale. As the primary inflationary scalar potential itself has $V(\phi) \ll 1$ this symmetry effectively takes the form of a shift symmetry $\phi \to \phi + c$. An exact shift symmetry forbids non-derivative couplings. We can now argue that breaking the shift symmetry *weakly and smoothly* by a large-field inflaton potential $V(\phi) \ll 1$ is sufficient to protect against the dangerous generic higher-dimension corrections discussed above. The arguably simplest large-field (and likely simplest overall) inflaton potential is a mass term 'citeLIN-DECHAOS

$$V(\phi) = \frac{1}{2}m^2\phi^2 \quad . \tag{29}$$

The ensuing argument works the same for a direct generalization to monomial potentials $V(\phi) = \mu^{4-p}\phi^p$, $p > 0$.

Firstly, we see that expanding the potential around a reference point ϕ_0 we get

$$V(\phi) = V(\phi_0)\left(1 + \frac{\delta\phi}{\phi_0}\right)^p = V(\phi_0) + \sum_n \binom{p}{n}\frac{\delta\phi^n}{\phi_0^{n-p}} \quad . \tag{30}$$

During inflation $\phi_0 \gg 1$ and so we see that the interaction terms $n \geq 3$ have effective self-couplings of the inflaton dying out as $1/\phi_0^{n-p}$ at large field displacements. We therefore expect the potential to be safe from dangerous radiative corrections induced by self-interactions at large field values.

Secondly, we can see this in field theory if we look at the relevant Feynman diagrams.[35,46] These daisy diagrams individually produce catastrophic-looking contribution, but their sum constitutes an *alternating* series, as e.g. for ϕ^4 theory They resum into a good-natured logarithmic correction

$$V(\phi) = \lambda\phi^4\left[1 + c\ln\left(\frac{\phi}{M_{\rm P}}\right)\right] \tag{31}$$

which corrects the numerical values of n_s, r a bit but is far from spoiling slow-roll.

$$\bigcirc + \bowtie + \cdots = (-1)^n\lambda\phi^4\left(\frac{\phi}{M_{\rm P}}\right)^{2n-4}$$

Finally, quantized general relativity as a low-energy limit of quantum gravity couples *only* to $T_{\mu\nu}$. This is sourced by $V(\phi)$ and $\partial_\phi^2 V$, but *not* by field displacements themselves. Consequently, graviton loops induce corrections to the effective scalar potential and Newton's constant of the form[46-48]

$$\delta V \sim \left(c\frac{V}{M_{\rm P}^4} + d\frac{\partial_\phi^2 V}{M_{\rm P}^2}\right)V(\phi) \quad , \quad \delta\mathcal{L}_{EH} \sim \sqrt{-g}\,\partial_\phi^2 V R \quad . \tag{32}$$

Large-field inflation requires $\mu \ll 1$ and hence $V, \partial_\phi^2 V \ll 1$ to realize a COBE-normalized CMB spectrum from the curvature perturbation. Hence, all quantum gravity corrections are highly suppressed in

V/M_P^4 , $\partial_\phi^2 V/M_P^2 \ll 1$ except at extremely large field displacements where V/M_P^4 , $\partial_\phi^2 V/M_P^2 \lesssim 1$.

The upshot of the whole discussion is that large-field slow-roll inflation is technically natural in 't Hooft's sense. If we succeed to generate the mass scale $\mu \ll M_P$ of a given large-field model dynamically, then it would become even natural in the Wilsonian sense.

2.4. Need for UV completion

From the above discussion it is clear that large-field inflation works perfectly fine in effective field theory. However, the role of a UV complete candidate theory of quantum gravity such as string theory becomes clear: Firstly, the theory has to describe how the shift symmetry of the inflaton arises in the first place. Secondly, the UV completion needs to describe how it generates the large-field inflaton potential including the large trans-Planckian field range. Finally, a fundamental description needs to describe how the coupling of the inflaton to heavy states such as the moduli of string theory participates in the breaking of the shift symmetry and backreacts on the dynamics of slow-roll.

We will discuss in detail below, how shift symmetries of axions in string theory arise as remnants of its higher-dimensional gauge symmetries. Hence, these p-form axions as natural large-field inflaton candidates in string theory. However, string theory seems to limit the periodicity field range of such axions to sub-Planckian values. We will see that monodromy induced by the presence of higher p-form fluxes and/or branes breaks the periodicity generating a large-field range monomial potential for the axions. Combined with moduli stabilization this can be done in a regime $V(\phi) \ll 1$ which provides radiative stability as seen above.

String theory provides the setting to discuss the interaction between the many moduli of its myriad compactifications to 4D and the inflationary dynamics of shift symmetry breaking from monodromy. However, we can see the basic effect of integrating out the heavy moduli already at this point by coupling a large-field inflaton to a heavy modulus field in field theory[49]

$$V(\phi, \chi) = \frac{1}{2}\phi^2\chi^2 + \frac{1}{2}M_\chi^2(\chi - \chi_0)^2 \quad . \tag{33}$$

If we choose $\chi_0 \sim m \ll 1$ and a heavy modulus $M_\chi \gg m$ we see that this generates $V \sim m^2\phi^2$ as the effective potential at small ϕ. However, at larger values of ϕ the full effective potential after integrating out χ reads

$$V_{eff.}(\phi) = M_\chi^2 m^2 \frac{\frac{1}{2}\phi^2}{\phi^2 + M_\chi^2} \quad . \tag{34}$$

Hence the presence of the heavy field leads to a *flattening* of the inflaton potential $V(\phi) \sim \phi^p$ compared to its naive tree-level power $V_0(\phi) \sim \phi^{p_0}$. This behavior is generic and occurs in many cases of large-field axion monodromy inflation after carefully including moduli stabilization.[49–51]

3. Ingredients of String Theory

The above discussion gives us a clear motivation to use string theory as a UV completion of inflation. We do not yet know if string theory is the correct description of our world. However, it is a highly successful candidate theory of quantum gravity while including a unification of SM-like gauge forces, chiral fermion matter. Strong supporting evidence for this claim arises from the various mathematical consistency checks and 'null' results obtained in the last 20 years or so. These include the web of dualities linking all five string theories and 11D supergravity, the AdS/CFT correspondence which provides a setting where a certain non-perturbative definition of quantum gravity is available, and the description of black hole entropy (even though for highly supersymmetric settings only so far).

We do not have a complete picture or non-perturbative description of string theory yet. Given that we have so far only various corners where perturbative control and/or the use of string dualities allow us calculational access, we cannot yet compare the theory as a whole with experimental data. At this stage we can only try to identify mechanisms in string theory which realizes certain types of dynamics which we need to describe our world. Certain mechanisms which realize inflation in controlled string theory settings will be our prime examples here. These mechanisms may then serve for comparisons with data, or they may inspire new model building and analysis for the bottom-up construction of inflation models in effective field theory as well.

The moving parts

Given this preamble, we start with collecting the necessary ingredients of string theory which guide our current understanding of the many vacua we get from string compactification to 4D. These ingredients provide the necessary dynamics for moduli stabilization, and building string inflation within the stabilized vacua of the string landscape. The 2D worldsheet action of superstrings gives rise to a 10D target spacetime if we demand the Weyl anomaly to vanish for a flat target spacetime. Note, that giving up flatness of the target space allows for string theories with $D > 10$ which have interesting consequences such as a spectrum with 2^D axions.[52–55] The maybe simplest exact solution then is 10D Minkowski space with either $\mathcal{N} = 1$ or 2 supersymmetry. Hence, in 10D the effective spacetime action is one of the five possible 10D supergravities (type I, heterotic $SO(32)/E_8 \times E_8$, type IIB, type IIA). We can write these 10D actions schematically as[56]

$$S = \frac{1}{2\alpha'^4} \int d^{10}x \sqrt{-G}\, e^{-2\phi} \left(R + 4(\partial_M \phi)^2 \right) + S_{matter}. \tag{35}$$

The matter part needs more detailed discussion. We can write its action, again schematically, as

$$S_{matter} = \int d^{10}x \sqrt{-G} \left[e^{-2\phi}|H_3|^2 + \sum_p |\tilde{F}_p|^2 + C.S. \right.$$
$$+ \sum_p \left(T_p^B \frac{\delta^{9-p}(x_\perp)}{\sqrt{-G_\perp}} - T_p^O \frac{\delta^{9-p}(x_\perp)}{\sqrt{-G_\perp}} \right) \tag{36}$$
$$\left. + higher\text{-}deriv. \right]$$

Here C.S. denotes Chern-Simons terms involving the various p-form gauge potentials and field strengths of string theory. The terms localized at the $9-p$ coordinates $x_\perp$ denote p-branes with tension T_p^B and orientifold planes with tension T_p^O. These different terms contribute various forms of potential energy to the effective action. The various branes labeled by T_p^B contribute effectively terms proportional to their tension times their worldvolume they wrap in the internal dimensions as potential energy. On a subclass of them, called D-branes, we can have open strings ending on them. This renders D-branes dynamical. Their positions in the orthogonal dimensions are scalar fields leaving on the worldvolume of a D-brane. Branes can carry charge under some of the higher-dimensional p-form gauge potentials of string theory, and we will make extensive use of that.

T_p^O labels objects with negative tension. These are so-called orientifold planes (O-planes) which arise from modding both the internal space and the spectrum with generalizations of Z_2 action. In general, the number and type of allowed O-planes follows from certain topological properties and quantum numbers of the compact six dimensions. Hence, the theory limits the amount of negative energy objects which is rather beneficial for vacuum stability.

The various terms $|H_3|^2$ and $|\tilde{F}_p|^2$ denote the kinetic and gradient energy contributions from the field strengths of higher-dimensional p-form gauge potentials in string theory. They also contribute was is known as quantized background fluxes to the effective action. Since we will employ these objects far and wide, we will start by discussing their properties and utilize their relation with ordinary electrodynamics as a theory of a 1-form gauge potential.

Our discussion will follow the lines of e.g. Ref. 57. The electromagnetic gauge potential $A_1 = A_{1,\mu}dx^\mu$ is a useful means of describing electromagnetism though containing redundancy. The field strength $F_2 = dA_1$, $F_{2,\mu\nu} = \partial_\mu A_{1,\nu} - \partial_\nu A_{1,\mu}$ is gauge invariant under the gauge transformation $A_1 \to A_1 + d\Lambda$. This is not the only gauge invariant object object, as there are for instance Wilson lines $e^{i\int A_1} = e^{i\oint A_{1,\mu}dx^\mu}$ around a compact extra dimension y. Let us look at the 5D metric with the y-direction compactified on a circle S^1 with radius $L_y/(2\pi)$ (so $y \to y+1$)

$$ds^2 = dt^2 - dx_3^2 - L_y^2 dy^2 \quad . \tag{37}$$

There is a flat gauge connection $A_1 = A_{1,y}dy$ constituting a Wilson line around the S^1 such that

$$A_1 = a(x)\omega_1 \quad , \quad \omega_1 = dy \quad , \quad a(x) \equiv \oint A_{1,y}dy \quad . \tag{38}$$

The quantity $a(x)$ with periodicity $a \to a + 1$ inherited from the S^1 is effectively an axion in 4D with a perturbatively exact shift symmetry. $a(x)$ is gauge invariant under $A_{1,y} \to A_{1,y} + \partial_y \Lambda$ since an S^1 has no boundary. Reducing out the gauge kinetic term for F_2 we get

$$S_{F_2} = \int d^4x\, dy \sqrt{-g}|F_{2,mn}|^2 = \int d^4x \sqrt{-g_4} \int dy \underbrace{\sqrt{-g_{yy}}}_{=1}\, g^{00} \underbrace{g^{yy}}_{L_y}\underbrace{\,}_{1/L_y^2} |F_{2,0y}|^2$$

$$= \int d^4x \sqrt{-g_4} L_y \frac{\dot{a}^2}{L_y^2} . \tag{39}$$

The axion kinetic term appears with an overall prefactor f^2, where we call f the axion decay constant which here is $f = 1/\sqrt{L_y}$, and thus inversely proportional to a power of the compact length scale. This is a generic property of all of the known string theory axions which arise by close analogy from the higher p-form gauge potentials in string theory in the presence of compact extra dimensions.

We now couple the electromagnetic field to a point charge. The interaction term

$$\int d^4x J^\mu A_{1,\mu} = \int_{worldline} A_1 \qquad (40)$$

effectively describes the Aharanov-Bohm effect: the worldline integral picks up a phase along a closed path around a region containing a B-field, i.e. magnetic flux $\int \mathbf{B}d\mathbf{A}$. This explains a second property of magnetic fluxes: on a compact space they are quantized, for instance $\int_{S^2} F_2 = N$. This should better be, as otherwise the Aharanov-Bohm phase picked up by a particle around a small loop on the compact space would not be a multiple of 2π producing a multi-valued wave function.

All of these properties generalize to higher dimensions with higher p-form gauge potentials, where the objects charged under the gauge fields generalize then as well from point charges to charged branes. The first basic generalization comes from the anti-symmetric 2-form gauge potential B_2 for which the Polykav action of the string allows a coupling on the string worldvolume Σ reading

$$S_{string} \supset \int_\Sigma B_2 = \frac{1}{\alpha'} \int d^2\sigma \sqrt{-\gamma}\epsilon^{ab} B_{2,MN}\partial_a X^M \partial_b X^N \quad . \qquad (41)$$

The 10D effective action for B_2 is the $|H_3|^2 = |dB_2|^2$ term in eq. (36) above, with the field strength of B_2 being $H_3 = dB_2$. The form of H_3 provides for gauge invariance under $B_2 \to B_2 + d\Lambda_1$. In close analogy with the electromagnetic case before this leads to axions

$$b_i(x) = \int_{\Sigma_2^i} B_2 \quad , \quad B_2 = b_i(x)\omega_2^i \qquad (42)$$

on non-trivial trivial 2-cycles Σ_2^i with their associated basis 2-forms ω_2^i. Again, the wavefunction of a string picks up an Aharonov-Bohm like phase on a loop around a region with magnetic flux H_3 due to the B_2 worldvolume coupling above. Consequently, there are quantized background fluxes $\int_{\Sigma_3^a} H_3 = N_a$ of H_3 on non-trivial 3-cycles Σ_3^a of the compact dimensions.

The first natural guess to get inflation from the b-axion would be to implement natural inflation.[58] The kinetic term for b-axion arises from reducing $|H_3|^2$, while a Euclidean string worldsheet wrapping an appropriate 2-cycle Σ_2 of the extra dimensions may provide a non-perturbative contribution to the scalar potential of b. We get

$$\mathcal{L} = f^2(\partial_\mu b)^2 - \Lambda^4 \left[1 - \cos(b)\right] \quad . \qquad (43)$$

Canonically normalizing b into $\phi = bf$ gives the potential

$$V = \Lambda^4 \left[1 - \cos(\phi/f) \right] \quad . \tag{44}$$

For large $f \gg M_{\mathrm{P}}$ the potential approximates $m^2 \phi^2$ around the minimum. We can now compute n_s and r from the slow-roll parameters. Then, we find that requiring $n_s > 0.945$ as required by the PLANCK 95% confidence limits implies $f \gtrsim 4.5 M_{\mathrm{P}}$.[37,59] However, the b-axion kinetic term gets a prefactor $f^2 \sim M_{\mathrm{P}}^2/L^q$ with $q > 1$ for a 2-cycle with length scale L in string units. This is in full analogy with the simple electromagnetic case before. Since a controlled string compactification requires all radii to be large, we have $L \gg 1$ and hence $f < M_{\mathrm{P}}$. This behavior is generic – all known string axions both from NSNS-sector or RR-sector p-form gauge fields have sub-Planckian axion decay constants.[60] Attempts to evade this were so far forced to resort to small radii or large string coupling.[61,62]

4. Axion Monodromy

Pushing further we note that in analogy with B_2 there are RR-sector p-form gauge potentials C_{p-1} and their field strengths $F_p = dC_{p-1}$. They arise from the type I/type II open string sector and have various branes as objects whose worldvolume couples to C_{p-1}. By the same arguments as above, RR-sector axions $c_k^{(p-1)}(x) = \int_{\Sigma_{p-1}^k} C_{p-1}$ arise from the invariance of F_p under $C_{p-1} \to C_{p-1} + d\Lambda_{p-2}$.

Finally, the full duality structure and set of gauge invariances of the various string theories forces generalization of the RR-sector p-form field strengths to include various Chern-Simons couplings

$$\tilde{F}_p = F_p + B_2 \wedge F_{p-2} \quad . \tag{45}$$

The corresponding kinetic terms $|\tilde{F}_p|^2$ are invariant under B_2 gauge transformation as well, if at the same time $C_{p-1} \to C_{p-1} - \Lambda_1 \wedge F_{p-2}$.

4.1. Axion monodromy inflation

We see that turning on RR-flux F_{p-2} provides C_{p-1} with a 'Stueckelberg' charge under the B_2 gauge transformation. C_{p-1} now shifts with Λ_1, while at the same time F_{p-2}-flux provides a mass term from $|\tilde{F}_p|^2$ for the B_2 gauge field and its associated axion $b(x)$.

This is in close analogy to the phenomenon of superconductivity: there, spontaneous symmetry breaking leads to the appearance of a 'Stueckelberg scalar' $\theta(x)$ with a coupling $(A_1 + d\theta)^2$ which provides the mass for the electromagnetic field inside a superconductor. The $A_1 \wedge d\theta$ coupling is the analogue of the $B_2 \wedge F_{p-2}$ coupling in higher dimensions. The coupling is necessary to preserve gauge invariance of the whole system even if the gauge field is massive, as now $A_1 \to A_1 + d\Lambda$ pairs with a shift $\theta \to \theta - \Lambda$.

Hence, turning on the lower-dimensional fluxes F_{p-2} provides a *non-periodic* potential for the b-axions of the form

$$V \sim \int d^6 y \sqrt{-g_6} |\tilde{F}_p|^2$$
$$\sim \int d^6 y \sqrt{-g_6} (F_p + B_2 \wedge F_{p-2})^2 \sim (N_p + b\, N_{p-2})^2 \quad . \tag{46}$$

Here N_p and N_{p-2} denote the flux quanta of F_p and F_{p-2} flux turned on. The field range of a given b-axion no longer shows periodicity and is a priori (kinematically) unbounded. This is the phenomenon of *axion monodromy* parametrically extending the axion field range along a non-periodic potential from fluxes.[34,35,49,63]

We recall here that the structure of the potential eq. (46) is quite similar to the 4D field theory version of axion monodromy in Refs. 35, 46, 64. There, we start from an axion and a 4-form field strength

$$\mathcal{L} \sim (\partial_\mu \phi)^2 + (F_{\mu\nu\rho\sigma})^2 + \frac{\mu}{\sqrt{-g}} \phi \epsilon^{\mu\nu\rho\sigma} F_{\mu\nu\rho\sigma} \quad . \tag{47}$$

Integrating out the 4-form while giving it q units of flux gives ϕ a potential $V \sim (q + \mu\phi)^2$. The underlying shift symmetry appears as a joint shift of both ϕ and q, but shifts in q are again mediated by exponentially suppressed brane nucleations. Hence picking q chooses a branch, giving the axion ϕ a non-periodic, a priori quadratic potential. The structure of this 4D theory is rather similar to the reduction of $|\tilde{F}_p|^2$ above.

Note, that we can compensate integer shifts of the axion b by appropriate integer changes in the F_{p-2} flux quanta. However, these changes are mediated by non-perturbative effects which are suppressed at weak string coupling and large volume. Hence, the full system displays a set of non-periodic potential *branches* for the b-axion.[34,35,46,64] The branches are labeled by the flux quanta of F_{p-2}. The periodicity of the full theory is now visible in the set of branches of non-periodic potentials when summing over branches. However, as in spontaneous symmetry breaking, once we pick a certain quantized flux F_{p2} which can only change by exponentially suppressed effects, we pick a certain branch along which the b-axion rolls in a non-periodic potential. Hence, axion monodromy clearly lends itself to realize large-field inflation in the context of string theory.

The generic structure of the axion effective action on one such branch picked by F_{p-2} flux looks like

$$\mathcal{L} = f^2(\chi)(\partial_\mu b)^2 + \mu(\chi)^{4-p_0} b^{p_0} + \Lambda^4(\chi)\cos(b) \quad . \tag{48}$$

χ summarily denote the moduli of a given string compactification, and the axion acquires a periodic contribution from non-perturbative contributions. However, we expect their scale Λ^4 to be exponentially suppressed in the radii of the extra dimensions and $1/g_s$. Hence, the generic axion potential is a large-field monomial with a priori power p_0 with tiny periodic modulations on top. The axion decay constant, however, is moduli dependent. As discussed above, in general $f \sim M_P/L^q, q > 0$. Hence, after moduli stabilization the backreaction of the moduli $\chi = \chi(b)$ due to the inflationary vacuum energy may very well change $f = f(\chi(b))$ and thus the canonically normalized inflaton field $\phi(b)$ we get from $d\phi(b) = f(\chi(b))db$.

Provided we realize this mechanism in a controlled setting with moduli stabilization at large volume and weak string coupling and with a sub-Planckian inflationary energy scale from axion monodromy, this mechanism will inherit all the properties of radiative stability discussed in the previous section. However, the presence of the moduli generically leads to backreaction effects as already dicussed above in field theory. The CS-couplings in the generalized flux kinetic terms lead to a priori axion potentials $V \sim b^{p_0}$ with powers $p_0 = 2, 3, 4$.[50,51,61,62,65-67] However, inclusion of the moduli will generically lead to *flattening* from backreaction of the moduli onto the axion potential[49,51,64]

$$V_{eff.} \sim \phi(b)^p \quad , \quad p < p_0 \quad . \tag{49}$$

The associated predictions for the inflationary observables are $n_s = 1 - (2 + p)/(2N_e)$ and $r = 4p/N_e \simeq 0.05 \ldots 0.24$ for $p \simeq 0.5 \ldots 4$ and $N_e = 50 \ldots 60$.

Finally, we can see the same basic effect of axion monodromy arising from the coupling of the p-form gauge potentials to branes. This must be the case, as the web of dualities in string theory in the ends relates compactification with certain sets of branes to dual geometries without branes but background fluxes describing the same physics.

For simplicity, we look at the effective action of Dp-branes wrapping $p + 1$ dimensions of space-time. To start, we look at the worldline action of a particle $S = \int dt\sqrt{1 - \dot{x}^2}$. If this particle is charged under the electromagnetic gauge potential, then

$$S = \int dt \sqrt{1 - \dot{x}^2} + q \int_{worldline} A_1. \tag{50}$$

This structure generalizes to higher dimensions, where Dp-branes by analogy have a worldvolume action of DBI form together with a CS term providing charge under the various RR-sector gauge potentials

$$S_{Dp} = T_{Dp}^B \int d^{p+1}\xi \sqrt{-\det(G + B_2) - \alpha' F_2} + \mu_{Dp}^B \int (C_{p+1} + \ldots) \tag{51}$$

$\det G$ here denotes the determinant of the induced metric on the worldvolume of the brane, while F_2 is the field strength of a $U(1)$ gauge field which a single brane can contain.

Now let us take a 5-brane living on $\mathcal{M}_4 \times T^2$. That is, the 5-brane fills all macroscopic 4D space-time, and wraps a small compact two-torus of the extra dimensions. Now we put a B_2 gauge field $B_{12} = -B_{21} = b(x)$ on the T^2, which in 4D is our b-axion. At same time there may be N_2 units of combined 2-form flux $N_2 = \int_{T^2} \mathcal{F}_2$, where $\mathcal{F}_2 \equiv = B_2 + \alpha' F_2$. Reducing the brane action plus the $|H_3|^2$ (for the b kinetic term) to 4D, we arrive at

$$\mathcal{L} = f^2 \dot{b}^2 - \sqrt{vol(T^2)^2 + (b + N_2)^2} \quad . \tag{52}$$

Again, we see the appearance of an infinite set of potential energy branches labeled by the flux integer N_2. However, once we pick a flux N_2, we are on one given branch. The b-axion now acquires a non-periodic scalar potential which is linear

$$V_{D5} \sim b \tag{53}$$

at large b on each branch.[34] Doing this on a D7-brane instead can produce $V_{D7} \sim b^2$ instead.[65] On each branch b will drive large-field slow-roll inflation, while relaxation to different branches with different N_2 is subject to generically exponentially suppressed brane nucleation tunneling events.[35,46,64] The general story is the same as before, as in fact it is related to the flux-induced version above by duality.

4.2. Inflating with several axions

Since axions seem to be ubiquitous in string compactifications (they comprise roughly $h^{1,1}/2$ of the CY moduli, and for supercritical strings there are $\mathcal{O}(2^D)$ axions present in 4D[52–55]), we might look for assistance effects of several axions helping each other towards slow-roll. The original example of this is N-flation,[68-70] where $N \ll 1$ axions are excited simultaneously from the minima of the non-perturbatively generated $V_i(\phi_i) = \Lambda^4[1 - \cos(\phi_i/f)]$ cosine-potentials. The Hubble parameter adds up the squares of the displacements in the quadratic approximation to the axion potentials $H \sim \sqrt{\sum_i V_i} \sim \sqrt{\sum_i m^2 \phi_i^2} \sim \sqrt{N}\sqrt{m^2\phi^2}$, $\phi_i \sim \phi \ \forall i$. This increases the friction term in the equation of motion for *each* axion by a factor $\sim \sqrt{N}$. However, the Planck mass renormalizes as well, yielding in total

$$H^2 \sim \frac{N \, m^2 \phi^2}{M_{\rm P}^2 + N \, M_*^2} \tag{54}$$

where M_* denotes the appropriate cut-off scale for the diagrams which renormalize the Planck scale. This renders parametric enhancement of H at large N difficult to achieve. However, we note here that describing the axion decay constants of an N-flation setup as a randomized ensemble enhances the relative likelihood of alignments or hierarchy among the decy constants. This can reduce the pressure towards large N for N-flation somewhat.[71,72]

The other option analyzed tries to realize natural inflation using a 2-axion system. We start with a 2-axion system which is aligned in the axion decay constants such that it has precisely one flat direction.[36] We begin with

$$\begin{aligned}\mathcal{L} =& f_r^2(\partial_\mu a_r)^2 + f_\theta^2(\partial_\mu a_\theta)^2 \\ & - \Lambda_1^4 \left[1 - \cos(p_1 a_r + p_2 a_\theta)\right] - \Lambda_2^4 \left[1 - \cos(q_1 a_r + q_2 a_\theta)\right] \end{aligned} \tag{55}$$

where the p_i, q_i denote coefficients in the Euclidean action (e.g. coxeter numbers for gaugino condensation) of the nonperturbative effects generating the axion potential. Canonically normalizing the axions gives

$$V(r,\theta) = \Lambda_1^4 \left[1 - \cos\left(p_1\frac{r}{f_r} + p_2\frac{\theta}{f_\theta}\right)\right] + \Lambda_2^4 \left[1 - \cos\left(q_1\frac{r}{f_r} + q_2\frac{\theta}{f_\theta}\right)\right] \tag{56}$$

Tuning an alignment $r_2/r_1 = q_2/q_1 \equiv \kappa$ produces a single flat direction.[36] We have can now generate a shallow long-range axion potential by slightly perturbing the alignment. There are two ways to do so. One way is to manifestly tune a finite small misalignment into the above condition, changing it to $r_2/r_1 = \kappa(1+\delta)$.[36] The flat direction lifts slightly producing an effective potential for the former flat direction

$$V_{eff.} \sim 1 - \cos(\phi_{eff.}/f_{eff.}) \quad , \quad f_{eff.} = \frac{\sqrt{f_\theta^2 + f_r^2\kappa^2}}{q_1\kappa}\frac{1}{\delta} \quad . \tag{57}$$

For $\delta \lesssim 0.1$ the approximately flat direction has a large-field potential with $f_{eff.} > 5M_{\rm P}$ even if $f_r, f_\theta \lesssim M_{\rm P}$.[36] The smaller the initial decay constants

f_r and f_θ are, the smaller we must tune δ to reach a desired f_{eff}.. For 2 axions the near-alignment typically constitutes at least 1-10% tuning. However, among a larger number of axions with a random decay constant distribution the likelihood for a random near-lignment of 2 axions with small enough δ among N axions can be rather sizable.[73]

Alternatively, we can leave the aligned situation $V_0 \sim 1 - \cos(r/f_r + \theta/f_\theta)$ by simply adding a potential such that $V = W(r) + V_0(r, \theta)$. $W(r)$ can originate from some brane or flux induced monodromy,[74] or can come from a non-perturbative effect itself $W(r) \sim \cos(r/f'_r)$.[75,76] Providing a hierarchy $f_r \ll f'_r, f_\theta$ is enough to give θ a long-range shallow potential with $f_{eff} > 5 M_P$ for sub-Planckian initial decay constants. The potential for the inflaton, which is approximately θ, sits like a set of terraces inside the potential slope for r.

4.3. General structure of moduli stabilization

String theory in its critical version lives in 10D. Describing four dimensional physics then typically employs compactification of the six extra dimensions on a small compact manifold. This process leaves various deformation modes as vacuum degeneracies which therefore describe massless and flat scalar fields in 4D, called the moduli. The presence of massless scalars is in contradiction with various experimental data (no 5th forces, light moduli screw up early Universe cosmology, etc.). The inclusion of background fluxes, branes, O-planes and internal curvature lead to a much improved understanding of moduli stabilization, that is giving mass to the various moduli.[77] Moduli stabilization also usually drives supersymmetry breaking. This enabled the first classes of de Sitter vacua in string theory, starting with the KKLT scenario.[30]

Inflation in string theory is the phenomenon which after embedding in string theory potentially lives closest to the string scale of all observable phenomena so far (a verification of BICEP2[21] would imply $V_{inf.}^{1/4} \sim M_{GUT} \sim 10^{16}$ GeV, as discussed before). Given its dependence on a slow-rolling scalar field, moduli stabilization is absolutely unavoidable before a meaningful discussion of string inflation becomes possible.

Hence we sketch here the basics of moduli stabilization as they are relevant for our discussion. The various fluxes, branes, and O-planes, as well as internal curvature contribute after reducing to 4D potential energy terms which scale as inverse powers of the radii and the volume $\mathcal{V}$ of the extra dimensions, and as different positive powers of the string coupling g_s.

These contributions fall into roughly three classes: i) certain branes, H_3-flux and negative internal curvature contribute positive terms of the form $V_1 \sim +ag_s^{r_1}/\mathcal{V}^{q_1}$. ii) Negative terms $V_2 \sim -bg_s^{r_2}/\mathcal{V}^{q_2}$ which fall slower ($q_2 < q_1$) with inverse volume arise from O-planes, positive internal curvature, and some quantum corrections. iii) Finally, a 2nd set of positive terms falling even slower with the inverse volume, $V_3 \sim +cg_s^{r_3}/\mathcal{V}^{q_3}$ arises from RR-sector fluxes. In total, these contributions add up to a scalar potential with a 3-term structure[78,79]

$$V = V_1 + V_2 + V_3 = a\frac{g_s^{r_1}}{\mathcal{V}^{q_1}} - b\frac{g_s^{r_2}}{\mathcal{V}^{q_2}} + c\frac{g_s^{r_3}}{\mathcal{V}^{q_3}} \quad . \tag{58}$$

Given the hierarchy $0 < q_3 < q_2 < q_1$ so that the negative middle term can produce a dip in an other positive potential, tuning the fluxes and the negative term from O-planes and/or positive curvature allows to realize exponentially many dS vacua with suppressed cosmological constant. Due to the large number of fluxes, the number of these flux vacua, called the 'landscape', can easily surpass scales like 10^{500}. This allows for environmential explanations of the observed amount of dark energy.[28,30,31,80]

In general, stabilizing all moduli along the lines sketched above implies supersymmetry breaking at the KK scales, as turning on the required fluxes generically breaks SUSY.[28]

However, we may insist on preservation of a single 4D supersymmetry to low energies for various phenomenological reasons (such as the gauge hierarchy problem). This case pretty much forces us to compactify critical 10D string theory on a Calabi-Yau (CY) manifold, or an orientifold thereof in type II settings. Preserving the Calabi-Yau structure restricts the types of background fluxes which are admissible. In particular, we may only use imaginary self-dual 3-form flux $G_3 = F_3 - \tau H_3$ in type IIB compactifications which probably provide still the best understood examples of moduli stabilization so far.[77] $\tau = C_0 + i/g_s = C_0 + ie^{-\phi}$ denotes the axio-dilaton, the complexified type IIB string coupling.

The moduli space of a CY compactification decomposes into a set of $h^{2,1}$ 3-cycle complex structure moduli U^a and $h_+^{1,1}$ Kähler or volume moduli T_i which measure 4-cycle volumina in the CY. The index "+" refers to those volume moduli which are even under the orienfold projection needed for 4D $\mathcal{N} = 1$ supersymmetry. A CY itself has $h^{1,1} \geq h_+^{1,1}$ volume moduli to start with. If $h^{1,1} > h_+^{1,1}$, the remaining $h_-^{1,1}$ moduli are orientifold-odd combinations of the B_2- and RR-sector C_2-form axions,[69,81] which in these settings are some of our natural large-field axion monodromy inflation candidates.[34] Turning on just the CY compatible 3-form fluxes results in

4D $\mathcal{N} = 1$ effective supergravity for the moduli sector described by a Kähler potential and a superpotential

$$K = -2\ln \mathcal{V}(T_i, \overline{T_i}) - \ln(-(\tau - \bar{\tau})/2i) - K_{c.s.}(U^a, \overline{U^a})$$

(59)

$$W = W_0(U^a, \tau) \ , \quad W_0(U^a, \tau) = \int_{CY} G_3 \wedge \Omega \ .$$

The 3-form fluxes induce a superpotential and corresponding scalar potential which serve to fix all the U^a and axio-dilaton τ.[28,77] However, this leaves the volume moduli T_i with a no-scale scalar potential $V(T_i) = 0 \forall T_i$.

It is this feature, which requires using a combination of perturbative α' and string loop corrections to the volume moduli Kähler potential[32,82–84]

$$K_{\text{Kähler}} = -2\ln(\mathcal{V}) \rightarrow K_{\text{Kähler}} = -2\ln(\mathcal{V} + \alpha'^3\xi) + \delta K_{g_s}$$

(60)

and non-perturbative corrections in the T_i from Euclidean brane instantons or gaugino condensation on D-brane stacks

$$W = W_0 \rightarrow W = W_0 + \sum_i A_i e^{-a_i T_i}$$

(61)

to stabilize the volume moduli.[28,30,32,33] These vacua are mostly AdS (except for those of Ref. 33). Reaching dS vacua involves either introducing manifestly SUSY breaking objects like anti-branes in warped regions (for control),[30] or effects generating F- and/or D-terms from the open string/matter sector.[85,86] In these CY based constructions, the scale of SUSY breaking associated with volume stabilization tends to be lower due to the underlying no-scale structure than in more generic settings away from CYs. Some CY schemes such as the Large Volume Scenario (LVS)[32] can reach TeV scale gravitino mass either by stabilizing at very large volume or by using sequestering to suppress the soft masses with respect to an intermediate scale gravitino mass.[87] We note, that in certain settings a racetrack combination of non-perturbative effects alone may suffice to stabilize all geometric moduli without the use of any flux. G_2 compactifications of M-theory provide one such an example.[88]

The underlying no-scale structure thus forces us to stabilize the volume moduli at least partly by combinations of exponentially small effects. This feature of CY compactification tends to render the volume moduli more susceptible to backreaction from the inflationary vacuum energy. In turn, once we require CY compactification, this more often requires us to separate the mass scale of the volume moduli from the scale of inflation, leading to more tuning in the moduli stabilization than necessary in the generic situation away from CYs.[34,49,50] In contrast, in more generic settings with KK scale supersymmetry breaking, and all sources of moduli potential energy available in 10D in play, the inflationary vacuum often *helps* with moduli stabilization.[51] This leads to mostly harmless backreaction driving flattening of the inflaton potential, replicating the spirit of the field theory discussion of the last section.

We finally note, that[78,79,89] used more general combinations of negative internal curvature, fluxes, branes and O-planes to generate new and more general classes of dS vacua by generalizing the basic flux and brane setup leading to the $AdS_5 \times X^5$ setup of the AdS/CFT pairs. These constructions lead to very promising paths towards a holographic description of dS vacua in terms of dS_d/dS_{d-1} pairs[90] or similar and potentially related FRW/CFT dual pairs.[91]

5. Where Do We Go from Here?

From the preceding discussion we clearly see that we have just begun to scratch the tip of an iceberg's worth of a rich set of models and relationships between various models of large-field inflation in string theory. We are just starting to map the underlying structure of axion monodromy which seems to be a widespread property of the axion sector of string compactifications. Given that large-field inflation with its GUT-scale inflaton potential is maybe our best hope to come close to string-scale physics using observations, we should understand as much as possible about the generic properties of the mechanism while continue to build more explicit sample construction. A synopsis of both, particular constructions and an understanding of the general properties of axion monodromy, might give us what we need to finally have estimates for the inflationary observables n_s and r beyond the level of a few lamp posts.

Acknowledgments

I wish to express my deep gratitude to a lot of deep and insightful conversations with many collaborators and friends which helped me most of the way towards the understanding expressed in these notes. This work was supported by the Impuls und Vernetzungsfond of the Helmholtz Association of German Research Centres under grant HZ-NG-603.

References

1. A. Borde, A. H. Guth, and A. Vilenkin, Inflationary space-times are incompletein past directions, *Phys.Rev.Lett.* **90**, 151301 (2003). doi: 10.1103/PhysRevLett.90.151301.
2. L. Cornalba and M. S. Costa, Time dependent orbifolds and string cosmology, *Fortsch.Phys.* **52**, 145–199 (2004). doi: 10.1002/prop.200310123.
3. B. Craps, Big Bang Models in String Theory, *Class.Quant.Grav.* **23**, S849–S881 (2006). doi: 10.1088/0264-9381/23/21/S01.
4. M. Berkooz and D. Reichmann, A Short Review of Time Dependent Solutions and Space-like Singularities in String Theory, *Nucl.Phys.Proc.Suppl.* **171**, 69–87 (2007). doi: 10.1016/j.nuclphysbps.2007.06.008.
5. W. Freedman et al., Final results from the Hubble Space Telescope key project to measure the Hubble constant, *Astrophys.J.* **553**, 47–72 (2001). doi: 10.1086/320638.
6. A. G. Riess et al., A 3% Solution: Determination of the Hubble Constant with the Hubble Space Telescope and Wide Field Camera 3, *Astrophys. J.* **730**, 119 (2011). doi: 10.1088/0004-637X/730/2/119.
7. A. G. Riess et al., Observational Evidence from Supernovae for an Accelerating Universe and a Cosmological Constant, *Astron. J.* **116**, 1009–1038 (1998). doi: 10.1086/300499.
8. S. Perlmutter et al., Measurements of Omega and Lambda from 42 High-Redshift Supernovae, *Astrophys. J.* **517**, 565–586 (1999). doi: 10.1086/307221.
9. G. Hinshaw et al., Nine-Year Wilkinson Microwave Anisotropy Probe (WMAP) Observations: Cosmological Parameter Results (2012).
10. P. Ade et al., Planck 2013 results. XVI. Cosmological parameters (2013).
11. P. Ade et al., Planck 2013 results. XXII. Constraints on inflation (2013).
12. J. L. Sievers et al., The Atacama Cosmology Telescope: Cosmological parameters from three seasons of data (2013).
13. K. T. Story et al., A Measurement of the Cosmic Microwave Background Damping Tail from the 2500-square-degree SPT-SZ survey (2012).
14. Z. Hou et al., Constraints on Cosmology from the Cosmic Microwave Background Power Spectrum of the 2500-square degree SPT-SZ Survey (2012).
15. A. H. Guth, The Inflationary Universe: A Possible Solution to the Horizon and Flatness Problems, *Phys.Rev.* **D23**, 347–356 (1981). doi: 10.1103/PhysRevD.23.347.
16. A. A. Starobinsky, A New Type of Isotropic Cosmological Models Without Singularity, *Phys.Lett.* **B91**, 99–102 (1980). doi: 10.1016/0370-2693(80)90670-X.
17. A. D. Linde, A New Inflationary Universe Scenario: A Possible Solution of the Horizon, Flatness, Homogeneity, Isotropy and Primordial Monopole Problems, *Phys.Lett.* **B108**, 389–393 (1982). doi: 10.1016/0370-2693(82)91219-9.
18. A. Albrecht and P. J. Steinhardt, Cosmology for Grand Unified Theories with Radiatively Induced Symmetry Breaking, *Phys.Rev.Lett.* **48**, 1220–1223 (1982). doi: 10.1103/PhysRevLett.48.1220.

19. D. Baumann, TASI Lectures on Inflation (2009).
20. P. Ade et al., Planck 2013 Results. XXIV. Constraints on primordial non-Gaussianity (2013).
21. P. Ade et al., Detection of B-Mode Polarization at Degree Angular Scales by BICEP2, *Phys.Rev.Lett.* **112**, 241101 (2014). doi: 10.1103/PhysRevLett. 112.241101.
22. M. J. Mortonson and U. Seljak, A joint analysis of Planck and BICEP2 B modes including dust polarization uncertainty (2014).
23. R. Flauger, J. C. Hill, and D. N. Spergel, Toward an Understanding of Foreground Emission in the BICEP2 Region (2014).
24. F. Boulanger. URL http://www.rssd.esa.int/SA/PLANCK/docs/eslab47/ Session04_Astrophysical_Results/47ESLAB_April_03_14_50_Boulanger. pdf (2014).
25. W. N. Colley and J. R. Gott, Genus Topology and Cross-Correlation of BICEP2 and Planck 353 GHz B-Modes: Further Evidence Favoring Gravity Wave Detection (2014).
26. M. Grana, Flux compactifications in string theory: A Comprehensive review, *Phys.Rept.* **423**, 91–158 (2006). doi: 10.1016/j.physrep.2005.10.008.
27. M. R. Douglas and S. Kachru, Flux compactification, *Rev. Mod. Phys.* **79**, 733–796 (2007). doi: 10.1103/RevModPhys.79.733.
28. F. Denef, Les Houches Lectures on Constructing String Vacua. pp. 483–610 (2008).
29. M. Cicoli, S. de Alwis, and A. Westphal, Heterotic Moduli Stabilization (2013).
30. S. Kachru, R. Kallosh, A. D. Linde, and S. P. Trivedi, De Sitter vacua in string theory, *Phys.Rev.* **D68**, 046005 (2003). doi: 10.1103/PhysRevD.68. 046005.
31. L. Susskind, The Anthropic landscape of string theory (2003).
32. V. Balasubramanian, P. Berglund, J. P. Conlon, and F. Quevedo, Systematics of moduli stabilisation in Calabi-Yau flux compactifications, *JHEP.* **0503**, 007 (2005). doi: 10.1088/1126-6708/2005/03/007.
33. J. Louis, M. Rummel, R. Valandro, and A. Westphal, Building an explicit de Sitter, *JHEP.* **1210**, 163 (2012). doi: 10.1007/JHEP10(2012)163.
34. L. McAllister, E. Silverstein, and A. Westphal, Gravity Waves and Linear Inflation from Axion Monodromy, *Phys.Rev.* **D82**, 046003 (2010). doi: 10. 1103/PhysRevD.82.046003.
35. N. Kaloper and L. Sorbo, A Natural Framework for Chaotic Inflation, *Phys.Rev.Lett.* **102**, 121301 (2009). doi: 10.1103/PhysRevLett.102.121301.
36. J. E. Kim, H. P. Nilles, and M. Peloso, Completing natural inflation, *JCAP.* **0501**, 005 (2005). doi: 10.1088/1475-7516/2005/01/005.
37. D. Baumann and L. McAllister, Inflation and String Theory (2014).
38. S. Kachru, R. Kallosh, A. D. Linde, J. M. Maldacena, L. P. McAllister, et al., Towards inflation in string theory, *JCAP.* **0310**, 013 (2003). doi: 10.1088/ 1475-7516/2003/10/013.
39. D. Baumann and L. McAllister, Advances in Inflation in String Theory, *Ann. Rev. Nucl. Part. Sci.* **59**, 67–94 (2009). doi: 10.1146/annurev.nucl.010909.

083524.

40. C. Armendariz-Picon, T. Damour, and V. F. Mukhanov, k - inflation, *Phys.Lett.* **B458**, 209–218 (1999). doi: 10.1016/S0370-2693(99)00603-6.

41. M. Alishahiha, E. Silverstein, and D. Tong, DBI in the sky, *Phys.Rev.* **D70**, 123505 (2004). doi: 10.1103/PhysRevD.70.123505.

42. V. F. Mukhanov and G. V. Chibisov, Quantum Fluctuation and Nonsingular Universe. (In Russian), *JETP Lett.* **33**, 532–535 (1981).

43. J. M. Maldacena, Non-Gaussian features of primordial fluctuations in single field inflationary models, *JHEP.* **0305**, 013 (2003). doi: 10.1088/1126-6708/2003/05/013.

44. C. Cheung, P. Creminelli, A. L. Fitzpatrick, J. Kaplan, and L. Senatore, The Effective Field Theory of Inflation, *JHEP.* **0803**, 014 (2008). doi: 10.1088/1126-6708/2008/03/014.

45. D. H. Lyth, What would we learn by detecting a gravitational wave signal in the cosmic microwave background anisotropy?, *Phys.Rev.Lett.* **78**, 1861–1863 (1997). doi: 10.1103/PhysRevLett.78.1861.

46. N. Kaloper, A. Lawrence, and L. Sorbo, An Ignoble Approach to Large Field Inflation, *JCAP.* **1103**, 023 (2011). doi: 10.1088/1475-7516/2011/03/023.

47. L. Smolin, Gravitational Radiative Corrections as the Origin of Spontaneous Symmetry Breaking!, *Phys.Lett.* **B93**, 95 (1980). doi: 10.1016/0370-2693(80)90103-3.

48. A. D. Linde, Particle physics and inflationary cosmology, *Contemp.Concepts Phys.* **5**, 1–362 (1990).

49. X. Dong, B. Horn, E. Silverstein, and A. Westphal, Simple exercises to flatten your potential, *Phys.Rev.* **D84**, 026011 (2011). doi: 10.1103/PhysRevD.84.026011.

50. A. Hebecker, S. C. Kraus, and L. T. Witkowski, D7-Brane Chaotic Inflation (2014).

51. L. McAllister, E. Silverstein, A. Westphal, and T. Wrase, The Powers of Monodromy (2014).

52. A. Maloney, E. Silverstein, and A. Strominger, De Sitter space in noncritical string theory. pp. 570–591 (2002).

53. A. Strominger, The Inverse Dimensional Expansion in Quantum Gravity, *Phys.Rev.* **D24**, 3082 (1981). doi: 10.1103/PhysRevD.24.3082.

54. R. Emparan, R. Suzuki, and K. Tanabe, The large D limit of General Relativity, *JHEP.* **1306**, 009 (2013). doi: 10.1007/JHEP06(2013)009.

55. S. Hellerman and I. Swanson, Cosmological solutions of supercritical string theory, *Phys.Rev.* **D77**, 126011 (2008). doi: 10.1103/PhysRevD.77.126011.

56. J. Polchinski, String theory. Vol. 2: Superstring theory and beyond (1998).

57. E. Silverstein, Les Houches lectures on inflationary observables and string theory (2013).

58. K. Freese, J. A. Frieman, and A. V. Olinto, Natural inflation with pseudo - Nambu-Goldstone bosons, *Phys.Rev.Lett.* **65**, 3233–3236 (1990). doi: 10.1103/PhysRevLett.65.3233.

59. E. Pajer and M. Peloso, A review of Axion Inflation in the era of Planck, *Class.Quant.Grav.* **30**, 214002 (2013). doi: 10.1088/0264-9381/30/

21/214002.

60. T. Banks, M. Dine, P. J. Fox, and E. Gorbatov, On the possibility of large axion decay constants, *JCAP*. **0306**, 001 (2003). doi: 10.1088/1475-7516/2003/06/001.

61. R. Blumenhagen and E. Plauschinn, Towards Universal Axion Inflation and Reheating in String Theory (2014).

62. T. W. Grimm, Axion Inflation in F-theory (2014).

63. R. Flauger, L. McAllister, E. Pajer, A. Westphal, and G. Xu, Oscillations in the CMB from Axion Monodromy Inflation, *JCAP*. **1006**, 009 (2010). doi: 10.1088/1475-7516/2010/06/009.

64. N. Kaloper and A. Lawrence, Natural Chaotic Inflation and UV Sensitivity, *Phys.Rev.* **D90**, 023506 (2014). doi: 10.1103/PhysRevD.90.023506.

65. E. Palti and T. Weigand, Towards large r from [p, q]-inflation, *JHEP*. **1404**, 155 (2014). doi: 10.1007/JHEP04(2014)155.

66. F. Marchesano, G. Shiu, and A. M. Uranga, F-term Axion Monodromy Inflation (2014).

67. S. Franco, D. Galloni, A. Retolaza, and A. Uranga, Axion Monodromy Inflation on Warped Throats (2014).

68. S. Dimopoulos, S. Kachru, J. McGreevy, and J. G. Wacker, N-flation, *JCAP*. **0808**, 003 (2008). doi: 10.1088/1475-7516/2008/08/003.

69. T. W. Grimm, Axion inflation in type II string theory, *Phys.Rev.* **D77**, 126007 (2008). doi: 10.1103/PhysRevD.77.126007.

70. M. Cicoli, K. Dutta, and A. Maharana, N-flation with Hierarchically Light Axions in String Compactifications (2014).

71. R. Easther and L. McAllister, Random matrices and the spectrum of N-flation, *JCAP*. **0605**, 018 (2006). doi: 10.1088/1475-7516/2006/05/018.

72. T. C. Bachlechner, M. Dias, J. Frazer, and L. McAllister, A New Angle on Chaotic Inflation (2014).

73. T. Higaki and F. Takahashi, Natural and Multi-Natural Inflation in Axion Landscape, *JHEP*. **1407**, 074 (2014). doi: 10.1007/JHEP07(2014)074.

74. M. Berg, E. Pajer, and S. Sjors, Dante's Inferno, *Phys.Rev.* **D81**, 103535 (2010). doi: 10.1103/PhysRevD.81.103535.

75. I. Ben-Dayan, F. G. Pedro, and A. Westphal, Hierarchical Axion Inflation (2014).

76. S. H. H. Tye and S. S. C. Wong, Helical Inflation and Cosmic Strings (2014).

77. S. B. Giddings, S. Kachru, and J. Polchinski, Hierarchies from fluxes in string compactifications, *Phys.Rev.* **D66**, 106006 (2002). doi: 10.1103/PhysRevD.66.106006.

78. E. Silverstein, Simple de Sitter Solutions, *Phys.Rev.* **D77**, 106006 (2008). doi: 10.1103/PhysRevD.77.106006.

79. X. Dong, B. Horn, S. Matsuura, E. Silverstein, and G. Torroba, FRW solutions and holography from uplifted AdS/CFT, *Phys.Rev.* **D85**, 104035 (2012). doi: 10.1103/PhysRevD.85.104035.

80. F. Denef and M. R. Douglas, Distributions of flux vacua, *JHEP*. **0405**, 072 (2004). doi: 10.1088/1126-6708/2004/05/072.

81. T. W. Grimm and J. Louis, The Effective action of N = 1 Calabi-Yau orien-

tifolds, *Nucl.Phys.* **B699**, 387–426 (2004). doi: 10.1016/j.nuclphysb.2004.08. 005.

82. K. Becker, M. Becker, M. Haack, and J. Louis, Supersymmetry breaking and alpha-prime corrections to flux induced potentials, *JHEP.* **0206**, 060 (2002).

83. M. Cicoli, J. P. Conlon, and F. Quevedo, Systematics of String Loop Corrections in Type IIB Calabi-Yau Flux Compactifications, *JHEP.* **0801**, 052 (2008). doi: 10.1088/1126-6708/2008/01/052.

84. M. Cicoli, J. P. Conlon, and F. Quevedo, General Analysis of LARGE Volume Scenarios with String Loop Moduli Stabilisation, *JHEP.* **0810**, 105 (2008). doi: 10.1088/1126-6708/2008/10/105.

85. C. Burgess, R. Kallosh, and F. Quevedo, De Sitter string vacua from supersymmetric D terms, *JHEP.* **0310**, 056 (2003).

86. M. Cicoli, S. Krippendorf, C. Mayrhofer, F. Quevedo, and R. Valandro, D-Branes at del Pezzo Singularities: Global Embedding and Moduli Stabilisation, *JHEP.* **1209**, 019 (2012). doi: 10.1007/JHEP09(2012)019.

87. R. Blumenhagen, J. Conlon, S. Krippendorf, S. Moster, and F. Quevedo, SUSY Breaking in Local String/F-Theory Models, *JHEP.* **0909**, 007 (2009). doi: 10.1088/1126-6708/2009/09/007.

88. B. S. Acharya, K. Bobkov, G. Kane, P. Kumar, and D. Vaman, An M theory Solution to the Hierarchy Problem, *Phys.Rev.Lett.* **97**, 191601 (2006). doi: 10.1103/PhysRevLett.97.191601.

89. M. Dodelson, X. Dong, E. Silverstein, and G. Torroba, New solutions with accelerated expansion in string theory (2013).

90. M. Alishahiha, A. Karch, E. Silverstein, and D. Tong, The dS/dS correspondence, *AIP Conf.Proc.* **743**, 393–409 (2005). doi: 10.1063/1.1848341.

91. B. Freivogel, Y. Sekino, L. Susskind, and C.-P. Yeh, A Holographic framework for eternal inflation, *Phys.Rev.* **D74**, 086003 (2006). doi: 10.1103/PhysRevD. 74.086003.

13

Dark Energy in String Theory

Brian Greene

Department of Physics and Department of Mathematics,
Columbia University, New York, NY 10027, USA

Gary Shiu

Department of Physics, University of Wisconsin,
Madison, WI 53706, USA
Center for Fundamental Physics & Institute for Advanced Study,
Hong Kong University of Science and Technology, Hong Kong

We review the tension between the observational evidence for dark energy and various theoretical considerations. This tension has motivated a reconsideration of the issue of naturalness, and spawned various exotic approaches toward an acceptable solution. We discuss attempts to realize dark energy in string theory, and the perspective on the string landscape that these results have suggested.

1. Introduction

1.1. *History of the cosmological constant*

The cosmological constant has a rich if checkered history. Its story began in the years following Einstein's completion of his General Theory of Relatiivy, in 1915. Shortly after this monumental achievement, Einstein applied the equations of General Relativity to the universe as a whole, and came upon an unexpected implication. The equations did not admit a solution in which the scale factor of the universe was constant. Instead, the scale factor would either have to increase or decrease over time, which would mean that space itself would be either stretching or contracting.

The overwhelming prejudice at the time was that on the largest of scales, the universe was static, eternal, and unchanging. All familiar astrophysical dynamics was deemed small scale motion, which on the largest of scales would average out to no change at all. Adhering to this perspective, Einstein modified the equations of General Relativity to allow for a static solution. From a technical standpoint, the modification itself was minimal and thoroughly reasonable. The Einstein field equations are nonlinear partial differential equations that involve terms that are first and second order differentials of the spacetime metric

$$R_{\mu\nu} - (1/2)g_{\mu\nu}R = 8\pi GT_{\mu\nu}. \tag{1}$$

Notice, however, that these equations leave out the sole term that would have the appropriate symmetries under spacetime differeomorphisms and be a zeroth differential of the metric–a term proportional to the metric itself, $g_{\mu\nu}$. Including this term, and calling the constant of proportionality Λ, the modified field equations become

$$R_{\mu\nu} - (1/2)g_{\mu\nu}R + \Lambda g_{\mu\nu} = 8\pi GT_{\mu\nu}. \tag{2}$$

Einstein called Λ the *cosmological member* but over time has acquired the moniker *cosmological constant*.

What is the physcial significance of Λ? There are two ways of answering, corresponding to the two ways of including Λ in the field equations: on the left hand side and on the right hand side. On the left hand side — the geometrical side of the equations — Λ is interpreted as a property of spacetime, a uniform intrinsic tension in the spacetime fabric. If written on the right hand side — the source side of the equations — Λ is interpreted as a homogenous contribution to the mass-energy whose equation of state is $p = -\rho$.

Of course, the physical implications of Λ are independent of the interpretation one chooses, and physicists routinely switch between the two, dictated by context and utility. For our purposes, the most useful equation involving Λ is the Friedmann equation for the scale factor of a Robertson-Walker universe with constant curvature spatial sections whose metric is ds^2:

$$ds^2 = -dt^2 + a(t)^2 ds^2. \tag{3}$$

The Friedman equations are then:

$$H^2 \equiv \left(\frac{\dot{a}}{a}\right)^2 = \frac{8\pi G}{3}\rho_{total} - \frac{k}{a^2} \tag{4}$$

$$\frac{\ddot{a}}{a} = -\frac{4\pi G}{3}\left(\rho_{total} + 3p_{total}\right) \tag{5}$$

where $k = 0, \pm 1$ correspond respectively to flat, positively and negatively curved spatial sections, and we have incorporated Λ on the right hand side–as a source–by defining

$$\rho_{total} = \rho + \frac{\Lambda}{8\pi G} \tag{6}$$

and

$$p_{total} = p - \frac{\Lambda}{8\pi G}. \tag{7}$$

From these equations we can immediately see Einstein's motivation for introducing Λ: If k = 1 (uniform constant positive curvature for the spatial sections, which can be taken to be three spheres, S_3), Λ allows the Friedmann equations to yield a static solution (with non-relativistic matter):

$$\rho = \frac{\Lambda}{4\pi G} , \qquad a(t) = \frac{1}{\sqrt{\Lambda}}. \tag{8}$$

This is Einstein's Static Universe proposal of 1917. Physically, we see that in the Friedmann equations positive Λ acts as an opposing gravitational force to that of ordinary matter and radiation, which is why Λ is sometimes colloquially described as generating "anti" or "repulsive" gravity. Einstein's static solution relies on a cancellation between these two flavors of gravitational force.

Most descriptions of Einstein's Static Universe note that in 1929 the astronomical observations of Edwin Hubble established that the universe is expanding, and with that Einstein was convinced that the static universe he sought was not relevant to reality. And so, he retracted the cosmological constant, citing it as one of his greatest blunders. However, there are two points worthy of emphasis.

First, even before the observations of Hubble, it was clear that the Static Universe had a fatal flaw. The balancing act between attractive and repulsive gravity is unstable. Perturbing the Friedmann equation around the static solution reveals this directly, but a simple physical argument shows it too. If the radius of the static universe is made a touch larger, attractive gravity will decrease since it is acting over larger distances, but the cosmological constant's repulsive gravity — being constant — will be unchanged. Thus, the repulsive push will now win out, causing the universe to expand unabated. If, on the other hand, the radius of the static universe is made a touch smaller, attractive gravity will increase since it is acting over shorter distances, but the cosmological constant's repulsive gravity —

being constant — will be unchanged. Thus, the repulsive push will now lose, causing the universe to contract unabated.

Second, even with such instabilities, and even with the observations of Hubble, there was no reason for Einstein to retract the cosmological constant. It no longer made any sense to invoke the cosmological constant to yield a static universe, but as the field equations involve first and second order derivatives of the metric, there is no *a priori* reason to not include the term proportional to the zeroth order derivative — the cosmological term. Indeed, Einstein himself was aware of this, noting in a letter that the true fate of the cosmological constant would await future astronomical observations. As is well known, and as we will note below, Einstein's evaluation of the cosmological constant was prescient. In 1998, observations of distant supernovae brought the cosmological constant back into mainstream physics.

1.2. *The cosmological constant problem*

Keeping to the historical timeline, by the early 1930s the need for a cosmological constant had vanished. Yet, as quantum mechanics and quantum field theory rapidly developed, it became clear that the cosmological constant would not be vanquished quite so easily.

The simplest quantum mechanical system, the harmonic oscillator, reveals the issue. Classically, the lowest energy state of a harmonic oscillator is the static configuration of the oscillator sitting at the lowest point in its potential. Quantum mechanically, the lowest energy state is quite different. Due to the Heisenberg Uncertainty principle, the oscillator can't have a definite location and a definite speed, and hence the classical zero energy configuration is simply not allowed. Instead, the lowest energy configuration has a non-zero energy, which is generally called the zero-point energy. A simple calculation shows that the value of the zero point energy is the famous factor for $\hbar\omega/2$ where ω is the oscillator's angular frequency. The import of this basic result for the cosmological constant becomes apparent when we generalize it to quantum field theory.

A (free) quantum field is nothing but an infinite collection of oscillators, one for each of the infinite possible wavelengths of the field fluctuations. Which means that the zero point energy of a quantum field is, formally, infinite. This yields a constant, but formally infinite energy density. The general expectation is that quantum field theory is an appropriate physical description on sufficiently long wavelengths, motivating a short distance cutoff which, to be definite, can be taken at the Plank scale, $l_p = 10^{-33}$ cm. With this cutoff, the constant energy density is on the order of 1 in Planck units, an enormous value compared with the observational limit of roughly 10^{-29} grams/cm^3, which is about 10^{-122} in Planck units. So, while astronomical observations led most everyone to anticipate that the cosmological constant was zero, quantum considerations suggested a vastly different from value.

The tension between observation and theoretical considerations becomes only more strained as we consider detailed models of particle physics. A key component of modern particle theory is symmetry breaking — a process in which the universe undergoes a phase transition that discontinuously changes the manifest symmetry group of the equations of motion. Such transitions are typically accompanied by a change in the energy of the local vacuum state the universe occupies, a process that in effect changes the value of the cosmological constant.

So, the cosmological constant we experience today arises from the combined influence of zero-point energies associated with each of nature's quantum fields, the effect of all phase transitions the universe undergoes, as well as whatever primordial (bare) cosmological constant the universe may have had at the outset. That these contributions, from disparate sources, should somehow cancel each other out, yielding a net value of zero for the cosmological constant, is the the puzzle that has become known as the "cosmological constant problem".

1.3. *Accelerated expansion*

For decades, physicists sought an explanation for why all the sources for a vacuum energy filling space would conspire to yield the combined value zero. The problem seemed difficult, but far from hopeless, because zero is a special value. One can at least imagine that some kind of symmetry principle or some kind of hidden relationship between the various sources would explain their cancellation. Many such proposals were put forward, although none was fully convincing, and hence the problem remained unresolved.

In 1998, the cosmological problem took on a decidedly different character. In that year, two teams of astronomers, who had been carefully measuring the rate at which the expansion of the universe was slowing over time announced a shocking result: the expansion is not slowing down. It is speeding up. The expansion is accelerating. In the years since, confidence in this result has only increased, requiring theorists to explain the outward force driving the accelerated growth of the scale factor. And the explanation which best aligns with the data is that the outward force is the repulsive push of a non-zero cosmological constant. In a sense, then, even Einstein's biggest "blunder" has turned out to be right.

Note however that while the resurrection of the cosmological constant affirms Einstein's conceptualization of this term in his equations, the value of the cosmological constant necessary to explain the accelerated expansion is far from the value Einstein invoked in developing his static model of the cosmos. In fact, the value of the cosmological constant suggested by the astronomical observations raises an even more acute version of the cosmological constant problem.

Namely, the data point to a value of Λ that in Planck units is about 10^{-122}. A fantastically tiny value when expressed in these natural units, but decidedly nonzero. Earlier we noted that explaining a value of zero for Λ posed a significant challenge, but one that theorists could attack with tools familiar from their standard mathematical arsenal – zero is value that in many other contexts emerges from and is protected by considerations of symmetry. But such an approach seems powerless to explain a fantastically tiny but non-zero value.

Which is not to say that over the past decade and a half theorists haven't tried. There is a vast literature of proposed explanations.[3] Some attempt to directly justify $\Lambda \sim 10^{-122}$. Others have been inspired by the fact that a vacuum energy of 10^{-122} is on par with the energy content of the universe's matter content, thus imagining dynamical values of vacuum energy that track the energy density of matter. It is fair to say that none of the proposals are sufficiently convincing to have risen to a point of prominence and consensus. Instead, there is a widespread opinion that explaining the origin of the accelerated expansion, in quantitative detail, ranks among the deepest problems faced by theoretical physics.

The uncertainty regarding the origin and identity of the invisible vacuum energy responsible for the accelerated expansion has led scientists to call it "dark energy," and the current failure to explain the dark energy and calculate its value, is known as the dark energy problem.

2. Anthropic Considerations

2.1. *Brief history of anthropic thinking*

When a problem in physics (and science more generally) stubbornly resists being solved, that often reflects that the problem is difficult. But sometimes it reflects that physicists are asking the wrong question, trying to solve the wrong problem. Centuries ago, Johannes Kepler sought a first principles explanation of the distance between the Earth and the Sun. Decades of work resulted in no progress. Did that mean Kepler was tackling an extremely difficult problem? In this case, the answer is no. Instead, the other explanation — that Kepler was asking the wrong question — comes to the fore.

Once Isaac Newton came on the scene of physics, it became clear that the distance between a planet and the star it orbits is not some fundamental constant of nature. Rather, from the standpoint of physics, planets can orbit their host star at essentially any distance. In any particular case, such as the Earth-Sun system, the distance reflects the historical development of the orbital system, and is thus not a question that is open to the kind of explanation Kepler sought. Which means that Kepler made no progress in explaining the Earth-Sun distance because he was asking the wrong question. Planets can be, and are, at a huge variety of distances from their host star. Given the intrinsic details of the Earth and the Sun, physics alone can not delineate the size of the orbit.

The question Kepler should have been asking is this: Of all the possible distances that planets might be from the Sun, why do we humans find ourselves on a planet that's at a distance of 93 million miles? And that's a question which has a sharp answer. At distances significantly different from 93 million miles, the temperature is either too hot or too cold for liquid water. And without liquid water, our form of life won't take hold.

The move from the question Kepler asked to the one we've just answered is a hallmark example of what's come to be known as "anthropic reasoning". Once one realizes that a feature of the natural world is not uniquely determined by physics (such as the radius of a planetary orbit) and once one realizes that the feature does in fact take on a wide variety of different values (planets orbit their host star at many different radii), we should shift our attention from a first principles analysis of the value we experience to describing it as a selection effect. For the feature in question, we experience the value we do because other values are incompatible with our existence.

Anthropic reasoning inspires heated debate, but most of the heat is hot air. When the condititions described in the previous paragraph hold, there can be no question that reasoning via selection effects is justified. This does not preclude the existence of other, non-selection effect explanations that may offer deeper or more satisfying insights. But anthropic reasoning — that is, taking account of selection effects — is beyond reproach.

To be clear, let's reiterate that anthropic reasoning requires that two conditions hold: (a) the physical feature which we seek to understand must assume a wide range of values across the landscape of reality and (b) most of the possible values (except for the value we experience) must be incompatible with our form of life. Attempts to use anthropic reasoning in circumstances for which either or both of these conditions do not hold (usually (a)) have led some to believe that anthropic reasoning is circular. But when applied appropriately, anthropic reasoning is unassailable.

2.2. Anthropic approaches to the cosmological constant

Long before the observations of accelerated expansion and the proposed explanation invoking a tiny presence of dark energy throughout space, the classic cosmological problem already posed a significant puzzle. As mentioned above, why would all the disparate contributions to the cosmological constant exactly cancel, yielding a net result of zero? In the mid-1980s, Steven Weinberg took a novel approach. Perhaps, he reasoned, the many years of failed attempts to solve the cosmological problem indicated that, much like Kepler, we've been asking the wrong question. Instead of a first principles explanation, perhaps an anthropic approach is called for.

How would that go? Well, first off, we would need to imagine that our universe, with its particular value of the cosmological constant, is but one of many universes which possess cosmological constants of differing values. And second, we'd need to be sure that most values of the cosmological constant would create conditions incompatible with our form of life. As Weinberg pointed out, this appears to be the case: most values of the cosmological constant are incompatible with the formation of galaxies, a prerequisite for having habitable planetary systems. The reason is that the clumping of dust into galactic seeds is easily disrupted by the repulsive force of a cosmological constant that is too large and positive; the longevity of a universe would be too short for galactic formation if it possessed a cosmological constant that is too large and negative.

Weinberg took this idea one step further and noted that these considerations don't favor a vanishing cosmological constant per se, but rather a cosmological constant whose value is as large as it could be without crossing into the uninhabitable zone. If the cosmological constant in our universe were far smaller than this distinguished value, anthropic reasoning would be ineffective at explaining its value: the mystery of why it was far smaller than it could be, given our own existence, would remain in full force. Weinberg worked out the distinguished value[4] and found it to be on the order of 10^{-121} (in Planck units). This calculation was in 1987, a decade before the observations of accelerated expansion gave the first evidence for dark energy, and also showed that the value was on par with Weinberg's prediction.

A number of authors have pointed out subtleties that afflict the power of Weinberg's argument (most notably, Weinberg considered universes distinguished solely by the difference in the cosmological constant values, but if other features can also vary, the conclusions are modified), but it launched an anthropic perspective on the cosmological contant that has been developed by many in the years since. One key point of the argument, though, is worth calling out. The number of universes with distinct values of the cosmological constant needs to be sufficiently large to ensure that the spectrum of values includes the one we've measured. How large? Focusing attention of cosmological constant values whose magnitude doesn't exceed one in Planck units, we'd need, say, at least 10^{121} distinct universes, with randomly distributed cosmological constant values, to ensure that a universe with our measured value would be represented.

Is there any natural way for a theory to generate a multiverse with such a vast assortment of constituent universes? According to string theory, the answer is yes.

3. String Theory and Dark Energy

The dark energy problem really only arises in the context of quantum gravity. This is because the zero of the vacuum energy can be chosen arbitrarily in the absence of gravity, and furthermore its value can be tuned at will classically. It is therefore natural to address this puzzle in a concrete framework of quantum gravity, such as string theory. In this section, we will review attempts to realize dark energy in string theory, and new twists on the anthropic reasoning that the string theory landscape might suggest.

3.1. Brief history of string compactifications

As it turns out, realizing dark energy in string theory also presents new hurdles. Overcoming these hurdles requires one to confront several vexing issues in string compactifications, such as moduli stabilization and supersymmetry breaking. It was not until recent years that approaches to address such central issues in string phenomenology[5] have been sufficiently developed to be carried out in a controllable and calculable way.

It is an old idea that unification of the fundamental forces in Nature may be rooted in the existence of extra dimensions. This idea which dated back to the works of Kaluza and Klein[1,2] reemerged around the 1970s-1980s with the discovery of string theory as a consistent quantum theory of gravity. Like the original Kaluza-Klein idea, string theory (at least in its known weak coupling limits) is formulated in higher dimensions. However, what distinguishes string theory from other Kaluza-Klein theories is that the number (and type) of extra dimensions is not arbitrary but rather dictated by quantum consistency.

While string theory helps limit the choice of extra dimensions, we are still faced with many possibilities. The size and shape of the extra dimensions can vary over four-dimensional space-time in a way consistent with the equations of motion. Thus, they describe four-dimensional fields known as *moduli*, so called because classically their potential energy function admits continuous families of solutions. Each solution describes a different universe with different fundamental parameters (e.g., the values of Newton's constant or gauge couplings). Without a mechanism for "lifting" these flat directions (i.e., giving the moduli a sufficiently heavy mass), these moduli lead to unacceptable phenomena such as new long range forces, or time dependence of fundamental parameters, in direct conflict with observations. Moduli stabilization is therefore a problem with two parts: how do we give masses to the moduli and why the fundamental parameters so obtained match the values we observe.

Another and related problem is supersymmetry (SUSY) breaking. While there are many phenomenological reasons to consider four-dimensional theories with low energy $\mathcal{N} = 1$ SUSY (to name a few, the Higgs hierarchy problem, gauge unification, and dark matter), this symmetry, if realized in nature, must be broken in the current state of our universe. The discovery of dark energy, as described above, adds another layer of complication: not only is our current universe in a SUSY breaking phase, observations require it to be a (meta)stable vacuum (or in the case of time varying dark energy, so-called quintessence) with a tiny positive vacuum energy. This is a tough order for the underlying string compactification and the mechanism used to stabilize the moduli.

In string theory, the way SUSY emerges has more to do with the theoretical control it provides. In compactifications that preserve $\mathcal{N} = 1$ supersymmetry at the Kaluza-Klein (KK) scale, the computation of the four dimensional effective Lagrangian is greatly simplified as powerful physical and mathematical tools can be brought to bear. Moreover, it is generally easier to prove metastability for a SUSY breaking state in a supersymmetric theory. These arguments for supersymmetry may appear to be more a matter of convenience, and indeed it has recently been advocated (see e.g., Ref. 6) that models with broken supersymmetry at the string scale can be equally brought under control as long as the vacuum is stabilized in the weak coupling regime. Nonetheless, much of the effort in string compactifications has been focussed on scenarios with unbroken $\mathcal{N} = 1$ SUSY at the KK scale, with SUSY broken by dynamical effects at low energy.

Since the mid 1980s, it has been known that string theory admits several weak coupling formulations. Four of these formulations are theories of closed strings, known as type IIA, type IIB, heterotic $E_8 \times E_8$, and heterotic $SO(32)$. In addition, type I string theory provides a formulation that involves both open and closed strings. The above five string theories are all formulated in ten dimensions, and admit low energy limits described by known supergravity theories (this fact will be important later on). The type II theories have 32 supercharges and do not at first sight include non-Abelian gauge sectors. One can obtain $\mathcal{N} = 1$ supersymmetric theories upon reducing the type II theories on non-geometric backgrounds (e.g., asymmetric orbifolds[7]), but worldsheet conformal field theory arguments demonstrated that the Standard Model cannot be obtained in this manner.[8] The other three string theories have 16 supercharges and include non-Abelian gauge sectors. Geometrical spaces that preserve $d = 4$, $\mathcal{N} = 1$ supersymmetry at the KK scale are manifolds of $SU(3)$ holonomy. Metric spaces with $SU(3)$ holonomy are known as Calabi-Yau manifolds, whose existence was proved in Ref. 9. Calabi-Yau manifolds are Ricci-flat and so serve as a good starting point for solving the supergravity equations of motion. Substantial advances have been made in constructing realistic vacua (i.e., those that can accommodate the Standard Model of particle physics and its extensions) from Calabi-Yau compactifications of heterotic string theories[a] (see e.g, Ref. 5 and also some reviews written in that pe-

---

[a]Relatively less effort was made on the Type I front in that period, see, e.g., Ref. 10 for a collection of papers on realistic string compactificaflons up to the late 1980s.

riod[11-13]).

3.2. Discovery of D-branes and flux compactifications

The above summarizes the state of affairs until the mid 1990s, when the second string theory revolution provided new insights into a non-perturbative formulation of string theory. The key observation is that string theory is not merely a theory of closed or open strings but also contains in its non-perturbative sector extended objects of higher dimensions, known as branes. Dirichlet branes (D-branes for short) defined as boundary conditions for open strings are a particularly simple description of these dynamical extended objects. They provide key links in the web of dualities between the aforementioned weak coupling descriptions of string theory as well as eleven dimensional supergravity. The different formulations of string theory are now understood as different "phases" of a fundamental theory known as M-theory.

D-branes are charged under certain massless fields appearing in the Ramond-Ramond (R-R) sector of the ten-dimensional type IIA/B string theories, and couple to those fields much as an electron couples to the photon. More specifically, a p-brane is an extended object with p space-like directions and one time-like direction and it couples to a $(p+1)$ form potential A_{p+1} as follows:

$$Q_p \int_{D_p} A_{p+1}. \tag{9}$$

The analogy with electromagnetism goes even further. If we compactify string theory on a manifold M with a non-trivial homology group $H_{p+2}(M)$ with Σ being a non-trivial element of homology, we can consider a configuration with a non-zero flux (satisfying the Dirac quantization condition):

$$\int_\Sigma F_{p+2} = n \in \mathbb{Z}. \tag{10}$$

As in electromagnetism, turning on a field strength costs energy, proportional to F^2. Because the fluxes are threading cycles in the compact geometry, this energetic cost will depend on the precise choice of metric on the internal space. To see this explicitly,

$$V = \int_M F_{p+2} \wedge (*F_{p+2}) \tag{11}$$

where the metric enters in the definition of $*$. Thus a potential on the moduli space is developed. Minimizing this potential (provided that it is sufficiently generic) will fix the moduli.

While this idea is general and applies to both type II theories, it is most explicitly expressed in the framework of type IIB string theory[14-17] (see Refs. 20–22 for some reviews) where a superpotential[23,24] $W = \int (F_3 - \tau H_3) \wedge \Omega$ is generated by the NS-NS 3-form H_3 and R-R 3-form F_3 flux background (τ is the axio-dilaton). There has been some success in constructing Standard Model like vacua in this context.[18,19] In type IIB theory, fluxes alone can only stabilize the complex structure moduli and the axio-dilaton, leaving the Kahler moduli unfixed (a result of the no-scale structure of the supergravity). Stabilization of the remaining moduli invokes additional quantum corrections as e.g., in Refs. 25, 26. In type IIA string theory, however, fluxes alone are sufficient for stabilizing all the closed string moduli.[27]

Moreover, depending on how the gauge and matter sectors are embedded, SUSY can be softly broken by the supergravity flux with soft terms generated in a calculable way.[28-33] In applying these results on flux-induced soft terms, one should ensure that there are no additional moduli stabilization contributions that could restabilize the vacuum to a supersymmetric minimum, or sizable effects from "uplifting" (to be defined later).

3.3. *No-go theorem for de Sitter space*

Moduli stabilization and SUSY breaking are necessary conditions for de Sitter vacua to arise from string compactifications but they are not sufficient conditions. In fact, under fairly general assumptions, one can prove that there are no non-singular de Sitter vacua in gravity theories.[34-36] Here we follow closely the arguments for the no-go theorem presented in Ref. 36.

Consider a D-dimensional ($D > 2$) gravity theory compactified to d dimensions, $ds_D^2 = \Omega^2(y) \left(dx_d^2 + \hat{g}_{mn} dy^m dy^n\right)$ satisfying the conditions:

(1) The gravity action does not contain higher curvature corrections.
(2) The potential is non-positive, $V \leq 0$[b].
(3) The theory contains massless fields with positive kinetic terms.
(4) The d-dimensional effective Newton constant is finite.

Taking the trace of the D-dimensional Einstein equations (which, due

---

[b]This condition is violated in massive IIA supergravity which has a positive cosmological constant. This case is therefore treated separately.

to (1), are the equations of motion):

$$\frac{1}{(D-2)\Omega^{D-2}}\nabla^2\Omega^{D-2} = R_d + \Omega^2(-T_\mu^\mu + \frac{d}{D-2}T_L^L) \tag{12}$$

where $\mu = 0, \ldots, d$ and $L = 0, \ldots, D$. (2) and (3) imply that the second term on the right hand side is non-negative. If $R_d \geq 0$, $\Omega^{D-2}\nabla^2\Omega^{D-2} \geq 0$. Integrating over a compact manifold (volume is finite because of (4)), we conclude that $\int \sqrt{g}(\hat{\nabla}\Omega^{D-2})^2 \leq 0$ which is only possible if Ω is constant and $R_d = 0$, thus no de Sitter solution.

3.4. Evading the no-go theorem

An implicit assumption that went into the above no-go theorem is the absence of localized sources. However, string theory is full of localized sources, e.g., D-branes, which can give additional contributions to the stress tensor. The no-go theorem can be evaded if the localized sources give (for the interesting case of $d = 4$, $D = 10$):

$$\left(T_m^m - T_\mu^\mu\right)^{loc} < 0 , \qquad \mu = 0, \ldots, 3 \text{ and } m = 4, \ldots, 10. \tag{13}$$

For example, in type IIB compactifications whose $4d$ $\mathcal{N} = 1$ SUSY is of the type preserved by $D3/D7$ branes, the following inequality

$$\frac{1}{4}\left(T_m^m - T_\mu^\mu\right)^{loc} \geq T_3\rho_3^{loc} \tag{14}$$

is *saturated* by D3 branes and O3-planes, and by D7-branes wrapping holomorphic cycles. ($\overline{D3}$ satisfies but not saturates it). Orientifold planes carry negative tension and thus provide a key ingredient to evade the no-go result.

Condition (1) is also expected to be violated in string theory. In fact, α' corrections have been employed in moduli stabilization scenarios, e.g., Ref. 26 though orientifold planes are also present in such constructions.

3.5. KKLT, LVS, classical de Sitter, and beyond

Since the discovery of dark energy, various scenarios to construct de Sitter vacua from string theory have been proposed. These scenarios can be broadly divided into two types: those that invoke quantum effects, and those within classical supergravity but which are supplemented by localized sources.

The prototypical scenarios of the first type include the so-called KKLT construction[37] and Large Volume Scenario (LVS).[26] The no-scale structure that leaves the Kahler moduli in type IIB supergravity unfixed is expected to be broken by quantum effects (in g_s and α'). The superpotential is only corrected non-perturbatively while the Kahler potential receives both perturbative and non-perturbative corrections. Schematically, we can write:

$$W = W_0 + W_{np} \tag{15}$$

$$K = K_0 + K_p + K_{np} \tag{16}$$

where W_0 is the classical superpotential from fluxes. The KKLT scenario corresponds to $W_0 \ll 1$ (in which case corrections to the Kahler potential are subdominant) whereas the LVS corresponds to $W_0 \approx 1$ (the gravitino mass is suppressed by large volume). Quantum corrections to the Kahler potential are difficult to compute. While some α' and g_s corrections to the Kahler potential of 4d $\mathcal{N} = 1$ type IIB compactifications have been derived in the last years (see, e.g., Refs. 38–42), a complete understanding is still lacking (though exact results were obtained, albeit for $\mathcal{N} = 2$ vacua, in[43]). It has been argued[26] that the leading corrections to the Kahler potential of relevance to the stabilization of Kahler moduli is that of Ref. 44, $K \supset -2\ln(\mathcal{V}_E - \frac{\chi(M)\zeta(3)}{4(2\pi)^3 g_s^{3/2}})$ where $\chi(M)$ is the Euler characteristic of the compactification.

Regardless of whether $W_0 \ll 1$ (KKLT) or $W_0 \approx 1$ (LVS), the quantum effects which stabilize the Kahler moduli generically lead to anti de Sitter minima. Thus, additional uplifting mechanisms are invoked to lift these minima to de Sitter vacua. In Ref. 37, an anti-brane is introduced at the tip of a highly warped throat within the compactification. The anti-brane provides a positive-definite contribution to the scalar potential and the warping can be adjusted to make the vacuum energy to tunably small and positive. It is worthwhile noting that finding explicit backreacted solutions of anti-branes in warped backgrounds (see e.g., Refs. 46–57) and deriving warping corrections to the effective action[58–65] are subtle issues that are currently being actively investigated. Other uplifting mechanism include D-term uplift,[66–68] Kahler uplift,[69–72] and the inclusion of dilaton-dependent non-perturbative effects on the superpotential.[73] The latter two mechanisms exploit the interplay of several quantum effects to attain metastable de Sitter minima.

Another class of de Sitter constructions are known as *classical de Sitter solutions*.[74–77] The motivation for considering such classical constructions is multifold. Quantum effects are difficult to compute explicitly in $\mathcal{N} = 1$ string compactifications. For example, the moduli stabilizing instanton effects which stabilize the Kahler moduli can in principle depend non-trivially on the complex structure as well as open string moduli. One has to invoke a hierarchy of scales^c in order to safely neglect the complicated moduli dependence. It is certainly desirable to find a class of string compactifications where the moduli stabilization potential can be explicitly computed. Moreover, the no-go theorem discussed above[34–36] does not exclude the possibility of classical de Sitter solutions (as long as appropriate localized sources are present). As stabilized anti de Sitter vacua were shown to arise in type IIA flux compactification,[27] it is natural to wonder if de Sitter vacua can be obtained in a similar context. As it turns out, there are further no-go theorems[75–79] that restrict the appearance of classical de Sitter solutions.

These additional no-go theorems follow from the scaling behavior of different scalar potential contributions with respect to the two orthogonal universal moduli $\rho = (\mathrm{vol}_6)^{1/3}$ and $\tau \equiv e^{-\phi/2}\sqrt{\mathrm{vol}_6}$ that occur in any perturbative and geometric string compactification. Here vol_6 and ϕ are the 6D internal volume in the string frame and ϕ is the 10D dilation. In the 4D Einstein frame, the contributions from the NS-NS flux H_3, RR flux F_p, O_q planes and the 6D Ricci scalar R_6 scale as:

$$V_H \sim \tau^{-2}\rho^{-3}, \quad V_{F_p} \sim \tau^{-4}\rho^{3-p} \quad V_{O_q} \sim \tau^{-3}\rho^{\frac{q-6}{2}}, \quad V_{R_6} \sim \tau^{-2}\rho^{-1}. \quad (17)$$

These scalings in particular forbid the existence of de Sitter critical point in IIA flux compactification on Calabi-Yau orientifolds[79] (where $V_{R_6} = 0$). More generally, one can derive inequalities of the form $(a\tau\partial_\tau + b\rho\partial_\rho)V \geq cV$ where a, b, c are real constants and $c > 0$. These inequalities point to the *minimal* ingredients for constructing classical de Sitter extrema.[75,76] Absence of tachyonic instabilities puts additional necessary conditions on the Hessian of the potential in the (ρ, τ) directions.[80] Taken together these arguments naturally lead one to seek for SUSY breaking states in compactifications on $SU(3)$ structure manifolds.[76,77] As homogeneous spaces are more amendable analytically, much of the studies on explicit $SU(3)$ structure manifolds have been focussed on those that can be realized as group manifolds and coset spaces (see e.g. Refs. 81–84). A rather extensive search for de Sitter vacua in orientifold compactifications of $SU(3)$ structure group/coset manifolds has been carried out in Ref. 77 though all the de Sitter critical points found there turn out to have tachyons. We will return to discuss why tachyonic instabilities seem ubiquitous in these constructions in Section 4.

---

^{c}E.g., in IIB flux compactifications, the complex structure moduli are assumed to be hierarchically heavier than the Kahler moduli so the former can be treated as constants.

3.6. *Fluxes and the landscape*

Not only are fluxes a key ingredient in moduli stabilization, they also suggest an interesting view of the string theory landscape. Without going into details of the string construction (e.g., the precise geometry, fluxes, branes involved, etc.), one can already see how the notion of a *discretuum* emerges.

Recall that the fluxes are quantized $\int_\Sigma F_{p+2} \in \mathbb{Z}$. In a compact space, these flux quanta need to satisfy some Gauss's law constraints which state that the total charge on the compactification manifold, including D-branes, orientifold planes and all other sources must vanish. In an open string picture, these constraints are related to anomalies and thus are an integral requirement of a consistent string compactification. The essential point is that these Gauss's law constraints provide an upper bound on the flux quanta. For example, in type IIB compactification with 3-form fluxes, the flux quanta are bounded by the number of orientifold 3-planes in the vacua. Therefore, even for a *fixed* internal space topology, there are a discrete set of string theory solutions labeled by the choices of flux quanta. The number of discrete states can be estimated as follows. Let N be the number of moduli (typically of $\mathcal{O}(100)$ to $\mathcal{O}(500)$, and F_{bound} be the upper bound on the flux quanta (which we can take for typical compactifications to be of $\mathcal{O}(10 - 100)$), then the number of discrete choices amounts to the number of discrete flux quanta in a N dimensional flux sphere, i.e., $\#_{\text{discrete}} \sim F_{bound}^N \sim 10^{100}$ to $\sim 10^{500}$. Each choice corresponds to a different string theory solutions (i.e., different D-brane content, volume of internal cycles, etc). Thus they give rise to universes with a priori different gauge groups, matter content, and fundamental constants.

Note that the number of theories–distinct possible universes–is what we need for a successful anthropic explanation of the cosmological constant. In this sense, string theory naturally offers an anthropic solution fo the dark energy problem.

The above reasoning also suggests a way to realize an old proposal[85] in a string theory context[86] (see also Ref. 87). It has been suggested in Ref. 85 that the cosmological constant could be explained in terms of a space-filling 4-form flux which could dynamically neutralize itself (in small steps) by membrane nucleation. At first sight, this mechanism does not seem to work in string theory since the supergravity fluxes are quantized and thus the steps in which the cosmological constant can be changed would exceed the current bounds. For example, the necessary flux can be obtained in M-theory compactifications to 4-dimensions by reducing the 7-form flux in 11-dimensions on non-trivial 3-cycles of the internal manifold (similar arguments can be made for other formulations of string theory). It was observed in Ref. 86 that in a typical compactification, there are numerous such 3-cycles in the internal space and hence many different 4-form fluxes in the four noncompact dimensions. The net cosmological constant is then:

$$\Lambda = \Lambda_0 + \frac{1}{2} \sum_i n_i^2 q_i^2 \qquad (18)$$

where Λ_0 is the bare (negative) cosmological constant, $n_i^2 q_i^2$ are the contributions from n_i quanta of each kind of 4-form (for the M-theory example, $q_i = M_{11}^{3/2} V_{3,i} / \sqrt{V_7}$. We can find cosmological constants within acceptable bounds $|\Lambda| < \epsilon$ if there exists a point within the n_i charge lattice that lie within a distance $\sim \epsilon$ of a sphere of radius Λ_0 centered at $n_i = 0$. For sufficiently large number of moduli, this can be achieved with $q_i \sim \mathcal{O}(1)$, i.e., the compactification scale can be comparable to the fundamental scale.

3.7. Anthropic thinking 2.0

The large number of flux vacua in the string theory landscape reminds us of statistical physics where the systems of interest involve a large number of degrees of freedom. As in statistical physics where such large numbers make studying individual states impractical, we can try to draw statistical statements from an ensemble of string theory solutions chosen to reflect generic features of the microscopic physics. This brings us to the notion of *vacuum statistics* and *stringy naturalness* to which we now turn.

Take the cosmological constant as an example. An implicit assumption that went into Weinberg's argument is that universes within the anthropically allowed window are realizable in the microscopic theory. The finely-spaced discretuum provides a plausible explanation for this assumption.

Going further, though, we'd like to not only be assured that vacua with small values of the cosmological constant exist within the landscape, we'd like to have a sense of the likelihood or abundance of such vacua. To that end, a useful concept is the vacuum counting distribution:

$$dN_{vac}(T) = \sum_i \delta(T - T_i) \qquad (19)$$

where T_i denotes collectively the discrete (e.g., gauge groups, matter representations, etc) and continuous (e.g., gauge and Yukawa couplings, cosmological constant, density perturbation, SUSY breaking scale, etc.) parameters of the theory. It is important to distinguish the vacuum counting distribution dN_{vac} from the probability distribution function

$$d\mu_P(T) = \sum_i P(i)\delta(T - T_i) \tag{20}$$

where we assign a "probability factor" for each vacuum. Thus, the vacuum counting distribution encodes information about the set of consistent string vacua, not the probability of their occurrence. A certain choice of parameters T_i can be said to be *stringy natural* if there is a large number of vacua dN_{vac} with $T = T_i$. See, e.g., Refs. 88–92 for a statistical analysis of the scale of SUSY breaking in the string landscape.

More generally, though, any delineation of "natural" or "unnatural" vacua requires commitment to a measure on the space of vacua, and such a measure need not align with vacuum counts. As is thoroughly familiar from quantum mechanics, a direct count of the number of ways a certain outcome might happen (e.g. the number of combinations of up and down spins that respect a given net spin value) need not have any relationship to the likelihood of that configuration being observed (a probability that is governed by the system's wavefunction). To date, though, finding a prefered, first principles measure on the space of string vacua, has remained beyond reach.

Thus, as of now, we can use the landscape of string theory to argue for the existence of hospitable universes, but we can't say anything about how common or natural such universes may be.

4. Stability and the Landscape

The counting of flux vacua provides a first step towards charting the landscape of string theory. And, as above, our understanding of the vacuum structure is still far from complete. Beyond the question of a measure on the landscape, we outline here some additional outstanding issues.

4.1. *A closer look at the landscape*

The argument presented in Section 3 suggested a large number of string theory solutions in the flux landscape. However, this argument alone does not guarantee that we have a similarly large number of (meta)stable vacua, especially those with positive vacuum energy. Even if we lift all the flat directions and hence the solutions are critical points of a dimensionally reduced $4d$ supergravity, such critical points can be a local minimum, maximum, or saddle point. The large number of moduli which gives rise to the landscape also makes it harder to find critical points with a positive definite Hessian. Furthermore, even if we find solutions that are perturbatively stable, they are subject to quantum tunneling. The large dimensionality of the moduli space again leads to multiple decay channels which enhance the tunneling rate. Here, we take a closer look at these issues in the landscape.

4.2. Of minima and stationary points

Among the stationary points of a scalar potential, only those with a positive definite Hessian matrix are minima. In the absence of supersymmetry, the scalar potential can easily develop instabilities along one of its many field space directions. The scalar superpartner of the Goldstino is thus a potentially unstable direction[93] (see e.g. Ref. 94 for an extensive study of this "sGoldstino" direction in supergravities). Explicit constructions point to further potential tachyonic instabilities along some universal directions[80] as well as semi-universal ones (e.g., orientifold volume).[95]

However, when the number of moduli is large, direct examination of the Hessian matrix at stationary points of an effective supergravity potential becomes impractical. One is naturally led to a statistical approach, in which the compactification data are taken to be random variables.[93] This random matrix approach was recently adopted to analyze the stability of classical de Sitter solutions[96] and random supergravity.[97] These results (improving an earlier estimate in Ref. 93) suggest that the probability for the full mass matrix to be positive definite is exponentially suppressed by a factor $\exp(-cN^p)$ (to leading order in N) where N is the number of moduli in the scalar potential and p lies between 1.3 to 2 depending on whether the soft masses are small compared with the supersymmetric masses. It appears that the large number of moduli which leads to the picture of the landscape (i.e., an exponentially large number of string theory solutions) also predicts its own demise.

There are however ways out of this conclusion. For example, the suppression due to stability requirements can be reduced if a large subset of the moduli is decoupled by gaining large supersymmetric masses. The number of critical points is determined by the total number of fields while the suppression factor only depends on the number of light fields.

4.3. Tunneling and quantum stability

Vacuum decay by quantum tunneling is well studied in quantum field theory. In Refs. 98, 99, it was shown that the decay proceeded by the nucleation of bubbles of a lower vacuum inside the original false vacuum. In the semi-classical approximation the nucleation rate per unit volume is governed by a bounce solution of the Euclidean field equations which can be written in the form $\Gamma = Ae^{-B}$, where A depends on the determinant of fluctuations around the bounce solution and B is the Euclidean action of the bounce. The analysis was extended to include gravitational effects by Coleman and De Lucia.[100] For decay from a de Sitter vacuum the gravitational corrections to the nucleation rate are typically small unless the potentials are Planckian in scale or the bubbles nucleate with a size comparable to the horizon length. With unusually flat potential barriers it can happen that there is no Coleman-De Luccia bounce, but in such cases there is always a Hawking-Moss solution[101] corresponding to a process in which an entire horizon volume fluctuates to the top of the potential barrier.

There are several ways in which this picture can be modified in string theory. First of all, different vacua of string theory differ not only in their vacuum energies, but they can have vastly different $4d$ properties (e.g., gauge groups, matter content etc). The aforementioned results on vacuum decay and quantum tunneling in quantum field theories may not be directly applicable. In fact, it has been argued that since different string theory solutions can lead to globally different space-times and thus appear to correspond to different quantum Hamiltonians, rather that different states of a single quantum theory.[102] Secondly, the presence of multiple vacua (resonant tunneling)[103,105] and stringy effects (tunneling via the DBI action)[104,105] were shown to enhance the usual Coleman-De Lucia tunneling. Finally, it has been suggested[105] that tunneling in the landscape can be rapid because of the multiple decay channels made accessible by the large dimensionality of the moduli space. Specifically, by studying theories with multiple scalar fields, Ref. 108 provided numerical evidence that the rate for tunneling out of a typical false vacuum grows rapidly as a function of the number of moduli fields. As a consequence, the fraction of vacua with tunneling rates low enough to maintain metastability appears to fall exponentially as a function of the moduli space dimension. If such results prove applicable to string theory, the landscape of metastable vacua may not contain sufficient diversity to offer a natural explanation of dark energy.

4.4. Open questions

Driven by the dark energy problem, remarkable progress has been made in charting the vacuum structure of string theory and its moduli dynamics.

Yet many open questions remain. Is our landscape picture guided by the low energy supergravity description of string theory a correct one? Could a combination of statistical studies and explicit constructions point us to a more promising region of the landscape? Besides introducing orientifold planes, are there other stringy mechanisms to evade the no-go theorems for de Sitter vacua in string theory? What can we learn from the increasingly explicit constructions of de Sitter solutions the microphysics of de Sitter entropy? Making progress on any of these fronts would undoubtedly propel our understanding of quantum gravity.

Acknowledgments

We would like to thank Bobby Acharya, Gordon Kane, and Piyush Kumar for their invitation to contribute to this volume. BG is supported in part by DOE grant DE-FG02-92ER40699. GS is supported in part by DOE grant DE-FG-02-95ER40896.

References

1. T. Kaluza, Sitzungsber. Preuss. Akad. Wiss. Berlin (Math. Phys.) **1921**, 966 (1921).
2. O. Klein, Z. Phys. **37**, 895 (1926) [Surveys High Energ. Phys. **5**, 241 (1986)].
3. S. Weinberg, Rev. Mod. Phys. **61**, 1 (1989).
4. S. Weinberg, Phys. Rev. Lett. **59**, 2607 (1987).
5. L. E. Ibanez and A. M. Uranga, "String theory and particle physics: An introduction to string phenomenology," Cambridge, UK: Univ. Pr. (2012).
6. E. Silverstein, hep-th/0405068.
7. K. S. Narain, M. H. Sarmadi and C. Vafa, Nucl. Phys. B **288**, 551 (1987).
8. L. J. Dixon, V. Kaplunovsky and C. Vafa, Nucl. Phys. B **294**, 43 (1987).
9. S. -T. Yau, Proc. Nat. Acad. Sci. **74**, 1798 (1977).
10. B. Schellekens, Amsterdam, Netherlands: North-Holland (1989) 514 p. (Current physics - sources and comments
11. M. Dine, AMSTERDAM, NETHERLANDS: NORTH-HOLLAND (1988) 476 P. (CURRENT PHYSICS - SOURCES AND COMMENTS; 1)
12. B. R. Greene, "Lectures on string theory in four-dimensions," CLNS-91-1046.
13. Z. Kakushadze, G. Shiu, S. H. H. Tye and Y. Vtorov-Karevsky, Int. J. Mod. Phys. A **13**, 2551 (1998) [hep-th/9710149].
14. K. Dasgupta, G. Rajesh and S. Sethi, JHEP **9908**, 023 (1999) [hep-th/9908088].
15. B. R. Greene, K. Schalm and G. Shiu, Nucl. Phys. B **584**, 480 (2000) [hep-th/0004103].

16. G. Curio, A. Klemm, D. Lust and S. Theisen, Nucl. Phys. B **609**, 3 (2001) [hep-th/0012213].
17. S. B. Giddings, S. Kachru and J. Polchinski, Phys. Rev. D **66**, 106006 (2002) [hep-th/0105097].
18. F. Marchesano and G. Shiu, Phys. Rev. D **71**, 011701 (2005) [hep-th/0408059].
19. F. Marchesano and G. Shiu, JHEP **0411**, 041 (2004) [hep-th/0409132].
20. M. Grana, Phys. Rept. **423**, 91 (2006) [hep-th/0509003].
21. M. R. Douglas and S. Kachru, Rev. Mod. Phys. **79**, 733 (2007) [hep-th/0610102].
22. R. Blumenhagen, B. Kors, D. Lust and S. Stieberger, Phys. Rept. **445**, 1 (2007) [hep-th/0610327].
23. S. Gukov, C. Vafa and E. Witten, Nucl. Phys. B **584**, 69 (2000) [Erratum-ibid. B **608**, 477 (2001)] [hep-th/9906070].
24. T. R. Taylor and C. Vafa, Phys. Lett. B **474**, 130 (2000) [hep-th/9912152].
25. S. Kachru, R. Kallosh, A. D. Linde and S. P. Trivedi, Phys. Rev. D **68**, 046005 (2003) [hep-th/0301240].
26. V. Balasubramanian, P. Berglund, J. P. Conlon and F. Quevedo, JHEP **0503**, 007 (2005) [hep-th/0502058].
27. O. DeWolfe, A. Giryavets, S. Kachru and W. Taylor, JHEP **0507**, 066 (2005) [hep-th/0505160].
28. P. G. Camara, L. E. Ibanez and A. M. Uranga, Nucl. Phys. B **689**, 195 (2004) [hep-th/0311241].
29. M. Grana, T. W. Grimm, H. Jockers and J. Louis, Nucl. Phys. B **690**, 21 (2004) [hep-th/0312232].
30. D. Lust, S. Reffert and S. Stieberger, Nucl. Phys. B **706**, 3 (2005) [hep-th/0406092].
31. D. Lust, S. Reffert and S. Stieberger, Nucl. Phys. B **727**, 264 (2005) [hep-th/0410074].
32. P. G. Camara, L. E. Ibanez and A. M. Uranga, Nucl. Phys. B **708**, 268 (2005) [hep-th/0408036].
33. F. Marchesano, G. Shiu and L. -T. Wang, Nucl. Phys. B **712**, 20 (2005) [hep-th/0411080].
34. G. W. Gibbons, three lectures given at GIFT Seminar on Theoretical Physics, San Feliu de Guixols, Spain, Jun 4-11, 1984.
35. B. de Wit, D. J. Smit and N. D. Hari Dass, Nucl. Phys. B **283**, 165 (1987).
36. J. M. Maldacena and C. Nunez, Int. J. Mod. Phys. A **16**, 822 (2001) [hep-th/0007018].
37. S. Kachru, R. Kallosh, A. D. Linde and S. P. Trivedi, Phys. Rev. D **68**, 046005 (2003) [hep-th/0301240].
38. M. Berg, M. Haack and B. Kors, Phys. Rev. D **71**, 026005 (2005) [hep-th/0404087].
39. M. Berg, M. Haack and E. Pajer, JHEP **0709**, 031 (2007) [arXiv:0704.0737 [hep-th]].
40. M. Berg, M. Haack and J. UKang, JHEP **1211**, 091 (2012) [arXiv:1112.5156 [hep-th]].

41. M. Cicoli, J. P. Conlon and F. Quevedo, JHEP **0801**, 052 (2008) [arXiv:0708.1873 [hep-th]].

42. D. Junghans and G. Shiu, arXiv:1407.0019 [hep-th].

43. I. Garcia-Etxebarria, H. Hayashi, R. Savelli and G. Shiu, JHEP **1303**, 005 (2013) [arXiv:1212.4831 [hep-th]].

44. K. Becker, M. Becker, M. Haack and J. Louis, JHEP **0206**, 060 (2002) [hep-th/0204254].

45. O. DeWolfe, S. Kachru and M. Mulligan, Phys. Rev. D **77**, 065011 (2008) [arXiv:0801.1520 [hep-th]].

46. P. McGuirk, G. Shiu and Y. Sumitomo, Nucl. Phys. B **842**, 383 (2011) [arXiv:0910.4581 [hep-th]].

47. I. Bena, M. Grana and N. Halmagyi, JHEP **1009**, 087 (2010) [arXiv:0912.3519 [hep-th]].

48. I. Bena, G. Giecold, M. Grana, N. Halmagyi and S. Massai, Class. Quant. Grav. **30**, 015003 (2013) [arXiv:1102.2403 [hep-th]].

49. J. Blaback, U. H. Danielsson, D. Junghans, T. Van Riet, T. Wrase and M. Zagermann, JHEP **1108**, 105 (2011) [arXiv:1105.4879 [hep-th]].

50. I. Bena, G. Giecold, M. Grana, N. Halmagyi and S. Massai, JHEP **1306**, 060 (2013) [arXiv:1106.6165 [hep-th]].

51. A. Dymarsky, JHEP **1105**, 053 (2011) [arXiv:1102.1734 [hep-th]].

52. I. Bena, M. Grana, S. Kuperstein and S. Massai, Phys. Rev. D **87**, no. 10, 106010 (2013) [arXiv:1206.6369 [hep-th]].

53. F. F. Gautason, D. Junghans and M. Zagermann, JHEP **1309**, 123 (2013) [arXiv:1301.5647 [hep-th]].

54. J. Blaback, U. H. Danielsson, D. Junghans, T. Van Riet, T. Wrase and M. Zagermann, JHEP **1202**, 025 (2012) [arXiv:1111.2605 [hep-th]].

55. J. Blaback, U. H. Danielsson and T. Van Riet, JHEP **1302**, 061 (2013) [arXiv:1202.1132 [hep-th]].

56. S. Massai, arXiv:1202.3789 [hep-th].

57. D. Junghans, arXiv:1309.5990 [hep-th].

58. S. B. Giddings and A. Maharana, Phys. Rev. D **73**, 126003 (2006) [hep-th/0507158].

59. G. Shiu, G. Torroba, B. Underwood and M. R. Douglas, JHEP **0806**, 024 (2008) [arXiv:0803.3068 [hep-th]].

60. M. R. Douglas and G. Torroba, JHEP **0905**, 013 (2009) [arXiv:0805.3700 [hep-th]].

61. A. R. Frey, G. Torroba, B. Underwood and M. R. Douglas, JHEP **0901**, 036 (2009) [arXiv:0810.5768 [hep-th]].

62. H. -Y. Chen, Y. Nakayama and G. Shiu, Int. J. Mod. Phys. A **25**, 2493 (2010) [arXiv:0905.4463 [hep-th]].

63. L. Martucci, JHEP **0905**, 027 (2009) [arXiv:0902.4031 [hep-th]].

64. F. Marchesano, P. McGuirk and G. Shiu, JHEP **0904**, 095 (2009) [arXiv:0812.2247 [hep-th]].

65. F. Marchesano, P. McGuirk and G. Shiu, JHEP **1105**, 090 (2011) [arXiv:1012.2759 [hep-th]].

66. C. P. Burgess, R. Kallosh and F. Quevedo, JHEP **0310**, 056 (2003) [hep-th/0309187].

67. D. Cremades, M. -P. Garcia del Moral, F. Quevedo and K. Suruliz, JHEP **0705**, 100 (2007) [hep-th/0701154].

68. S. Krippendorf and F. Quevedo, JHEP **0911**, 039 (2009) [arXiv:0901.0683 [hep-th]].

69. V. Balasubramanian and P. Berglund, JHEP **0411**, 085 (2004) [hep-th/0408054].

70. A. Westphal, JHEP **0703**, 102 (2007) [hep-th/0611332].

71. M. Rummel and A. Westphal, JHEP **1201**, 020 (2012) [arXiv:1107.2115 [hep-th]].

72. J. Louis, M. Rummel, R. Valandro and A. Westphal, JHEP **1210**, 163 (2012) [arXiv:1208.3208 [hep-th]].

73. M. Cicoli, A. Maharana, F. Quevedo and C. P. Burgess, JHEP **1206**, 011 (2012) [arXiv:1203.1750 [hep-th]].

74. E. Silverstein, Phys. Rev. D **77**, 106006 (2008) [arXiv:0712.1196 [hep-th]].

75. S. S. Haque, G. Shiu, B. Underwood and T. Van Riet, Phys. Rev. D **79**, 086005 (2009) [arXiv:0810.5328 [hep-th]].

76. U. H. Danielsson, S. S. Haque, G. Shiu and T. Van Riet, JHEP **0909**, 114 (2009) [arXiv:0907.2041 [hep-th]].

77. U. H. Danielsson, S. S. Haque, P. Koerber, G. Shiu, T. Van Riet and T. Wrase, Fortsch. Phys. **59**, 897 (2011) [arXiv:1103.4858 [hep-th]].

78. T. Wrase and M. Zagermann, Fortsch. Phys. **58**, 906 (2010) [arXiv:1003.0029 [hep-th]].

79. M. P. Hertzberg, S. Kachru, W. Taylor and M. Tegmark, JHEP **0712**, 095 (2007) [arXiv:0711.2512 [hep-th]].

80. G. Shiu and Y. Sumitomo, JHEP **1109**, 052 (2011) [arXiv:1107.2925 [hep-th]].

81. C. Caviezel, P. Koerber, S. Kors, D. Lust, D. Tsimpis and M. Zagermann, Class. Quant. Grav. **26**, 025014 (2009) [arXiv:0806.3458 [hep-th]].

82. C. Caviezel, P. Koerber, S. Kors, D. Lust, T. Wrase and M. Zagermann, JHEP **0904**, 010 (2009) [arXiv:0812.3551 [hep-th]].

83. R. Flauger, S. Paban, D. Robbins and T. Wrase, Phys. Rev. D **79**, 086011 (2009) [arXiv:0812.3886 [hep-th]].

84. U. H. Danielsson, P. Koerber and T. Van Riet, JHEP **1005**, 090 (2010) [arXiv:1003.3590 [hep-th]].

85. J. D. Brown and C. Teitelboim, Nucl. Phys. B **297**, 787 (1988). J. D. Brown and C. Teitelboim, Phys. Lett. B **195**, 177 (1987).

86. R. Bousso and J. Polchinski, JHEP **0006**, 006 (2000) [hep-th/0004134].

87. J. L. Feng, J. March-Russell, S. Sethi and F. Wilczek, Nucl. Phys. B **602**, 307 (2001) [hep-th/0005276].

88. L. Susskind, In *Shifman, M. (ed.) et al.: From fields to strings, vol. 3* 1745-1749 [hep-th/0405189].

89. M. R. Douglas, hep-th/0405279.

90. M. Dine, hep-th/0410201.

91. M. Dine, E. Gorbatov and S. D. Thomas, JHEP **0808**, 098 (2008) [hep-th/0407043].

92. M. Dine, D. O'Neil and Z. Sun, JHEP **0507**, 014 (2005) [hep-th/0501214].

93. F. Denef and M. R. Douglas, JHEP **0503**, 061 (2005) [hep-th/0411183].

94. L. Covi, M. Gomez-Reino, C. Gross, J. Louis, G. A. Palma and C. A. Scrucca, JHEP **0806**, 057 (2008) [arXiv:0804.1073 [hep-th]].

95. U. H. Danielsson, G. Shiu, T. Van Riet and T. Wrase, JHEP **1303**, 138 (2013) [arXiv:1212.5178 [hep-th]].

96. X. Chen, G. Shiu, Y. Sumitomo and S. H. H. Tye, JHEP **1204**, 026 (2012) [arXiv:1112.3338 [hep-th]].

97. D. Marsh, L. McAllister and T. Wrase, JHEP **1203**, 102 (2012) [arXiv:1112.3034 [hep-th]].

98. S. R. Coleman, Phys. Rev. D **15**, 2929 (1977) [Erratum-ibid. D **16**, 1248 (1977)].

99. C. G. Callan, Jr. and S. R. Coleman, Phys. Rev. D **16**, 1762 (1977).

100. S. R. Coleman and F. De Luccia, Phys. Rev. D **21**, 3305 (1980).

101. S. W. Hawking and I. G. Moss, Phys. Lett. B **110**, 35 (1982).

102. T. Banks, arXiv:1208.5715 [hep-th].

103. S. -H. Henry Tye, hep-th/0611148.

104. A. R. Brown, S. Sarangi, B. Shlaer and A. Weltman, Phys. Rev. Lett. **99**, 161601 (2007) [arXiv:0706.0485 [hep-th]].

105. S. Sarangi, G. Shiu and B. Shlaer, Int. J. Mod. Phys. A **24**, 741 (2009) [arXiv:0708.4375 [hep-th]].

106. P. Ahlqvist, B. R. Greene, D. Kagan, E. A. Lim, S. Sarangi and I-S. Yang, JHEP **1103**, 119 (2011) [arXiv:1011.6588 [hep-th]].

107. P. Ahlqvist, B. R. Greene and D. Kagan, JHEP **1207**, 066 (2012) [arXiv:1202.3172 [hep-th]].

108. B. Greene, D. Kagan, A. Masoumi, D. Mehta, E. J. Weinberg and X. Xiao, Phys. Rev. D **88**, 026005 (2013) [arXiv:1303.4428 [hep-th]].

14

Cosmological SUSY Breaking and the Pyramid Scheme

Tom Banks

Department of Physics and SCIPP
University of California, Santa Cruz, CA 95064, USA
and
Department of Physics and NHETC
Rutgers University, Piscataway, NJ 08854, USA

I review the ideas of holographic space-time (HST), Cosmological SUSY breaking (CSB), and the Pyramid Schemes, which are the only known models of Tera-scale physics consistent with CSB, current particle data, and gauge coupling unification. There is considerable uncertainty in the estimate of the masses of supersymmetric partners of the standard model particles, but the model predicts that the gluino is probably out of reach of the LHC, squarks may be in reach, and the NLSP is a right handed slepton, which should be discovered soon.

1. Introduction

All known consistent string theory models, in asymptotically flat space-time, are exactly supersymmetric. All well established examples of the AdS/QFT correspondence, with radius that can be parametrically large in string units, give flat space limits that are exactly super-symmetric.

The world around us is not supersymmetric. If a theory of quantum gravity is to have relevance to the world, it must explain to us how SUSY is broken, and extant string theory models do not do this. The rest of the papers in this volume treat SUSY breaking in "string theory" by taking a low energy quantum field theory derived as an approximation to a supersymmetric string model, and adding supersymmetry breaking terms to it, which are *at best* plausibly connected to excitations in string theory. I have criticized this procedure extensively [1]. I believe instead that SUSY breaking is connected to the asymptotic structure of space-time via degrees of freedom that are thrown away in effective field theory. These are not high energy DOF, but rather very low energy excitations, which decouple from particles localized in the bulk because they are localized on the causal horizon. In flat and AdS space, the horizon actually recedes to infinity, and the horizon DOF need not be included in the Hilbert space, but in de Sitter (dS) space, the finite horizon leads to a finite amount of SUSY breaking.

The theory of Holographic Space-time explains the "empirical facts" of supersymmetry in string theory by choosing (in Minkowski space) the variables of quantum gravity to be the cut-off generators of the (generalized) super-BMS algebra. The super-BMS algebras are defined on the lightcone in 4 dimensional[a] Lorentzian momentum space. It has two components, corresponding to the top and bottom of the cone. The $P_0 > 0$ component has generators $\psi_\alpha^+(P, a)$, satisfying

$$[\psi_\alpha^+(P, a), \bar{\psi}_{\dot{\beta}}^+(Q, b)]_+ = Z_{ab} M_\mu(P, Q) \sigma_{\alpha \dot{\beta}}^\mu \delta(P \cdot Q),$$

and

$$\psi_\alpha(P, a) \sigma_{\alpha \dot{\beta}}^\mu P_\mu = 0.$$

The delta function is non-zero only when P and Q are collinear, and so both positive multiples of $(1, \boldsymbol{\Omega})$, where $\boldsymbol{\Omega}$ is a point on the unit 2-sphere. M^μ points in the same direction and chooses the minimum of the two positive multipliers. The constraint on ψ says that it lies in the holomorphic spinor bundle over the two sphere, while it's conjugate lies in the anti-holomorphic bundle. The local super-algebra, at fixed P, completed by the commutators of Z_{ab} with $psi(P, a)$, and with themselves, has a finite dimensional unitary representation generated by the action of the fermionic generators on a single state. One should think of the index a as labelling the eigen-functions of the Dirac operator on some compact manifold, with a cutoff that fixes the volume of that manifold in Planck units [3]. Other geometric properties of the compact space are encoded in the super-algebra. For the purposes of this note, we can just set the Z_{ab} to be a c-number times δ_{ab} and let a, b run over the 16 components of a spinor in 7 dimensions. This is the quantum theory of eleven dimensional supergravity compactified on a 7 torus of Planck size. For fixed P, the super-BMS algebra is just

---

[a]We stick to four dimensions for simplicity of exposition. The formalism generalizes to higher dimensions.

the algebra of a single eleven dimensional graviton, and its superpartners, compactified on a Planck sized torus.

Scattering theory, at least in theories of gravity, is considered to be a map between the past and future representations of the super-BMS algebra[b].

$$\psi_\alpha^{(+)} = S^\dagger \psi_\alpha^{(-)} S.$$

Scattering states are states in which the operator valued measures $\psi_\alpha(P, a)$ vanish outside of the endcaps of a finite number of Sterman-Weinberg cones, and at $P = 0$.

For finite causal diamonds, these singular measures are replaced by a sum over a finite set of spinor spherical harmonics. The $P = 0$ modes are the degrees of freedom responsible for the entropy of horizons. For finite horizons they contribute finite terms to the Hamiltonian, but they decouple in the infinite horizon limit.

Cosmological SUSY breaking (CSB) is an attempt to implement the consequences of these abstract ideas in low energy effective field theory, and use them to guess at the correct model of Tera-scale physics. It leads to a quite restrictive set of models. The phenomenological analysis of these models is difficult because they must contain a new strongly coupled sector at the TeV scale, but a recent breakthrough has allowed me to make a lot of qualitative predictions for the spectrum of standard model super-partners. The models, which are called Pyramid Schemes [4], have a mechanism that produces large mixing between gauginos and composite adjoint chiral superfields. As a consequence, they predict heavy gauginos, squarks and right handed sleptons that should be in reach of the LHC, and a very complicated Higgs sector, whose properties are hard to extract from the brown muck of the new strong interactions. One can certainly get a 125 GeV Higgs, but it is not clear that its interactions are close enough to the standard model to fit the data. Questions of whether the weak scale is fine tuned in these models are beset by similar strong interaction obscurities.

The Pyramid Schemes also provide a novel solution to the strong CP problem, a novel dark matter candidate, a possible connection between the dark matter density and baryogenesis, and a possible pathway to explaining extra dark radiation (if the data indicating dark radiation improve to the point where it *needs* explanation) . They explain the absence of all

---

[b]For the BMS sub-algebra, Strominger has interpreted this equation as an expression of spontaneous breakdown of the BMS symmetry, and shown that Weinberg's graviton low energy theorems follow from this equation.

dimension 4 and 5 operators which could mediate unobserved baryon and lepton violation, while permitting the dimension 5 seesaw operator, which gives rise to neutrino masses.

2. Cosmological SUSY Breaking

The zero energy generators of the super-BMS algebra provide a huge set of degrees of freedom, localized on the horizon (the conformal boundary) of Minkowski space that are not incorporated in quantum field theory. They decouple from the S matrix of particles in Minkowski space. The basic idea of CSB is that the coupling between these states and particles remains finite, in the finite causal diamond of a single geodesic in de Sitter (dS) space. We view the radius of dS space as a tunable parameter[c] and ask how the coupling between particles and the horizon DOF leads to SUSY breaking.

To proceed we will have to understand a bit more about the geometry of dS space. The most important fact about dS space is that even a hypothetical observer, who lives for an infinite amount of time, can only see a finite distance R away. The entire history of the universe takes place inside a sphere whose radius can never be bigger than R. That sphere is actually running away from the observer at the speed of light. What is peculiar about the dS universe is that the expanding sphere describing where the backward light-cone from time T meets the forward light-cone from time $-T$, has a finite radius, even as T goes to infinity. This remains true if the universe is not exactly dS space, but began a finite time in the past, and becomes dS space in the asymptotic future. If observations on the acceleration of the universe are given their simplest interpretation in terms of a positive value for Einstein's cosmological constant (c.c.), then this is what is going on in the universe we live in. The radius of our *cosmological horizon* is about 10^{61} Planck units[d].

Now we want to think about implementing the idea that SUSY breaking comes from interactions with the cosmological horizon in effective field theory (EFT), the framework into which all physics below the Planck energy

---

[c]The question of what determines this radius in the real world goes beyond the bounds of this review, but has been addressed in [5]. Briefly, that cosmological model provides a multi-verse of possibilities for the dS radius, and the choice is made by invoking the anthropic principle.

[d]The Planck distance scale is $L_P = 10^{-33}$ cm. . In units where $\hbar = c = 1$ that corresponds to about $T_P \sim 10^{-44}$ seconds, and $M_P = 10^{19}$ GeV. In these units, Einstein's c.c. is about $10^{-123} M_P^4$.

scale has been assumed to fit. In effective field theory, SUSY is a gauge symmetry and can only be broken spontaneously. Gauge symmetries have space-time dependent parameters $\epsilon(x)$. We can do a gauge transformation on *any* Lagrangian. If the original Lagrangian was not gauge invariant, the result is a new Lagrangian, where $\epsilon(x)$ is a new field. This new Lagrangian *is* gauge invariant, if we let both the original fields and $\epsilon(x)$ transform under the gauge transformation[e]. The gauge potential is a field $A_\mu^a(x)$. If the original Lagrangian was not gauge invariant, the semi-classical expansion reveals a *massive* excitation of the gauge field, and we say that the Schwinger-Anderson-Higgs-Brout-Englert-Guralnik-Hagen-Kibble phenomenon has taken place. In the case of supersymmetry, the gauge parameters are a fermionic spinor field $\epsilon_\alpha(x)$. The massive excitation is that of the gravitino field $\psi_{\mu\alpha}(x)$ and has spin $3/2$. It's mass is denoted $m_{3/2}$.

The mass of the gauge excitation is proportional to the gauge coupling. In the limit that the gauge coupling goes to zero, the Higgs phenomenon morphs into the phenomenon of Nambu-Goldstone spontaneous symmetry breaking (whence the somewhat inaccurate name spontaneous breaking of gauge symmetry for the Higgs phenomenon). The longitudinal part of the massive gauge field becomes the Nambu-Goldstone particle associated with symmetry breaking. The mass is the product of the gauge coupling, and the value of an order parameter, F, whose size depends on the energy scale at which the symmetry is broken in the limit of zero gauge coupling. In the case of supersymmetry, the gravitino is the symmetry partner of the graviton, and the relevant coupling is $\frac{1}{m_P}$[f] The order parameter F has dimensions of squared mass, and determines the typical difference in squared masses between bosons and their supersymmetric partner fermions, in non-gravitational supermultiplets. The gravitino mass is

$$m_{3/2} \sim \frac{F}{m_P}.$$

Super-Poincare invariance appears naturally in $N = 1$ SUGRA only in the presence of a gauged discrete complex R symmetry, which sets the constant in the super-potential to zero. Indeed, super-symmetry is compatible with general negative values of the cosmological constant. This has been abundantly confirmed by the AdS/CFT correspondence, in which the

---

[e]This is straightforward for a gauge group with one parameter. If there are multiple parameter $\epsilon_a(x)$ and the different transformations do not commute with each other the formalism is more complicated, but the results are the same [9] .

[f]$m_P = \frac{M_P}{\sqrt{8\pi}} \sim 2 \times 10^{18}$ GeV, is called the *reduced Planck mass.*.

quantum theory of many space-times of the form

$$AdS_d \times \mathcal{K},$$

where $\mathcal{K}$ is a compact manifold, and Anti-deSitter space is the maximally symmetric space with negative c.c.. SUSY is incompatible with positive c.c.. The formula for the c.c. in SUGRA is

$$\Lambda = e^K (|F|^2 - \frac{3}{m_P^2}|W|^2).$$

F is the SUSY breaking order parameter, and W is the order parameter for R symmetry. R symmetry is the subgroup of the discrete symmetry group of the model, which acts on the generators of the supersymmetry algebra[g].

CSB depends on the hypothesis that the interactions with the horizon generate R violating terms in the effective action, which in turn lead to spontaneous SUSY breaking. It is also assumed that the gravitino is the lightest particle carrying R charge.

Then the leading order diagrams that could lead to R violating interactions coming from the horizon, have a gravitino line going out to the horizon and another coming back, violating R symmetry by two units . These diagrams are all proportional to

$$e^{-2m_{3/2}R_{dS}}C.$$

We can think of this as a term in second order perturbation theory in the interaction via which the horizon emits and absorbs gravitinos, so that

$$C = \sum_s \langle out|V^\dagger \frac{1}{E-H}|s\rangle\langle s|V|in\rangle,$$

where the sum is over all states of the horizon with which the gravitino can interact. The energy denominators are all of order $\frac{1}{R_{dS}}$ because the $e^{\pi(R_{dS}M_P)^2}$ states of the horizon, live in a band of this size.

The horizon is a null surface and the gravitino can propagate on it for a proper time of order $\frac{1}{m_{3/2}}$. According to conventional Feynman diagrams, written in the Fock-Schwinger proper time parametrization, it performs a random walk on the surface, with a step size in proper time given by the UV cutoff. If, as we will assume, the theory has extra dimensions large in 10 or 11 dimensional Planck units, then the step size is given by the higher dimensional Planck scale. Following Witten [6] we will assume that this

---

[g]More precisely: R symmetry is the coset of the discrete symmetry group G by the subgroup H, which leaves the SUSY generators invariant. It can be shown that H is a normal subgroup of G, so that G/H is a group, called the discrete R symmetry.

is the scale of coupling unification which is 2×10^{16} GeV. The entropy of the horizon states with which the gravitino interacts is proportional to the area in 4 dimensional Planck units that it covers in proper time $\frac{1}{m_{3/2}}$, and is $c\frac{1}{M_U m_{3/2}}$. Thus the full amplitude is proportional to

$$e^{-2m_{3/2}R}e^{c\frac{M_P^2}{M_U m_{3/2}}}.$$

Assuming that $m_{3/2}$ goes to zero as a power of R_{dS}^{-1}, we find a contradiction unless the power is precisely $R^{-\frac{1}{2}}$. Indeed, if the mass is assumed to go to zero more rapidly than this, the formula for R breaking interactions blows up exponentially as $R_{dS} \to \infty$, while if it goes to zero more slowly the strength of these interactions vanishes exponentially.

Using the reduced Planck mass $M_P^2 = 8\pi m_P^2$ and the relation $m_P R_{dS}^{-1} = \sqrt{\Lambda/3}$, we get

$$m_{3/2}^2 = \frac{4\pi c}{\sqrt{3}}\frac{m_P}{M_U}\sqrt{\Lambda}.$$

For a unification scale $M_U = 2 \times 10^{16}$ GeV, this gives

$$m_{3/2} = K10^{-2}\text{eV},$$

with K a constant of order one. The SUSY breaking order parameter is thus

$$F = 2K \times 10^7 (\text{GeV})^2.$$

This is a remarkably low value for the SUSY breaking scale, a fact which drives much of the analysis below.

The structure of diagrams contributing to the R violating terms in the Lagrangian implies that these terms *do not* satisfy the usual constraints of technical naturalness, familiar from QFT and perturbative string theory. Any diagram with more than two gravitinos, or with heavier R charged particle states, mediating between the local vertex and the horizon, will be exponentially suppressed. Diagrams involving R neutral exchanges with the horizon give contributions which have a finite limit as $R_{dS} \to \infty$, and are already incorporated in the $\Lambda = 0$ effective Lagrangian.

As a consequence, apart from the R symmetry itself, symmetries, or approximate symmetries of the $\Lambda = 0$ model, are also preserved by the R-violating terms. We exploit this in the following way: we choose the R symmetry to forbid all terms of dimension 4 or 5 in the $\Lambda = 0$ model, which violate B and L, apart from the dimension 5 superpotential

$$W_\nu = \frac{b}{M_{seesaw}}(H_u L)^2,$$

which generates neutrino masses of roughly the right order of magnitude[h]. Insertions of higher dimension B and L violating operators into a diagram with a pair of gravitino lines going out the horizon cannot generate the lower dimension operators, because the extra gravitino loop is cut-off at the SUSY breaking scale or below, by its space-time structure. We will see later that, with one extra mild assumption, this mechanism for R violation also provides a novel solution to the strong CP problem.

3. The Pyramid Schemes

We now want to build an effective field theory, compatible with current experiment, and with the mechanism of CSB. It must contain the MSSM, as well as an uncontrained goldstino superfield X, and must preserve SUSY and a discrete R symmetry, but spontaneously violate SUSY when R breaking terms are added. It must be consistent with the bounds on super-partner masses.

If the only dynamically generated scale in the theory is the QCD scale, it is impossible to do this. The most general renormalizable Lagrangian is that of the Minimal Supersymemetric Standard Model (MSSM), with the additional superpotential[i]

$$W_X = g_X X H_u H_d + C(X),$$

where C is a cubic polynomial. This always has supersymmetric solutions[j]. Even if we could generate a non-zero F_X the gluino mass generated by this model would be too small to be compatible with experimental bounds. Non-renormalizable corrections to this Lagrangian would be suppressed by powers of M_U or m_P and cannot help with these problems.

---

[h]We do not attempt to explain why M_{seesaw} is an order of magnitude or so less than M_U. This is a high energy problem.

[i]The superpotential is an analytic function of the complex fields, whose value,W_0 at the minimum is the order parameter for spontaneous R symmetry breaking. In the $\Lambda = 0$ limit, $W_0 = 0$, and the SUSY breaking order parameter is given by the gradient of W w.r.t. the fields, and must vanish.

[j]A meta-stable SUSY violating solution can be acceptable only if the probability for transitions into a Big Crunch by tunneling into the basin of attraction of the SUSic minimum is the inverse recurrence time for dS space [10] [11]. There are not enough parameters in the model to engineer this.

To remedy the gluino mass problem, we must include a strongly coupled hidden sector, some of whose fields carry color, in order to generate a coupling between X and the QCD field strength W_α^a, which can give a large enough gluino mass. This is the only way to generate a new low energy scale in a natural manner. The scale Λ_3 of this new strongly coupled sector has to be close to the SUSY breaking scale, since $\frac{F_X}{\Lambda_3}$ will be the natural scale that enters into the formula for the gluino mass, and F_X is so low. The Pyramid Scheme models we propose, have a natural explanation for this coincidence of scales.

The necessity for new colored particles is potentially problematic, if we wish to preserve coupling unification. The obvious solution to this is to include complete multiplets of some unified gauge group, but one must also be sure that the gauge couplings have no Landau poles[k] below the unification scale[l]. This puts restrictions on the size of the new strong gauge group.

Seiberg's general analysis of the IR behavior of asymptotically free SUSY gauge theories [12] enables us to rule out many possibilities. Initially, I was led to the $N_F = N_C = 5$ theory as the unique possibility that could preserve $SU(5)$ unification, but a careful analysis of two loop effects showed that the model had Landau poles below the unification scale. I cannot claim to have made an exhaustive survey, but at the moment the only class of models that survives all of these simple tests are the Pyramid Schemes [4].

The Pyramid Schemes utilize Glashow's Trinification [13], an $SU(3)^3 \ltimes Z_3$ subgroup of E_6, with 3 generations of chiral fields in the $(1,3,\bar{3}) \oplus (\bar{3},1,3) \oplus (3,\bar{3},1)$ representation. The Z_3 permutes the three $SU(3)$ groups and ensures equality of couplings in the symmetry limit. I will have to assume the reader is familiar with this, and refer to the ith subgroup as $SU_i(3)$. Color is embedded in $SU_3(3)$ and the weak $SU(2)$ in $SU_2(3)$. 10 chiral fields of each generation are assumed to obtain mass at the unification scale. The Higgs fields $H_{u,d}$ also belong to an incomplete multiplet, but we

---

[k]Renormalization group running of *marginally irrelevant* parameters in QFT, makes them grow at high energy. A postulated value at low energy, chosen to fit phenomenology, makes the coupling so strong at some scale, that one no longer knows how to analyze the model. This is called a Landau pole.

[l]Some authors like to preserve coupling unification to two loop accuracy. Two loop corrections are of the same size as one loop threshold corrections at the unification scale, so I have never been very impressed by the "better fit" given by the two loop results. I will try only to preserve the one loop results and the fact that the two loop corrections are small.

do not specify what it is. More generally, we make no attempt to explain details of physics at the unification scale.

The quiver diagram[m] of Trinification is a chiral triangle. The simplest extension of it answering our needs extends the triangle to a Pyramid with triangular base (a tetrahedron). The fields connecting the apex of the Pyramid to the base transform in the vector like representation $T_i \oplus \bar{T}_i \in (F, \bar{3}_i) \oplus (\bar{F}, 3_i)$ and are called *trianons*. F is some representation of the *Pyramid Group* G_P. Both the group and the representation must be fairly small, to preserve standard model coupling unification.

While I don't pretend to have made an exhaustive search, the only examples I've found that work are $G_P = SU_P(k)$ with $k = 3, 4$ and F the fundamental representation. The case $k = 3$ is more attractive in a number of ways. The minimal R symmetry group that works for $k = 3$ is Z_8, compared to Z_{14} for $k = 4$. Furthermore, there's a natural explanation for the coincidence between Λ_3 and the scale of SUSY breaking for $k = 3$, and an interesting dark matter candidate. So far, there are no analogous advantages for $k = 4$.

3.1. The singlet sector

The R symmetry is chosen [4] to forbid all relevant super-potential terms, which would otherwise be expected to be of order M_U or greater. Another way to say this is that we insist that the $\Lambda = 0$ theory be technically natural. The super-potential is thus

$$W_3 = \sum \kappa_i T_i^3 + \tilde{\kappa}_i (\tilde{T})_i^3 + W_{std},$$

where the second term is the familiar standard model super-potential, with no μ term, and no terms violating B or L. The expression T_i^3 refers to the cubic invariant of the $(3_P, \bar{3}_i)$ representation. The unified group would set all these couplings equal, but unification scale symmetry breaking could easily change that, without ruining the success of one loop gauge coupling unification.

The R breaking diagrams with gravitinos propagating out to the horizon can induce terms of the form

$$W_{1, \not R} = m_i T_i \tilde{T}_i + \mu H_u H_d + W_0.$$

---

[m]Quiver diagrams are schematic representations of the interactions in gauge theories, with products of $SU(N)$ gauge groups. They are geometric figures whose vertices are labelled by the different gauge groups, and they have arrows going between the i and jth pair of gauge groups, if there are chiral fermions transforming in the $(F_i, \bar{F}_j)$ representation of $SU(N_i) \otimes SU(N_j)$.

The model has a SUSY preserving minimum both with and without the extra terms, and so does not satisfy the requirements of CSB. This indicates the necessity of introducing other low energy fields.

The simplest way to do this, and perhaps the only one, which doesn't disturb the running of the standard model couplings, is to add singlets under the full gauge group. It seems that the minimal number is 3 fields S^i. There is no reason to assume that the index here transforms under the Z_3 of the trinification group, though it is suggestive of interesting structure at the unification scale. The R symmetry action on the S_i can be chosen so that the trilinear couplings

$$W_S = \frac{1}{6} C_{ijk} S^i S^j S^k + \alpha_i^j S^i T_j \tilde{T}_j + \beta_i S^i H_u H_d,$$

are allowed. However, we will also impose an additional discrete symmetry, which does not act on the supersymmetry generators, which ensures that the matrix $C_{ij} = C_{ijk} S^k$ has a zero eigenvalue for any choice of S^k. The full $\Lambda = 0$ super-potential, $W_S + W_3 + W_{std}$ has a SUSic, R symmetric minimum when all fields vanish. Although the cubic super-potential has flat directions, these are all lifted by non-renormalizable R symmetric corrections to the Kahler and super-potentials, scaled by the unification or Planck scales. Finally, with the given field content, the gauge couplings remain small at low energy so that the model preserves both SUSY and R symmetry. Thus, this low energy model is consistent with being the low energy limit of the $\Lambda \to 0$ limit of a model of stable dS space.

We now add

$$W_{\not R} = m_i T_i \tilde{T}_i + \mu H_u H_d + \frac{1}{2} m_{ij} S^i S^j + F_i S^i + W_0,$$

to the superpotential. The equations for a supersymmetric point become

$$(\mu + \beta_i S^i) H_{u\ (d)} = 0,$$

$$\sum_j \alpha_i^j T_j \tilde{T}_j + \frac{1}{2} C_{ijk} S^j S^k + \sum_j m_{ij} S^j + F_i + \beta_i H_u H_d = 0,$$

$$(\sum_i S^i \alpha_i^j + m_j) T_j + \tilde{g}_j (\tilde{T})_j^2 = 0,$$

$$(\sum_i S^i \alpha_i^j + m_j) \tilde{T}_j + g_j (T)_j^2 = 0.$$

$(\tilde{T})_j^2$ is the bilinear obtained by differentiating the trilinear invariant w.r.t. $\tilde{T}_j$.

As noted, we can choose the R symmetry, plus another discrete symmetry which does not act on the supersymmetry generators, to ensure that the matrix

$$C_{ij} = C_{ijk}S^k,$$

is not invertible for any S^i. We further assume that the coefficients in $W_{I\!\!R}$ are chosen so that μ_{ij} shares the zero modes of C_{ij} and that the S^i independent terms in $\frac{\partial W}{\partial S^i}$ have components in the zero mode subspace. In this case, there can be no SUSic minimum. The constraints on $W_{I\!\!R}$, do not follow from symmetries, but these terms arise from a very special class of diagrams. It's only by imposing these constraints that we obtain a low energy model compatible with an underlying gravitational model that breaks SUSY.

To understand fully the dynamics of SUSY breaking in this model, we must first make sure that the $SU_P(3)$ gauge theory indeed becomes strongly coupled, and find the relation between its dynamical scale Λ_3 and the CSB SUSY breaking scale. The $SU(3)_P$ Lagrangian at high energies is SUSY QCD, with $N_F = 3N_C$. Its one loop beta function vanishes, but in the absence of other couplings the two loop beta function is IR free. However, if the couplings g_i and $\tilde{g}_i$ are all equal to $\sqrt{4/3}\times$ the gauge coupling, then we have a line of fixed points. This line is attractive. We imagine that, at the unification scale M_U the effective theory lies in the domain of attraction of this line and is rapidly sent to a point where the coupling is relatively strong, but barely in the perturbative regime[n]. The couplings then remain fixed until we reach the highest mass threshold of the trianon fields. That mass scale is set by the parameters m_i, which come from interactions with the horizon. These three parameters are of comparable order of magnitude and all vanish like $\Lambda^{1/8}m_P^{1/2}$, when the c.c. is sent to zero. We do not know how to calculate them more accurately than that. For phenomenological reasons, we will assume that the lightest mass is m_3, the mass of the colored trianon.

Below the first two trianon thresholds, the lagrangian has $N_F = N_C$ and is asymptotically free. We have assumed that the fixed line value of the gauge coupling is fairly large, so the confinement scale Λ_3 is slightly below the masses $m_{1,2}$ of the two colorless trianons. We can think of the

---

[n]The last restriction is imposed in order to be able to do the analysis. It is likely that even stronger couplings will also work, but it is hard to calculate in that regime.

relations between these scales as roughly analogous to that between the charmed quark mass and the QCD scale, $m_i \approx 4\pi\Lambda_3$. We will also assume that m_3 is of order Λ_3, somewhat analogous to the strange quark mass in QCD. This means that we can analyze the low energy dynamics in terms of chiral perturbation theory, which in this case means Seiberg's effective Lagrangian [12].

The colored trianons are confined into a three by three matrix **M** of *pyrmesons*, which transform as an octet and singlet of color, and singlet *pyrmabaryon*, **B**, and *anti-pyrmabaryon*, $\tilde{B}$, fields. The effective super-potential on the moduli space is

$$W_{mod} = \Lambda_3^3[L(B\tilde{B} - 1 - \det \mathbf{M}) + (m_3 + \alpha_j^3 S^j)\text{tr }\mathbf{M}$$

$$+\kappa_3 B_{\tilde{\kappa}_3}\tilde{B} + (\beta_j S^j + \mu)H_u H_d + C(S)] + W_{std} + W_0.$$

W_0 is a constant added in order to tune the c.c. to its observed value. It does not affect the low energy dynamics, which is independent of the Planck mass to first approximation, once we fix the relevant couplings m_i, μ, and F_i. In this formula we've rescaled all fields and parameters by powers of Λ_3, to make them dimensionless. Apart from this rescaling, $C(S)$ is the cubic polynomial in the singlets, that appeared in the super-potential above the scale Λ_3. L is a Lagrange multiplier field.

Before analyzing the predictions of this model, we note that something very similar results if we set $kappa_i = ka\tilde{p}pa_i = 0$ for $i = 1$ or $i = 2$. The UV model no longer has a fixed line, but the couplings vary slowly. In particular, although the gauge coupling is now IR free, we can still have a strong coupling scale Λ_3 without producing a Landau pole below M_U, as long as $\Lambda_3 < 2$ TeV [14][0]. This is interesting because a model that preserves one of the pyrma-baryon symmetries at the renormalizable level, allows the lightest particle carrying this quantum number to be a dark matter candidate if an appropriate asymmetry is generated in the very early universe. We will discuss this further below.

4. Crude Estimates of MSSM Super-partner Masses

The first order of business is to estimate the masses of supersymmetric partners of standard model particles in this model. Here I've recently been

---

[0]In [14] we were trying to preserve predictions of the original Pyramid Scheme, which had explained various lepton excesses, now considered to be due to pulsars. We preferred schemes where the conserved pyrma-baryon number was carried by colored trianons. The current scenario will not produce light PNGBs.

surprised. I'd initially thought that the Pyramid scheme was a form of direct gauge mediation [15]. In fact, for the gluino at least, the gauge mediated masses are much smaller that a contribution from mixing between the gaugino and the pyrmeson octet. Write

$$\mathbf{M} = M + \lambda_a M^a,$$

where the fields are dimensionless a λ_a represent the Gell-Mann matrices. These fields, are the pyrmesons. Consider an operator

$$\int d^4\theta \; f(M, \bar{M}) D_\alpha M^a W_a^\alpha + h.c.$$

$$= g_3 f_{MM^*} \frac{F_M \bar{F}_M}{\Lambda_3^3} \psi_\alpha^a \lambda_a^\alpha + \text{h.c.} + \text{fermions} + \text{derivatives}.$$

The function f and its derivatives have a factor of the QCD coupling g_3, but no loop factors. They are analogs of hadron magnetic moments in QCD, with the insertion of one weakly coupled field into an effective Lagrangian for composites. The QCD fine structure constant $\sim .1$ gives $g_3 \sim 1.4$, so this is nominally of the same size as contributions to the Majorana mass

$$m_3 \int d^2\theta \; M_a M^a.$$

(These formulae must be rescaled by the Kahler potential to get physical masses), because F_M, Λ_3 and m_3 are all in the TeV range.

The result is a pair of octet Majorana fermions, whose masses and mass splitting are all of order $\frac{|F_M|^2}{\Lambda_3^3}$. Given the rules of CSB, the numerator is bounded by about $10^{14}(\text{Gev})^4$. RG running gives an *upper* bound on $\Lambda_3 < 2 \times 10^3$ GeV. The lower bound is harder to determine but is related to the fact that we haven't seen any of the pyrma-hadrons and so is probably about 1 TeV. Thus, the lightest mass eigenstate with the quantum numbers of the gluino is between $10 - 100$ TeV. It is a mixture, with order one mixing angle, of the gluino interaction eigenstate and the composite octet fermion in the N^a super-multiplet. The conventional gauge mediated Majorana gluino mass is suppressed by a factor $\frac{1}{16\pi^2}$ relative to these masses [16].

The dominant contribution to squark squared masses comes from a *one QCD loop*, convergent, diagram, as in super-soft models [17]. The squark masses are universal and are of order

$$m_{\tilde{q}} \sim m_{gluino} \sqrt{\frac{g_3^2}{16\pi^2}} \sim 900 - 9000 \text{ GeV}.$$

The lower reaches of these estimates mean that squarks but not gluinos will be within the reach of the LHC, while the upper values bode ill for near term experimental detection of these particles.

A similar formula is also appropriate for the other gauginos, though here the argument is more complicated. Basically, using arguments analogous to those invoked when discussing mixing between the photon and strongly interacting vector mesons, one counts factors of g_i, and 4π, with everything else determined by dimensional analysis and the scale Λ_3. In this kind of bilinear mixing, there are no loop factors. Then one argues that when $g_i = 0$ there is a stable adjoint fermion, with a Majorana mass of order $2m_2$ for the $SU(2)$ triplet, and the lighter of $2m_1$ and $2m_2$ for the $U(1)$ adjoint (since both of the colorless trianons have $U(1)$ couplings). Mixing between the gauginos and these states is a seesaw mechanism [18], giving Majorana gaugino masses of order

$$m_{1/2}^{(i)} \sim \frac{g_i^2 |F_Y|^4}{2\Lambda_3^6 m_i}.$$

It's not completely clear which F terms will give the dominant contribution here. The masses are probably less than a TeV, though there's considerable uncertainty in these estimates. Slepton masses will be down from this by the square root of a loop factor

$$\sqrt{\frac{\alpha_i}{4\pi}}.$$

One of the right handed sleptons is thus the NLSP. If the bino weighs a TeV this crude estimate gives right handed slepton masses of order 30 GeV, which is already ruled out. Indeed, for the decay topology of slepton going to Goldstino, sleptons are ruled out up to about 260 GeV [19] [20] and the next run of the LHC might explore another 100 GeV in mass[p]. If we took our estimates seriously, this would push the bounds on bino and chargino masses up to about 9 and 27 TeV. However the strong $SU_P(3)$ uncertainties do not warrant such drastic conclusions. I cannot emphasize too strongly how much uncertainty there is in these estimates, but they lead us to expect the discovery of right handed sleptons in the near future.

There is one caveat to the claim that a RH slepton is the NLSP, since we have not studied the masses of all the states in the singlet sector. However, the diagrams contributing to slepton masses are lepton flavor blind. The coupling of sleptons to the singlet sector is mediated by the Higgs boson

---

[p]I'd like to thank Scott Thomas and Patrick Draper for explaining the LHC bounds on right handed sleptons to me.

and we know that the Higgs couplings of the leptons are small, ranging from 10^{-5} to 10^{-2}. Decays of a slepton into a lepton and a hypothetical light singlino, would occur outside the LHC detectors. Thus, even if it turns out that the NLSP is a singlino, the light right handed sleptons predicted by the model should be observed.

5. Pyramidal Cosmology

5.1. *Dark matter*

In the original paper on the Pyramid Scheme, Fortin and I got caught up in the excitement surrounding positron excesses and other dark matter signatures. The majority opinion seems to be that these excesses are no longer considered to be relevant to dark matter. Since then I've returned to the simple idea [22] that dark matter is one of the pyrma-baryons of the strongly coupled Pyramid sector. This requires that we omit one pair of trilinear couplings from the underlying Lagrangian, and one can choose the R symmetry properties of the model to make this natural. We have seen that the attractive RG structure of the model is preserved when we do this, as long as the scale $\Lambda_3 < 2$ TeV.

We've seen above, that we want to keep the trilinear couplings of the colored trianon. This implies that (if the dark matter is the fermion in the supermultiplet) dark matter has a magnetic moment. This is an old idea, which goes back to technicolor [23] and has potential observational consequences [21].

In order to get the right dark matter density, we need to postulate an asymmetry generated in the very early universe. This is very easy to do, but has no predictive power. However, it opens the door to a connection between the dark matter density and the baryon asymmetry of the universe. The standard model couplings of the trianons lead to a coupling

$$\frac{\alpha_2^2}{\Lambda_3^2} J_{PB}^{\mu} J_B^{\mu},$$

which implies that an asymmetry in one quantum number will give rise to a chemical potential for the other. If this chemical potential is substantial when the interactions that violate the corresponding quantum number go out of equilibrium, then spontaneous (pyrma) baryogenesis will occur [24] [25], thus connecting the dark matter and baryon densities of the universe.

There may also be a possible dark matter candidate in the singlet sector of the model, about which I understand too little to make a definitive statement. Presumably, if it exists, it would be much more like a WIMP. If this is the dark matter, we can restore the possibility of UV equality of all the trianon trilinear couplings, which is somewhat more elegant.

5.2. *Dark radiation*

The gravitinos in any model implementing CSB are very light and were certainly relativistic at the eras where the CMB and structure formation may indicate the need for more relativistic species. Standard estimates [26] indicate that such light gravitinos decouple before the electroweak phase transition and contribute much less than a neutrino species to the evolution of the universe. However, non-thermal repopulation of the gravitinos by late decaying NLSPs, could generate the required excess. This could only occur if the NLSP was part of the singlet sector, because our bounds on light MSSM super-partners rule out such late decays.

6. The Strong CP Problem

As pointed out in [27] the Pyramid Scheme provides a novel solution to the strong CP problem. When $\Lambda = 0$ the model has many $U(1)$ symmetries at the renormalizable level, which allow us to rotate away all CP violating phases except the CKM phase. This would lead to an axion with a decay constant that has been ruled out be experiment. However, the R symmetry violating interactions coming from the horizon break all these symmetries and give the axion a large mass. Normally, when we try to do this in QUEFT, the $U(1)$ breaking terms re-introduce CP violating phases.

In the Pyramid scheme, these terms come from a very special class of diagrams, where two gravitinos are exchanged with the horizon. The part of these diagrams localized near the origin has all the symmetries of the $\Lambda = 0$ theory, and the CP violating θ_{QCD} induced through the CKM matrix is tiny. The other end of the gravitino lines is more mysterious, but since it lies on the horizon it is at a very high local temperature, of order the unification scale. Thus, if the fundamental origin of CP violation is spontaneous breakdown, at scales $\ll M_U$ there will be no CP violation near the horizon. Thus, the phases in all the R breaking diagrams are small, without either fine tuning or an axion.

7. The Higgs Potential and the Electroweak Scale

Neglecting loops involving standard model fields, the Higgs potential in the Pyramid Scheme is

$$K^{i\bar{i}}(\beta_i H_u H_d + \alpha_i^3 M + F_i)(\bar{\beta}_{\bar{i}} \bar{H}_u \bar{H}_d + \bar{\alpha}_{\bar{i}}^3 \bar{M} + \bar{F}_{\bar{i}}) + |\beta_i S^i + \mu|^2 (|H_u|^2 + |H_d|^2).$$

The Kahler potential depends on the singlet fields through the combinations $\alpha_i^k S^i$, for $k = 1, 2$. This comes from integrating out the colorless trianons. In [27] this part of the Kahler potential was calculated in zeroth order perturbation theory in the Pyramid coupling. This approximation is not really justified because the masses of the trianons are just a few times Λ_3. The parameters F_i and μ come from interactions with the horizon. We expect them to be of order a few TeV, but do not have a way to calculate them with any precision.

One should also include contributions to the Higgs potential from stop loops, and, given the size of $SU(2) \times U(1)$ preserving gaugino masses that we have estimated, loops of TeV scale gauginos. We will also want to choose the couplings β_i to be fairly large, which means that loops of singlets will also be important in determining the Higgs potential.

In [27] we included some, but not all of these effects, many of which push in opposite directions. We found that we could fit the LHC bounds but that this required a few percent tuning. Given our new insights into gaugino masses, and the singlet loops, which we simply forgot in [27] , the problem becomes more complicated. In addition, the large β_i present us with the possibility of large mixing between singlets and the lightest Higgs. Neglect of the complicated dependence of the Kahler potential on the S^i was unjustified.

Note that the tuning in the Pyramid scheme is not really the same as the oft discussed little hierarchy problem of the MSSM. It really comes from the fact that the Higgs potential above contains a number of relevant parameters whose natural scale in CSB is multiple TeV. The dimensionless parameters are bounded from above in order to avoid Landau poles below the unification scale. On the other hand, we have a rather complicated function of 6 complex variables (the neutral Higgs fields $h_{u,d}$, the singlet pyrmeson M, and the three S^i fields) to minimize, so it seems premature to conclude that a tuning of one part in a hundred is unnatural. It is, at any rate, too complex to attempt here.

8. Conclusions

The Pyramid Schemes are the only low energy effective field theories compatible with both the very low scale of SUSY breaking required by CSB, extant experimental data, and standard model gauge coupling unification. They contain a new strong coupling gauge theory, with fields carrying standard model quantum numbers. The most attractive candidate so far has an $SU_P(3)$ gauge group.

The strong interactions complicate the computation of the Higgs potential and parts of the spectrum, but terms that give rise to TeV Dirac masses for gluinos (and probably the electroweak gauginos as well) enable us to make a few robust predictions

- The MSSM spectrum can be characterized as "flipped mini-split SUSY", with squarks and sleptons systematically lighter than gauginos. Gluinos will probably not be detected at the LHC, but squarks should show up in the next run, with production and decay modes characteristic of the gluino decoupling limit. The entire Higgs sector is complicated by mixing with the singlet fields in the low energy model. This spectrum is predicted by the model. It's realized more generally in any model in which there are adjoint chiral superfields, Dirac masses comparable to the supersymmetric adjoint mass term, and small SUSY breaking Majorana terms for the gaugino. Models with adjoint fields that are elementary up to scales much larger than the SUSY breaking scale, will have problems with gauge coupling unification.

- The NLSP is either a right handed slepton, or something from the singlet sector, but in any case the right handed sleptons are "detector stable" and should be seen soon at the LHC, since they decay to leptons and very light gravitino LSPs. The crudest calculations put their masses 9 times lower than the LHC lower bound. The simplest way to solve this problem is to assume that the bino and charged winos are at 9 and 27 TeV, but there is so much uncertainty in these estimates from hidden sector strong interactions that one should not take these drastic values that seriously. Anyone who has followed my work on the Pentagon and Pyramid schemes will know that I'd previously estimated that the bino was the NLSP and that charginos should be found at the LHC. The recent discovery of operators that give Dirac gaugino masses has changed everything in a dramatic way.

Apart from that, the Pyramid Schemes retain the flavor structure of gauge mediation [15]. The only violation of rotation symmetries among the generations comes from Standard Model Yukawa couplings, and the mechanism determining the pattern of those is assumed to operate at very high scale. Dimension 4 and 5 baryon and lepton number violation is eliminated by a combination of the discrete R symmetry of the $\Lambda = 0$ model and the special properties of the R breaking operators coming from the horizon. A similar conspiracy solves the strong CP problem. The discrete R symmetry imposes an accidental $U_{PQ}(1)$ Peccei-Quinn symmetry on the renormalizable terms of the $\Lambda = 0$ theory, and the special nature of discrete R violation, combined with the assumption that CP is spontaneously broken at a scale below the unification scale, guarantee that the would be axion is lifted to a high mass, without introducing new phases into low energy couplings.

The Pyramid Schemes also have interesting implications for cosmology. If we assume one of the pyrma-baryon symmetries is preserved at the renormalizable level, then the dark matter candidate is a standard model singlet fermion, with a mass of 10s of TeV and a commensurate magnetic dipole moment. The correct dark matter density is obtained by assuming an appropriate primordially generated asymmetry, and there is a potential connection between the dark matter density and the ordinary baryon asymmetry via a form of spontaneous baryogenesis [25].

On the other hand it is possible, though not guaranteed, that there can be a light state in the singlet sector that could serve as dark matter. In this case we would be able to have an elegant and symmetric theory at the unification scale, which would explain the coincidence of scales between Λ_3 and SUSY breaking. The model with only two of the three renormalizable PB violating couplings does the same job, but is less elegant.

If the singlet dark matter candidate were sufficiently light, it could be the NLSP, and its stability only due to R parity. Then it could also be a form of late decaying dark matter, which would produce a dark radiation density in the form of non-thermal gravitinos. It may be that cosmological data will eventually require us to explain such a density of dark radiation. The very light gravitinos of the Pyramid Scheme are hard to detect, but beg to be used as dark radiation. Much more investigation along these lines is needed.

Acknowledgments

I would like to thank M.Dine, W. Shepherd, P. Draper, S. Thomas, D. Shih, L. Carpenter, J-F Fortin, and H. Haber, for numerous conversations about the topics covered in this review. This work was supported in part by the Department of Energy.

References

[1] T. Banks, "A Critique of pure string theory: Heterodox opinions of diverse dimensions," hep-th/0306074. T. Banks, "Strings in a Landscape,"; T. Banks, "Landskepticism or why effective potentials don't count string models," hep-th/0412129; "The Top 10^{500} Reasons Not to Believe in the Landscape," arXiv:1208.5715 [hep-th].

[2] Z. Komargodski and N. Seiberg, "Comments on the Fayet-Iliopoulos Term in Field Theory and Supergravity," JHEP **0906**, 007 (2009) [arXiv:0904.1159 [hep-th]].

[3] T. Banks and J. Kehayias, "Fuzzy Geometry via the Spinor Bundle, with Applications to Holographic Space-time and Matrix Theory," Phys. Rev. D **84**, 086008 (2011) [arXiv:1106.1179 [hep-th]].

[4] T. Banks and J. -F. Fortin, "A Pyramid Scheme for Particle Physics," JHEP **0907**, 046 (2009) [arXiv:0901.3578 [hep-ph]]; T. Banks and J. -F. Fortin, "Tunneling Constraints on Effective Theories of Stable de Sitter Space," Phys. Rev. D **80**, 075002 (2009) [arXiv:0906.3714 [hep-th]]; T. Banks, J. -F. Fortin and S. Kathrein, "Landau pole in the pyramid scheme," Phys. Rev. D **82**, 115015 (2010) [arXiv:0912.1313 [hep-ph]].

[5] T. Banks and W. Fischler, "An Holographic cosmology," hep-th/0111142; T. Banks, W. Fischler and S. Paban, "Recurrent nightmares? Measurement theory in de Sitter space," JHEP **0212**, 062 (2002) [hep-th/0210160]; T. Banks and W. Fischler, "Holographic cosmology 3.0," Phys. Scripta T **117**, 56 (2005) [hep-th/0310288]; T. Banks, W. Fischler and L. Mannelli, "Microscopic quantum mechanics of the p = rho universe," Phys. Rev. D **71**, 123514 (2005) [hep-th/0408076]; T. Banks and W. Fischler, "Holographic Theories of Inflation and Fluctuations," arXiv:1111.4948 [hep-th]; T. Banks, "Holographic Space-Time: The Takeaway," arXiv:1109.2435 [hep-th]; T. Banks and J. Kehayias, "Fuzzy Geometry via the Spinor Bundle, with Applications to Holographic Space-time and Matrix Theory," Phys. Rev. D **84**, 086008 (2011) [arXiv:1106.1179 [hep-th]]; T. Banks, "TASI Lectures on Holographic Space-Time, SUSY and Gravitational Effective Field Theory," arXiv:1007.4001 [hep-th];
T. Banks, "Holographic Space-time from the Big Bang to the de Sitter era," J. Phys. A A **42**, 304002 (2009) [arXiv:0809.3951 [hep-th]].

[6] E. Witten, "Strong coupling expansion of Calabi-Yau compactification," Nucl. Phys. B **471**, 135 (1996) [hep-th/9602070].

[7] T. Banks, W. Fischler, "Holographic Theory of Accelerated Observers, the

S-matrix, and the Emergence of Effective Field Theory", arXiv:1301.5924 [hep-th].

[8] M. A. Awada, G. W. Gibbons and W. T. Shaw, "Conformal Supergravity, Twistors And The Super Bms Group," Annals Phys.171, 52 (1986); T. Banks, "The Super BMS Algebra, Holographic Space-Time and Infra-red Divergences", to appear.

[9] T. Banks, *Modern Quantum Field Theory: A Concise Introduction*, Cambridge University Press, New York, 2008.

[10] A. Aguirre, T. Banks and M. Johnson, "Regulating eternal inflation. II. The Great divide," JHEP **0608**, 065 (2006) [hep-th/0603107].

[11] T. Banks and J. -F. Fortin, "Tunneling Constraints on Effective Theories of Stable de Sitter Space," Phys. Rev. D **80**, 075002 (2009) [arXiv:0906.3714 [hep-th]];

[12] N. Seiberg, "Exact results on the space of vacua of four-dimensional SUSY gauge theories," Phys. Rev. D **49**, 6857 (1994) [hep-th/9402044]. N. Seiberg, "Electric - magnetic duality in supersymmetric nonAbelian gauge theories," Nucl. Phys. B **435**, 129 (1995) [hep-th/9411149].

[13] S. L. Glashow, "Trinification Of All Elementary Particle Forces," Print-84-0577 (BOSTON).

[14] T. Banks, J. -F. Fortin and S. Kathrein, "Landau pole in the pyramid scheme," Phys. Rev. D **82**, 115015 (2010) [arXiv:0912.1313 [hep-ph]].

[15] M. Dine and W. Fischler, "A Phenomenological Model of Particle Physics Based on Supersymmetry," Phys. Lett. B **110**, 227 (1982); M. Dine and A. E. Nelson, "Dynamical supersymmetry breaking at low-energies," Phys. Rev. D **48**, 1277 (1993) [hep-ph/9303230]; M. Dine, A. E. Nelson and Y. Shirman, "Low-energy dynamical supersymmetry breaking simplified," Phys. Rev. D **51**, 1362 (1995) [hep-ph/9408384]; C. Csaki, Y. Shirman and J. Terning, "A Simple Model of Low-scale Direct Gauge Mediation," JHEP **0705**, 099 (2007) [hep-ph/0612241]; K. Agashe, "An Improved model of direct gauge mediation," Phys. Lett. B **435**, 83 (1998) [hep-ph/9804450]; T. Han and R. -J. Zhang, "Direct messenger - matter interactions in gauge - mediated supersymmetry breaking models," Phys. Lett. B **428**, 120 (1998) [hep-ph/9802422].

[16] T. Banks, "Dirac Gluinos in the Pyramid Scheme," arXiv:1311.4410 [hep-ph].

[17] P. J. Fox, A. E. Nelson and N. Weiner, "Dirac gaugino masses and supersoft supersymmetry breaking," JHEP **0208**, 035 (2002) [hep-ph/0206096].

[18] M. Gell-Mann, P. Ramond and R. Slansky, "Complex Spinors and Unified Theories," Conf. Proc. C **790927**, 315 (1979) [arXiv:1306.4669 [hep-th]].

[19] CMS-PAS-SUS-13-06 http://cds.cern.ch/record/1563142/files/SUS-13-006-pas.pdf ; B. Fuks, M. Klasen, D. R. Lamprea and M. Rothering, "Revisiting slepton pair production at the Large Hadron Collider," arXiv:1310.2621 [hep-ph].

[20] http://xxx.lanl.gov/pdf/1208.2884v2.pdf

[21] J. Bagnasco, M. Dine and S. D. Thomas, "Detecting technibaryon dark matter," Phys. Lett. B **320**, 99 (1994) [hep-ph/9310290]; T. Banks, J. -

F. Fortin and S. Thomas, "Direct Detection of Dark Matter Electromagnetic Dipole Moments," arXiv:1007.5515 [hep-ph].

[22] T. Banks, J. D. Mason and D. O'Neil, "A Dark matter candidate with new strong interactions," Phys. Rev. D **72**, 043530 (2005) [hep-ph/0506015].

[23] S. Nussinov, "Technocosmology: Could A Technibaryon Excess Provide A 'natural' Missing Mass Candidate?," Phys. Lett. B **165**, 55 (1985); R. S. Chivukula and T. P. Walker, "Technicolor Cosmology," Nucl. Phys. B **329**, 445 (1990); S. M. Barr, R. S. Chivukula and E. Farhi, "Electroweak Fermion Number Violation and the Production of Stable Particles in the Early Universe," Phys. Lett. B **241**, 387 (1990); D. B. Kaplan, "A Single explanation for both the baryon and dark matter densities," Phys. Rev. Lett. **68**, 741 (1992).

[24] A. G. Cohen and D. B. Kaplan, "Spontaneous Baryogenesis," Nucl. Phys. B **308**, 913 (1988); A. G. Cohen, D. B. Kaplan and A. E. Nelson, "Spontaneous baryogenesis at the weak phase transition," Phys. Lett. B **263**, 86 (1991).

[25] T. Banks, S. Echols and J. L. Jones, "Baryogenesis, dark matter and the Pentagon," JHEP **0611**, 046 (2006) [hep-ph/0608104].

[26] H. Pagels and J. R. Primack, "Supersymmetry, Cosmology and New TeV Physics," Phys. Rev. Lett. **48**, 223 (1982).

[27] T. Banks and T. J. Torres, "Update on the Pyramid Scheme," Eur. Phys. J. C **72**, 2185 (2012) [arXiv:1205.3073 [hep-ph]]; T. Banks and T. J. Torres, "Approximate Particle Spectra in the Pyramid Scheme," Phys. Rev. D **86**, 115015 (2012) [arXiv:1207.5096 [hep-ph]].

Printed in the United States
By Bookmasters